'To create a future where humans and other animals are free, we need inspiring visions to guide us. After all, we can only create what we can imagine. Over 14 chapters, this collection offers much-needed examples and visions of more liberatory ways of seeing and relating to our animal cousins, providing glimpses of what this promises and helping us uncover pathways to create these alternative futures.'

Dr Laila Kassam, *Animal Think Tank, UK*

'Taking us across sanctuaries, neighbourhood trees, dovecotes, and design projects, Arcari's curation of "heterotopia" shows *why* and *how* noticings and imaginings of possibilities otherwise are imperative for liberatory politics to be conceivable and hence actionable. Each chapter exuding care, this groundbreaking collection heralds a momentous generational shift in animal studies, composing a conceptual insurrection, a new political grammar, and critically, a tangible cartography for radicalising human-to-animal alliances.'

Associate Professor Yamini Narayanan, *Deakin University, Australia*

Heterotopia, Radical Imagination, and Shattering Orders

This volume takes ending the oppression of other animals seriously and confronts the question 'What would happen to all the animals?' by showcasing real, promissory, and imagined counter-sites or *heterotopia*, where animals 'happen' in different ways, free of anthropocentric orders of value and purpose.

Rejecting persistent understandings of the oppression of nonhuman animals, across the entire breadth of the Animal-Industrial Complex (A-IC), as either non-existent, unproblematic, and/or fundamentally unalterable – open to merely being reduced in scale or made less harmful – the collection offers readers a variety of pathways towards radically 'disordered' ways of thinking about and relating to other animals. Over 14 chapters, authors describe more liberatory relational reconfigurations playing out in the present and undertake conceptual, imaginative, and embodied explorations of liberatory futures. The chapters are united by a common commitment to heterotopic disturbance – to contesting and subverting the anthropo-capitalo-centric space in which we live. Each chapter approaches this subversion in its own way, using prefiguration, restorying, speculation, radical imagination, and combinations thereof, to disturb or shatter orders, explore the kinds of liberation and resistance their disturbance demonstrates, demands, or embodies, and ultimately illustrate exactly what *would* or *could* happen to all the animals.

Heterotopia, Radical Imagination, and Shattering Orders will appeal to scholars, students, and individuals interested not only in challenging normalised binaries, hierarchies, and orders of value, both human and nonhuman, but in creating and realising liberatory alternatives. Scholar-activists, activists, professionals working in animal advocacy, and anyone undertaking activities aimed at radically changing how other animals are understood and used will also find inspiration, new insights, and information that enhance their current methods and approaches. Some readers may also find simply confirmation and comfort in the knowledge that so many others are working in solidarity with the 'disordered' belief that shattering the A-IC is possible.

Paula Arcari is an independent scholar living in Melbourne Australia, and a former Leverhulme Early Career Fellow (2019–2022) hosted by the Centre for Human Animal Studies (CfHAS), Edge Hill University, UK. She is the author of *Making Sense of 'Food' Animals: A Critical Exploration of the Persistence of 'Meat'* published in 2019.

Routledge Human – Animal Studies Series
In Memory of Professor Henry Buller, founding series editor.

The new *Routledge Human – Animal Studies Series* offers a much-needed forum for original, innovative and cutting-edge research and analysis to explore human – animal relations across the social sciences and humanities. Titles within the series are empirically and/or theoretically informed and explore a range of dynamic, captivating and highly relevant topics, drawing across the humanities and social sciences in an avowedly interdisciplinary perspective. This series will encourage new theoretical perspectives and highlight ground-breaking research that reflects the dynamism and vibrancy of current animal studies. The series is aimed at upper-level undergraduates, researchers and research students as well as academics and policy-makers across a wide range of social science and humanities disciplines.

Winged Worlds
Common Spaces of Avian-Human Lives
Edited by Olga Petri and Michael Guida

Methods in Human-Animal Studies
Engaging with animals through the social sciences
Edited by Annalisa Colombino and Heide K. Bruckner

What We Owe to Nonhuman Animals
The Historical Pretensions of Reason and the Ideal of Felt Kinship
Gary Steiner

Horses, Power, and Place
A More-Than-Human Geography of Equine Britain
Neil Ward

Heterotopia, Radical Imagination, and Shattering Orders
Manifesting a Future of Liberated Animals
Edited by Paula Arcari

For more information about this series, please visit: www.routledge.com/
Routledge-Human-Animal-Studies-Series/book-series/RASS

Heterotopia, Radical Imagination, and Shattering Orders

Manifesting a Future of Liberated Animals

Edited by Paula Arcari

LONDON AND NEW YORK

First published 2025
by Routledge
4 Park Square, Milton Park, Abingdon, Oxon OX14 4RN

and by Routledge
605 Third Avenue, New York, NY 10158

Routledge is an imprint of the Taylor & Francis Group, an informa business

© 2025 selection and editorial matter, Paula Arcari; individual chapters, the contributors

The right of Paula Arcari to be identified as the author of the editorial material, and of the authors for their individual chapters, has been asserted in accordance with sections 77 and 78 of the Copyright, Designs and Patents Act 1988.

With the exception of Chapter 1 no part of this book may be reprinted or reproduced or utilised in any form or by any electronic, mechanical, or other means, now known or hereafter invented, including photocopying and recording, or in any information storage or retrieval system, without permission in writing from the publishers.

Chapter 1 of this book is available for free in PDF format as Open Access at www.taylorfrancis.com. It has been made available under a Creative Commons Attribution-NonCommercial-NoDerivatives (CC-BY-NC-ND) 4.0 International license.

Trademark notice: Product or corporate names may be trademarks or registered trademarks, and are used only for identification and explanation without intent to infringe.

British Library Cataloguing-in-Publication Data
A catalogue record for this book is available from the British Library

ISBN: 978-1-032-43300-4 (hbk)
ISBN: 978-1-032-43303-5 (pbk)
ISBN: 978-1-003-36670-6 (ebk)

DOI: 10.4324/9781003366706

The Open Access version of Chapter 1 was funded by Julius Rosenwald Fund (The University of Chicago).

For Product Safety Concerns and Information please contact our EU representative:
GPSR@taylorandfrancis.com
Taylor & Francis Verlag GmbH, Kaufingerstraße 24, 80331 München, Germany

This book is dedicated to all humans working to shatter orders and all the other animals helping to show
us how.
*100% of the authors' proceeds from this book are donated to vegan animal sanctuaries.

Contents

Figures

Tables

About the Editor

Paula Arcari is an independent scholar living in Melbourne, Australia, and a former Leverhulme Early Career Fellow (2019–2022) hosted by the Centre for Human Animal Studies (CfHAS), Edge Hill University, UK. She is the author of *Making Sense of 'Food' Animals: A Critical Exploration of the Persistence of 'Meat'* published in 2019.

Contributors

Amy Dover is an experienced fine artist, illustrator, and researcher, currently working on a practice-based PhD in fine art at the School of Arts and Culture, Newcastle University. Her work has been exhibited internationally, and she has collaborated with international animal charity organisations. Dover also works as a programme director at Edinburgh College of Art, Edinburgh University, as well as an educator for other organisations and institutions.

Amina Grunewald obtained her PhD on native self-representations in the works of Sherman Alexie at the Humboldt University of Berlin. Her postdoctoral studies focus on environmental and animal justice in American cultural productions while broadening her horizon on critical animal studies, critical museum studies, and blues/jazz studies.

Richard Healey is a postdoctoral researcher as part of the 'Not in My Name' project at the London School of Economics (LSE). He works in moral, political, and legal philosophy. His current research focuses on normative powers, interpersonal relationships, and animal rights.

Scott Hurley is Associate Professor of Religion and Identity Studies at Luther College, Iowa. His research interests focus on the critical analysis of racism, sexism, speciesism, white body supremacy, and white privilege in US policy, law, and traditions; embodied social justice; Buddhist and Daoist ideology and practice from the nineteenth century to the present; Black and Asian American Buddhism; and the intersections of martial arts, religion, and identity. He is co-editor of a Critical Animal Studies Series for a leading publisher.

Alexandra Isfahani-Hammond is Associate Professor Emerita of Comparative Literature and Luso-Brazilian Studies at the University of California, San Diego. She works at the intersection of critical animal studies, decolonial studies, and comparative race and slavery studies. Her current book project, *Home Sick*, blends theory with creative nonfiction and poetry to meditate on grief, the end of life, and the commodification of human and more-than-human animals.

Katya Krylova is a doctoral candidate in cultural studies at the University of Canterbury. Her research addresses ethical issues related to the cute economy and the simulation of animal speech on social media. Katya's book *The Market of Convenient Animals*, exploring the roles imposed upon companion animals in neoliberal cultures, was published in 2023 in Russian. She teaches workshops on posthumanist architecture for practising architects and serves as a guest editor for *Pulse: Journal of Science and Culture*.

Charlotte A. Kunkel is Professor of Sociology and Identity Studies at Luther College, Iowa. Her research and teaching centre around stratification and systems of power, including the sociology of gender, constructs of race, and wealth and poverty. She is passionate about imagining systemic social justice, incorporating both visual sociology and social psychology. Her publications include articles in *Deviant Behavior*, *The Sociological Quarterly*, and *The Peace Studies Journal*. She serves as the inaugural Department Head of Identity Studies and Chair of the Sociology Program at Luther.

Marie Leth-Espensen is a sociologist, qualitative researcher, and member of the Lund University Critical Animal Studies Network. Her research interests centre on human–animal relations, the ethics and politics of food and eating, and the possibility of achieving a non-anthropocentric foundation for care.

Maria Martelli is an independent scholar who writes and researches mixing antispeciesist and queer-feminist theories. She has an MA in sociology and is slowly working on an independent activist research on the liberatory potential of animal sanctuaries and multispecies imaginaries. She is part of the antispeciesist collective *just wondering . . .*, of the Queer Vegan Community in Romania (CVQ) and of the autonomous literary group Cenaclul X.

Angie Pepper is Senior Lecturer in Philosophy at the University of Roehampton. Angie is interested in what we owe to other animals as a matter of justice and has written on various topics, including multispecies climate justice, ethics in animal shelters, and animals' privacy rights. Angie's recent work focuses on the normative significance of nonhuman animal agency; in other words, she is interested in what other animals do and why it matters morally, socially, and politically.

Sue Hall Pyke teaches creative writing and Indigenous studies at the University of Melbourne and is part of the editorial collective for *Swamphen*. She writes from the snake-friendly Stony Rises in Djargurd Wurrong Country, a territory governed by the Eastern Maar Nation. Sue's published thinking around snakes includes 'Snake Church' (*Animal Studies Journal*, 2022), 'Meeting Selena' (*Meanjin*, 2022), 'Tiger Snake Co-Existence 101' (*Feral Feminisms*, 2022), and 'Citizen Snake' (*The Materiality of Love*, 2017). Her monograph *Animal Visions* (2019) considers posthumanist literary representations. Sue also collaborates with animal studies scholars on questions of animal welfare. Details of her publications can be found at suehallpyke.com.

Sutirtho Roy graduated with first class from Calcutta University. He ranked third in a state-wide essay contest regarding drug abuse and earned the first position in a writing contest nationally. He has also served as a panellist at international workshops pertaining to post-humanism. His growing interest in environmental humanities and critical animal studies has helped him present several papers at international conferences. He has immensely enjoyed the experience of delving into a wide gamut of literature and exploring the potential of storytelling and proper linguistic frameworks in conservation.

Jennifer Rebecca Schauer is a human animal studies scholar, teacher, and activist. Her work centres on human coexistence with animals and the earth. She mentors and collaborates with students in her Sentience Collective at San Jose State University (SJSE). Her work can be read in journals such as *Animalia, Nature and Culture, Society and Animals,* and *Trace.*

Elizabeth Tavella is a scholar of critical animal studies and environmental humanities at the University of Chicago whose work builds on critical theory, decolonial methods, and queer ecology to investigate interlinked systems of oppression and liberation histories from an artistic and cultural perspective. Elizabeth is the creative writing and arts editor of *Ecozon@: European Journal of Literature, Culture and Environment* and co-editor of the multimedia digital publication *Animated Wor(l)ds: Language and Relationality for Multispecies Kinship.*

Daniel Vandersommers is Assistant Professor of Environmental History at the University of Dayton, Ohio. He is the author of *Entangled Encounters at the National Zoo: Stories from the Animal Archive* (2023), which was a finalist for the ASEH George Perkins Marsh Prize. He is the lead editor of the new journal *Animal History* and co-editor of the volume *Zoo Studies: A New Humanities* (2019), which was chosen as a *Choice* 'Outstanding Academic Title.' Dan is currently working on projects concerning racism and zoology in the nineteenth century, global history and animal displays, and the history of goldfish breeding.

Michelle Westerlaken is a postdoctoral researcher at the University of Cambridge with a PhD in interaction design from Malmö University. Her research and design work investigates possibilities for more relational – multispecies – ways of living on this planet. She collaborated with humans, cats, dogs, ants, penguins, forests, and various interactive technologies.

Foreword

Richard Twine

How do we change the world? Is that not the question all critical animal scholars, activists, and 'progressives' are asking everywhere, almost to the extent now that, in these times of escalating urgency, it becomes a deafening, slightly tortured crescendo? How do we intervene in an inherently dynamic and unpredictable world so that it might be fairer, caring, and just? How to make such concepts speak to an inescapably posthumanist and global world? How do we upscale the efforts and creativity of our own lives and communities to make a dent in the tragically aloof and cowardly phallocentric formal political sphere? How to make meaningful interventions in the practices of the status quo that seem determined to perpetuate the irrationality of reducing everything (education, healthcare, work, the food system, humans, the more-than-human, etc.) to monetary instrumental value?

Just do. Do practices that embody a counter-hegemonic set of meanings. Create space, create practices, create meanings, weave them together, and demonstrate their qualities. Practice radical acts of kindly decommodification. Focus practice making on contesting the persistent overlapping systems of domination and exclusion in the contemporary Capitalocene. You may know them by now: classism, sexism, racism, colonialism, anthropocentrism, ableism, ageism, heteronormativity, transphobia. Well, this is but *one* answer to some of these very difficult questions posed above.

What heterotopias have you participated in, researched, and helped to create? What, you might be asking, is this concept? A little posthumous gift from Michel Foucault? I will leave the definitional work to Paula Arcari's introduction, and it was Paula who brought it to my attention. I have employed it recently, finding appeal in the notion of already existing counter-sites that might be linked together in, for example, a nascent vegan and decommodified food heterotopology, spaces, and practices which are in a sense inciting, mocking, and exceeding the prevailing order. Futural fragments awaiting your enrolment in political struggle. One can conduct a familial conversation with its conceptual kin: prefiguration, killjoying, and multiple other siblings across the social change and transition literature and practice space. Heterotopias will not be politically pure spaces but sites of experimentation where persistent relations of power can be brought out into the open and tackled in process.

The creative and imaginative process of heterotopia also provides some riposte to the persistent claim that critical theory fails to offer alternative imaginaries and ways of living. However, it should not be seen as a call to neglect attention to those discourses and spaces that maintain oppressive, overlapping orders. Just as attempts to intervene in narratives and framings around animal advocacy gain insights from prevailing meanings and stereotypes, hypertopias – what Paula Arcari describes as sites that magnify hegemonic orders – may be points of instability and generative for heterotopology. From butchers becoming vegan butchers to farms becoming sanctuaries? Or in my town a shop that used to sell guns and hunting paraphernalia is now a vegan café. The slaughterhouse, for example, perhaps a paradigmatic hypertopia, is a space now so out of step with prevailing views on violence and 'civility' that it is arguably precarious and shameful. It is no wonder some countries and US states legislate via so-called ag-gag laws, attempts to stifle undercover investigations.

Speaking of unsustainable incongruity, how confounding must it be to be an otherwise progressive scholar or activist to then discover that one's everyday consumerism places oneself within the nexus of classist, racist, sexist, and speciesist exploitation that is the contemporary slaughterhouse? I am not claiming that those of us living differently may not also find ourselves unwittingly perpetuating relations of privilege, but unfortunately the point stands that a block on the counter-anthropocentric heterotopology coming to constitute a new social fabric is not merely the hapless right but also most of the progressive left. A humanist political frame alone just won't cut it in the twenty-first century.

If that is you, I ask you not to be offended. I am also speaking to my former self, and neither are intersectional critical animal scholars and activists considering themselves somehow the politically perfect subject. The 'individual' after all is constrained by operating within larger infrastructures of practice that shape what they can do. The financial might and ability of, for example, capitalism to shape practices at scale remain one of the biggest obstacles to simply forging a new social fabric and dismantling a significant dimension of that (dis)order: the animal-industrial complex. It is true that the 'just' of 'Just do' is doing a lot of heavy lifting. The burden of our own positionality and the dominance of the status quo constrain heterotopia, but as Foucault might have said if he had lived to become an aged eco-radical (nice image, right?): resistance is fertile, and another world is possible.

Inescapably we *are* living in a climate and biodiversity crisis, and if you are resisting your own enrolment into practices that contest this, you are making a specific contribution to the systemic problem. These are crises that are not just about an imagined human 'us,' but they are threatening to unravel conditions of earthly survival for all species. On the other hand, heterotopias offer a positive galvanising direction in the face of such weighty issues. This volume reminds us of the rewarding possibilities of human–animal relations beyond that of commodification. An antidote to the contemporary malaise is that self and social transformation add joy and meaning to lives, taking us in unexpected directions.

This volume is overflowing with the very ideas, spaces, imagination, and creativity to inspire change. The contributors have successfully given voice to a multitude of nonhuman animal species. Understandably here you will find a focus on topics such as domestication and on sites such as sanctuaries. To work your way through this volume also takes you on a diverse journey through subversive spaces, multi-species world-making, anarcho-geography, gothic enviro-toons, posthumanist humour, productive bewilderment, hospitality with snakes, pedagogical transformation, non-anthropo-capitalo-centric futures, feminist science fiction, and black liberation theology, to mention just a few.

A big thank you to Paula Arcari and all the contributors.

Read this, and then pass it on.

Preface
Once There Was a Bear Here

Amy Dover

My present doctoral inquiry explores the themes of loss, disappearance, and the traces of ghosts of non-human animals. The disappearance of these animals from our natural world and their confinement, captivity, and exploitation for cultural representations have reduced them to being the hapless slaves of human society, thereby rendering them mere products in a world increasingly estranged from nature. This has led to a world where wild mammals make up only 4% of all mammals on the planet (Bar-On et al., 2018); humans are ever disconnected from our wild relatives. I seek to explore whether a fine-art practice can not only aid in the conservation of non-human animals but also re-wild humans and develop a greater connection with nature.

Non-human animals were the first drawings created by humans, and they have been portrayed in art for centuries (Berger, 2009). Animals are often a symbol representing cultural beliefs and values. However, with the growing loss and disappearance of non-human animals, this representation has taken on a new significance. The bear, once the revered king of beasts in Europe (Pastoureau, 2011), inhabiting the land in substantial numbers, was hunted to extinction in Britain 1,500 years ago (O'Regan, 2018). Consequently, the bear was stripped of its title and became a villain with the rise of Christianity's association with the Lion, and the bear's association with paganism. This pattern of loss is still happening, with an estimated 150–200 species going extinct globally every day (WWF, 2022). There is now a growing call for re-wilding, re-establishing lost keystone species. Several animal species have been successfully reintroduced in Britain, such as the beaver and the Eurasian bison, leading to an increase in biodiversity (Tree, 2003). Nonetheless, the bear continues to be a controversial species, primarily portrayed as a terrifying man-eater that perpetuates negative perceptions. This inquiry seeks to challenge such representations and engage the public with empathy and respect.

Presenting statistical evidence, scientific studies, and protesting have had limited impact so far on conservation outcomes. In some cases, protests can even have a negative impact on public engagement (Polanco, 2022). Art, however, can have the capacity to 'disarm' speciesism and have a greater impact. As artist in residence for Natural England's Natural Futures, which seeks to re-wild Britain and develop

Figure 0.1 Bear, at Newcastle contemporary arts gallery.

humans' connection with nature, I developed this piece. A large pencil portrait drawing of a bear who faces out at the viewer, every hair is visible, tactile, and almost lifelike (Figure 0.1). I filmed the process over a number of weeks as he developed while an audience watched. He was installed in the gallery at Newcastle Contemporary Art (NCA). Visitors were then invited to wipe him out with a rubber eraser (Figure 0.2). The audience was very reluctant to cause harm to a piece of art, and it made them very uncomfortable. However, in the written feedback people said the experience made them feel a greater connection to nature and a need to conserve it.

In his book *Dark Ecology*, Timothy Morton (2018) discusses how small acts that seem inconsequential in the Anthropocene contribute collectively to the destruction of nature. The emotional response elicited by the act of erasing the bear drawing had an impact, bringing the consequences of visitors' individual actions to the forefront of their minds and centring their own responsibility. The video of its creation and destruction has since been transformed into a video piece that has been shown in exhibitions and shared on social media. Currently, it has reached over 800,000 people, garnered more than 150,000 likes, and attracted hundreds of comments and contemporary discourse.[1] The drawing has a crucial role to play in raising awareness of the human-made sixth mass extinction and inspiring a greater connection between humans and non-human animals, ultimately leading to greater conservation efforts.

Natural Futures is a collaborative project between Natural England, Tees Valley Nature Partnership, Borderlands, MIMA Gallery, and Teesside University: natural-futures.co.uk.

Figure 0.2 Erasure.

Note

1 www.tiktok.com/@amy_dover/video/7103643853441289477

References

Bar-On, Y. M., Phillips, R., & Milo, R. (2018). The Biomass Distribution on Earth. *Proceedings of the National Academy of Sciences, 115*(25), 6506–6511. https://doi.org/10.1073/pnas.1711842115.Â
Berger, J. (2009). *Why Look at Animals?* Penguin Books.
Morton, T. (2018). *A Dark Ecology: For a Logic of Future Coexistence*. Columbia University Press.
O'Regan, H. J. (2018). The Presence of the Brown Bear *Ursus arctos* in Holocene Britain: A Review of the Evidence. *Mammal Review, 48*, 229–244.
Pastoureau, M. (2011). *The Bear: History of a Fallen King*. The Belknap Press of Harvard University Press.

Polanco, A. (2022). The Challenges of Researching Animal Advocacy Protests. *Faunalytics*, June 22. https://faunalytics.org/the-challenges-of-researching-animal-advocacy-protests/

Tree, I. (2003). *The Book of Wilding : A Practical Guide to Rewilding Big and Small.* Bloomsbury.

WWF. (2022). *Living Planet Report 2022 – Building a Nature-Positive Society* (R. E. A. Almond, M. Grooten, D. Juffe Bignoli, & T. Petersen, Eds.). WWF, Gland, Switzerland.

Acknowledgements

The symposium that led to this collection, and the research presented in Chapter 4, were supported by a Leverhulme Trust Early Career Fellowship (2019–2022) and by Edge Hill University, UK. I thank Richard Twine and Claire Parkinson, co-Directors of Edge Hill University's Centre for Human Animal Studies (CfHAS), for endorsing my Fellowship and hosting me through CfHAS. The networks, conversations, and events I was fortunate to be part of during my time with CfHAS helped shape the thinking that underpins this collection. In particular, I thank Richard Twine, again, for his ongoing support and guidance, and everyone who attended the Heterotopia Symposium as part of the 7th Biennial Conference of the European Association for Critical Animal Studies (EACAS, 2021), hosted by CfHAS. This collection was only possible thanks to their enthusiastic engagement and the generous commitment of all subsequent contributing authors. Thank you to Shae Hunter and Iris Bergmann for always being willing to gift their time, intellect, and editorial skills to my scholarly endeavours. Finally, I thank the scholars, friends, family, and nonhumans who have shaped and guided my thinking, intentionally or otherwise, and most of all my sister Nicki Arcari and brother Antony Arcari for their life-long encouragement and support.

Introduction

Paula Arcari

Is shattering orders possible?

It is an aspiration that shapes the hopes, ideals, and missions of many movements aimed at social and global change, so it must be at least conceivable.

In his 1966 publication, *Order of Things*, Michel Foucault describes heterotopias as a discursive device, capable of undermining language, shattering common names, and destroying the 'syntax' that causes words and things to 'hold together' (1989 [1966], p. xx). Heterotopias, he says, "desiccate speech, stop words in their tracks, contest the very possibility of grammar at its source; they dissolve our myths and sterilize the lyricism of our sentences" (ibid).

Foucault is articulating here the uneasy, distressing effect of disordered language (specifically a passage from Argentine writer Jorge Luis Borges) on normalised understandings of the arrangement of things; how it can upset the coherence of cultural schemas of perception, exchange, value, and hierarchies of practice. In the moment of reading Borges' text, Foucault's (our) familiar landmarks of thought were shattered, "breaking up all the ordered surfaces and all the planes with which we are accustomed to tame the wild profusion of existing things" (ibid., p. xvi). Effectively, reading Borges was, for Foucault, a heterotopic encounter that showed how it was possible to disassemble "the categories that make it possible for us to name, speak, and think," to "collapse our age-old distinction between the Same and the Other," and to realise that "these orders are perhaps not the only possible ones or the best ones" (ibid., pp. xvi–xxii).

A year later, as part of a public lecture published as *Of Other Spaces* (1967), Foucault describes heterotopias in more explicitly spatial terms, as counter-sites where "all the other real sites that can be found within the culture, are simultaneously represented, contested, and inverted" (3). These *may* be real sites, but Foucault's key point, in my interpretation, is that these 'other places,' like Borges' text, enact a "simultaneously mythic and real contestation of the space in which we live" (ibid. p. 4). This 'space' in which we live, encompassing all its dimensions – discursive, spatial, virtual, cognitive, affective, emotional, and sensory – is one permeated by orders that draw countless distinctions between Same and Other.

Many of us may live in more equal societies whose achievements in challenging long-standing discriminatory orders, particularly regarding gender and 'race,'

DOI: 10.4324/9781003366706-1

are rightly celebrated. However, many more of us do not. Furthermore, achievements constitute partial and sometimes temporary advances; unstable progress towards, rather than the realisation of, the ideals of a post-sexist and post-racist future (Littler, 2018).

Political and economic regimes rise and fall, but the social hierarchies on which they depend, and work to maintain, persist. These hierarchies rest on arbitrary, intersecting, and mutually compounding orders of value determined by attributions relating not only to a person's gender, 'race,' ethnicity, indigeneity, and colour but also to their sexuality, socio-economic status, age, ability, religious beliefs, and others. Hierarchies arise because there is, in each case, a designation considered to have primacy over *all* others, creating a primary dualism. The most forceful of these in relation to the world's current crises include *male* vs female, *white* vs nonwhite, *rich* vs poor, *culture* vs nature, *civilised* vs 'wild,' *human* vs animal, and *reason* vs emotion (Plumwood, 1993; Gaard, 2004). Against these italicised ideals, all bodies are measured, ordered, and endure consequent limits, exclusions, harms, persecutions, and violences that are legitimated, more and less explicitly, by their social location(s).

For centuries, people have been trying to dismantle these orders and hierarchies. It should theoretically be possible. They may have endured so long as to profoundly constitute the world's prevailing social, economic, political, and juridical systems and practices, forming foundations of Western thought that have "become so entrenched in our worldview that we simply take them for granted" (Lent, 2021, p. 4). Yet, they emerged, coalesced, and intensified at particular points in history, suppressing alternate worldviews in the process and making us "heirs of an operation of cultural and social eradication" (Apffel-Marglin, 2020). Though still dominant, these orders and the worldview they nurture have, Lent observes, expired, and more viable futures are being envisioned through the recovery, (re)assertion, and integration of wisdoms confined for so long to the 'wrong' (right-hand) side of the aforementioned equations of self-bestowed human entitlement.

It is impossible to identify the spatially and temporally dispersed events, movements, thoughts, and other happenings that contribute to the kinds of rumbling social shifts that gather broader momentum. I refer specifically to those that generate (and indicate) a growing sense of un-ease with what is, awareness that 'it' (e.g., racism, sexism, capitalism, or anthropocentrism) has run its course or expired, and resonance with what could be – the mobilisation or stirring of a collective imagination. As per Haraway (1992) and Barad (2007), we cannot ultimately know which patterns of diffraction make *the* difference. The shift is generally only recognisable after lots of happenings have been happening for a while, until connections are made, patterns recognised, networks formed, and agency built.

As an aside, I would note that it is often possible to predict which happenings are likely to have questionable and even detrimental impacts on disordering ambitions. In relation to the nonhuman subjects of this collection, I would argue that the alternative meat/plant-based movements, interwoven with discourses of health and environmental sustainability, and movements for the rights of sentient beings, are two such happenings. The former, critiqued also by Miller (2012), Sexton (2016), and Daly (2020), reinforce animals' utilitarian over inherent value through a neo-liberal, depoliticised, and unstable strategy of 'behaviour trading' that only extends to some nonhuman animals (Arcari, 2024). The latter reinforce hierarchies of value based on a human-determined notion of sentience[1] – hierarchies that, as with humans, bestow limited legal rights that can further solidify rather than disturb dominant orders (see Wise, 2013; Blattner, 2019). The recognition of animals' sentience (often within existing *welfare* bills) does not preclude their being used and harmed. It simply opens avenues to further cast their treatment as ethical under renewed perceptions of it being done while recognising and respecting their sentience.[2]

Returning to how patterns of diffraction can generate broader shifts, it is perhaps in this way that oppressive orders of human supremacy may conceivably be shattered – through the multiplication of heterotopic counter-sites of unreserved resistance, dissent, and alternative/revolutionary practice. An expanding heterotopology or counter-map of such sites increases points of contact between ordered and dis-ordered understandings, relations, and practices, thereby increasing opportunities to weaken, undermine, and ultimately dismantle or even shatter expired ideologies. And so to the themes of this volume's title.

Heterotopia

Scholarly applications of heterotopia often favour one reading over another: textual vs spatial (Knight, 2017; Palladino & Miller, 2015; Johnson, 2013; Dehaene & De Cautier, 2008; Soja, 1996), real sites vs imaginary (Saldanha, 2008), and also intentional/purposeful vs incidental/emergent (Beckett et al., 2017). The concept has also attracted staunch criticism, as Johnson details. However, in light of imperatives for social and global change, I emphasise the fundamental and potentially anarchic elements of Foucault's theorisation in terms of contesting and inverting normalised orders of perception and hierarchies of practices. As Soja asserts, heterotopias "are meant to detonate, to deconstruct, not to be comfortably poured back into old containers" (1996, p. 76). Given Foucault's original use of the term in a volume whose central aim is to explore the effects of order on knowledge, including "the knowledge of living beings, the knowledge of the laws of language, and the knowledge of economic facts" (1989 [1966], p. x), and considering his later works on knowledge and power, this appears to me a natural reading. It is also one that echoes Hook (2007), who describes the capacity of heterotopias "to represent a point of destabilisation for current socio-political or discursive orders of power" (p. 184).

On this basis, and admittedly quite presumptuously, I would recast many of the heterotopias of Foucault's 1967 piece, including brothels, prisons, and colonies rather as *hypertopias*, understood conversely as places of excess where the majority of practices are organised around and towards the magnification and solidification of hegemonic orders so that they take on an exaggerated form (Arcari, 2023). Where heterotopias enact the disruption of orders, hypertopias demonstrate and mobilise their intensification. With this, I suggest that concerns over the precise definition and theoretical integrity of Foucault's concept miss a key point in his related work, which is what heterotopias *do*. What happens to bodies occupying these permanent or temporary 'sites' – real, virtual, or imagined? How are their lives and experiences affected? What is different? What happens to a person, or more accurately, the socially constituted understandings and practices they carry/represent, when they encounter the disordering effects of certain arrangements, narratives, words, images, sights, sounds, and sensations? What happens when the dominant orders and hierarchies by which humans name, speak, think, *and* act are disturbed, inverted, and possibly shattered?

From this perspective, I suggest that heterotopias signal two phenomenological experiences: 1. that of creating and holding an 'other place,' and 2. that of being somewhere different that is, importantly, not just unfamiliar or uncommon but hierarchically subversive and thereby uncomfortable (or more comfortable) on some level. Not every order or hierarchy need be completely inverted in heterotopias – only to the extent that the work required to maintain/defend its integrity or make sense of a discomfiting encounter is felt physically, emotionally, sensorially, and/or cognitively.

Hence, rather than the 'site' itself, it is these 'happenings' that define a heterotopia, that, by virtue of their doings – contesting, undermining, shattering, destroying, desiccating, disassembling, dissolving, inverting – constitute it as an 'other place.' In this respect, I agree with Johnson (2013, p. 800) that "heterotopias have no axe to grind, just scissors to cut" but take a more inclusive view on the 'spaces' of imagination, invention, and diversification in which these cuts can occur. To acknowledge the diversity of interpretations and applications, I could perhaps distinguish this reading with a suffix such as counter-hegemonic heterotopias (Annist, 2013) or anti-oppression heterotopias (Arcari, 2022).

The history of resistance, dissent, and revolutionary effort is characterised by challenging the artifice of hierarchies and orders. Foucault's concept has thus drawn favour across a range of critical applications, including anarchism (Shantz, 2012), civil disobedience (Anfinson, 2020), 'race,' gender, and queer studies (Bailey & Shabazz, 2014), social movement studies (Beckett et al., 2017), and climate activism (Edwards & Bulkeley, 2017). However, the extent to which bodies that carry Othered designations (including but not limited to gender and 'race') can fully escape their ascribed meanings and consequences is uncertain. Exactly 200 years after Mary Robinson's 1799 feminist polemic[3] in which she entreats, "Is not woman a human being?" (as cited in Botting, 2020, p. 6), feminist and legal scholar Catharine MacKinnon (2007 [1999]) asserts the continued relevance of this question in '*Are Women Human?*,' highlighting the restrictions and violations to which women around the world are still subjected. And over 75 years since the Universal Declaration of Human Rights, ratified by 192 member states, Morehouse et al. (2023) show how racial/ethnic groups perceived as less human are more associated with animal attributes.

Perhaps, as many scholars have argued (e.g., Wolfe, 2003; Stanescu & Twine, 2012; Freeman, 2020), the more fundamental problem is the obstinate duality of the human vs the animal, which makes ascriptions of human-ness and animal-ness available as mechanisms to elevate and exclude. As Heise (2013, p. 640) argues, "humans' willingness to assume a categorical difference between themselves and other kinds of animals enables discrimination against other persons"; and also Ko (2017, p. 46): "as long as these notions of 'the animal' and 'the human' are intact, white supremacy remains intact." The human/animal dualism supports the idea of an essential human-ness, but the duality only exists and can do this because humans also recognise that they are animals.[4] This has historically been, and continues to be, a widespread source of discomfort and tension (Yampell, 2008). Humans strive and strain to reconcile undeniable features of their animality with learned narratives of entitlement, exceptionalism, dominion, and superiority, constructing in the process a hierarchy of human-ness dependent on particular representations, or orders of knowledge, and their ongoing duplication (Foucault, 1989 [1966], p. 257). Hence, as individuals and groups push ever harder to break structural, institutional, epistemic, and discursive representations that order, limit, contain, denigrate, and exclude human bodies, the extent to which their efforts are hindered by excluding nonhuman bodies is insufficiently recognised, if at all.

This collection is intended to contribute to the growing volume of critical work aimed at bringing about a wholesale corrective to this state of affairs. It began, unknowingly at the time, with the idealised concept of a vegan pantopia, which I constructed as an analytical vantage point from which to observe and interrogate cartographies of meat (Arcari, 2019). I suggested the seeds of this pantopia were to be found in vegan heterotopias – diffuse sites of resistance that reject normalised constitutions of meat consumption and 'food' animals. Then, I was focused on better understanding these dominant constitutions to offer more effective ways of troubling habitual ways of thinking and acting involving 'food' animals. However, among other conclusions, I returned to my hope for the expansion of vegan hetero-topias, proposing, to that end, "an attentiveness to caesuras, ruptures, paradoxical spaces, and heterotopia" (p. 298) and, particularly the entitlement of hegemonic orders and hierarchies they foreground.

Radical Imagination

Dovetailing with these thoughts over the following two or more years, and help-ing to define the parameters for this volume, is my longstanding attraction to sci-ence fiction. Specifically, the kind that constructs detailed, immersive worlds filled with wonderfully subversive characters, and even whole societies, who understand themselves and others differently, who value things differently, and who question their actions with respect to a broader sense of (in)justice. These worlds can still be hierarchical, but *differently* so, with sufficient inversion and detonation to offer critical nourishment coupled with a welcome and sometimes joyful escape from the orders, constrictions, and senseless harms of this earthly world. The fact that many of the protagonists are not recognisably human – being variations of post- and hybrid humans, or are very identifiably nonhuman animals, as in Xenofiction (fictional stories told from a nonhuman animal's point of view) – makes them even more satisfying.

And yet, so often, stories I largely enjoy and appreciate for their critical other-placeness betray an enduring anthropocentrism in the ways animals continue to be used as food, transport, labour, or other kinds of resource.[5] If they are bestowed with any kind of subjectivity, as opposed to being purely figurative or symbolic, they are typically used as human enhancements or extensions, or as narrative devices to add depth to human characters or interrogate the nature of humanity (Bellour, 2015, as cited in Radner & Fox, 2018; Mills, 2017). They can also be entirely absent, and not in any critical or reflective way. Hungry for enrichment, I still scour reviews in search of fiction that promises respite from the exhausting, inescapable repetition of anthropocentric orders but can often be disappointed a few pages in and cast it aside. It has become increasingly apparent that there are few 'caesuras, ruptures, paradoxical spaces, and heterotopia' relating to nonhuman animals to which to give attention. Where sexism, racism, classism, heterosexism, ableism, ageism, and other discriminatory constructs, including even anthropocentrism, are standard critical fare for genres of science fiction and fantasy literature that speculate alternate worlds, speciesism and human exceptionalism in relation to nonhuman animals (rather than other forms of nonhuman) are less so, and are more typically reinforced.

For Anthony Dunne and Fiona Raby (citing Clark, 2011), the role of speculation is to "unsettle the present rather than predict the future" (2013, p. 88). However, speculation provides a fertile medium for heterotopias precisely because it disturbs the present while also providing alternative imaginaries to help anchor disordered schemas and steer them along particular pathways. Whether understood as utopian thinking, situated speculation, radical imagination, counter-conduct, critical subversion, radical resignification, prefiguration or many other terminologies, any imaginary that unsettles the present creates a heterotopian ripple extending outwards, spatially and temporally. However, critical reflection is also required on the intentions and effects of this unsettling. This relates to my earlier point regarding heterotopias being about more than just difference. It is fine to unsettle and disturb but one must ask critically to what end? Speculation always emerges from and is immersed in the world (Halewood, 2017). Commitments to being critical, radical, and transformative should therefore be central. We must, Halewood contends, "stay faithful to this ground, to what enables us to jump, to speculate, at the very moment that we make our jump. We must also consider where we are going to land" (2017, p. 58).

Naturally, art, literature, and other creative media provide especially rich speculative forums to radically bend, contort, and upend normalised realities; explore new realities; and play with the implications. Steve Cooke writes that art and literature "act as moral educators, revealing truths by making new ways of thinking available and opening up new possibilities" (2017, p. 15). Speculative nonhuman heterotopias are less common, but literature and art do offer a number of examples.

To name just a few, within the genre of Xenofiction, there is Adam Roberts' *Bête* (2014) in which animal rights activists insert microchips into animals that allow them to speak, creating escalating tensions between humans and animals and, ultimately, war. In *The Wildings* Nilanjana Roy (2016) creates a more other-immersive world that invites readers to embody non-human and, furthermore, 'stray' ways of being. Finally, Deborah Levy's *Diary of a Steak* (1997) is narrated from the perspective of a disembodied piece of 'meat' lying in a butcher's shop and charts the disintegration of a 'mind that has been symbolically culled.' Others that have stirred and/or delighted my imagination, or discombobulated even my heterotopic expectations, include *The Hollow Kingdom* by Kira Jane Buxton (2019), *The Awareness* by Gene Stone and Jon Doyle (2014), *Fifteen Dogs* by Andre Alexis (2015),[6] *The Animals in that Country* by Laura Jean McKay, *The White Bone* by Barbara Gowdy (2000), and works by Jeff VanderMeer.[7]

Visual media offers the 2017 mockumentary *Carnage*, with which many readers are likely familiar. Set in a vegan future, the film's characters look back with incredulity on today's meat-centred culture. Bong Joon-ho's *Okja*, Darron Aronofksy's *Noah*, and Jordan Peele's *Get Out* also variously mobilise the disruption of the human/animal and other binaries. Taking a similar time-warping approach to *Carnage*, Australian artist Lynn Mowson's *Museum of Forgotten Objects* depicts a future museum displaying an array of objects that have lost their meaning, among them meat claws, bone-handled butcher's knives, bird cages, and egg cups in the shape of tiger, rabbit, and ferret heads. Visitors are imagined pondering the purpose of these testaments to fossilised practices (Shove & Pantzar, 2007).

These attempts to unsettle the present and to mobilise present futures for nonhuman animals raise a mirror to our present arrangements of life, creating a virtual distance and a concrete space of freedom – of anticipatory illumination (Allen, 2015). Among these examples, I must also include works of non-fiction such as George Monbiot's *Feral* (2014) and other eco-literature that offer alternate imaginaries for free-living nonhumans. However, these imaginaries are far from mainstream, as evidenced when reading or watching more popular representations of possible futures or fictional worlds, whether utopian, dystopian, or apocalyptic. Furthermore, and perhaps more concerning, is the similar dearth of nonhuman animals in more formalised practices of forecasting, backcasting, scenario development, and speculative design, widely used by governments, the military, international organisations, private think tanks, corporations, and independent research institutes to prepare for altered and compromised future conditions (politically, economically, socially, and environmentally).

These imaginaries are heavily guided by political and economic interests, and most commonly focus on what are widely constituted as the three key pillars of society: energy, transport, and communication. The environment features primarily in relation to the sustainable use of natural resources (encompassing models for green and blue economies), the preservation of ecosystem services, biodiversity management, safeguarding food supplies, and the insurance of human and planetary health. In these constructs, animals (if they appear at all) are aggregated accordingly as 'nature,' 'wildlife,' or 'livestock.' More holistic models that integrate the social sciences and arts, while capable of generating important, thought-provoking, and inspiring visions, are still often quietly devoid of nonhumans. In fact, thus far, I am aware of only one scenario-planning exercise by Richard Twine (2024) that takes seriously the notion of winding back and dismantling not only an unsustainable, climate-ruining, animal-based food system but the entire animal-industrial complex (A-IC). No others were found that even posit such a scenario, let alone explore the practical implications and opportunities it would generate, and/or provide alternate visions of how humans could live more ethically and carefully with, alongside, and apart from other animals. If they do exist, I apologise for my oversight, though my point regarding their scarcity would still stand.

This visionary vacuum, I argue, is a significant part of the reason why that hackneyed question, some version of 'What would happen to the animals[8] if everyone went vegan?' continues to be trotted out in response to arguments calling for an end to their use. In many ways, it is understandable that alternative ways of conceiving, relating to, and living with nonhumans are so unthinkable. As children and adults, in everything we encounter, we are presented with only one way. Other ways are arrived at often sidelong or by accident – a heterotopic encounter that manages to cause a seismic shift in the solid ground on which we thought we stood. The artifice of normalised and naturalised orders that maintains a wall against the unthinkable might be easier to perceive if there were more tangible and relatable examples that show what being part of a future of liberated human *and* nonhuman animals means – what it looks and feels like, what it involves, requires, demands, and also what it promises. Then, by making other animals "more real and more credible as objects of policy and activism" (Gibson-Graham, 2008, p. 613), it may also be possible to define the features of a map that will help bring this future into being.

CAS and some animal studies scholars are making inroads in this regard (HARNEC, 2015; Mummery & Rodan, 2022; Hodge et al., 2022), their work contrasting with enduringly anthropocentric conceptions of multispecies futures being advanced in a number of fields. New research directions, fields of study, guiding principles, and values are being proposed to underpin explicitly liberatory visions (Freeman, 2020; Oliver et al., 2023; Narayanan, 2023; Meijer, 2023). More tangibly, frameworks are being developed by which other animals might become part of human political and moral communities in ways that acknowledge current injustices and recognise their intrinsic value and individual subjectivities (Donaldson & Kymlicka, 2011; Dunayer, 2013; Blattner et al., 2020). As yet, however, detailed descriptions, prefigurations, and visions of different ways of understanding and relating to all nonhumans in our everyday practices (not only those relating to food) are still underdeveloped but are essential to articulating pathways to their broader realisation.

Shattering Orders

In addition to my imaginings of a vegan pantopia and the proliferation of heterotopias needed to achieve it, and my dissatisfaction with the representation of alternate futures involving nonhumans, a third, more personal and embodied impulse provided the practical framework for the research from which this volume emerged. As is the case for many researchers working in spaces that involve confronting interminable cruelty, violence, pain, suffering, and gross injustices, with no immediate way to stop or even mitigate their effects, the 'work' is not confined to certain hours and can exact a mental and emotional toll. I felt a deep need to insert myself into more positive spaces, to focus for a while on the very happenings I had only speculated about, on the people and animals who were already living in disordered arrangements – to energetically imbibe hope. Somewhat like my search for escapist fiction, I wanted to experience a reprieve from negative thoughts and feelings about humankind, to have these tempered by spending time with those who also rejected (certain) orders and hierarchies, and to grow my own resilience with increased knowledge of what differences were being enacted here and now, and importantly, how.

All the aforementioned were realised, but I had not fully acknowledged the fact that it is where resistance is strongest that mechanisms of power (and sometimes ultimately domination) are also more visible (Foucault, 1994; Palmer, 2001). I was therefore also learning and seeing how certain heterotopias came about, what it took to maintain their integrity, and what can happen at the interface between their resistance and the dominant power mechanisms that surround them.

Humankind has amassed a rich storehouse of critical perspectives on power, its mechanisms, and its unequal effects covering a wide range of historical, social, and cultural perspectives. Coupled with this is a growing record of the physical, psychic, generational, and ecological impacts of power, both in its coercive, disciplinary aspects and as an overt force mobilised in relations of violence (Foucault, 1982). In whatever forms power is exercised and its effects are felt – whether through war, genocide, imperialism, human supremacy,[9] institutional and societal norms, dominant discourses, orders of knowledge, the pursuits of profit, 'progress,' and/or pleasure, environmental degradation and destruction, 'natural' disasters, and of course climate change[10] – its mechanisms vary in their spatial and temporal scope, and in their degree of visibility or recognition (Wadiwel, 2015; Arcari, 2023). Indeed, power permeates our lives and becoming aware of it can be a visceral, embodied experience, whether felt in the numbing resignation or simmering anger generated by repeated encounters with the normalised everyday, in the moment of realisation that you are the subject of power, or the understanding of the power bestowed in you by your social location, and which you are thereby entitled to exercise regardless of the implications for Others.

As explained, Foucault argues that relations of power always contain the potential for domination, where the possibility of the Other's resistance is removed in order to achieve the desired outcome. The rhizomatic roots of human oppression are well theorised – how bodies are shaped by their aforementioned social locations, constituting them as more and less subject to exploitation, marginalisation, disempowerment, cultural imperialism, and violence (Young, 1990). Animalisation permeates these social locations, reinforcing understandings of entitlement (to oppress), and legitimating ideologies of supremacy. The integration of nonhuman animals into speculative scenarios and visions for more just, liveable, and sustainable futures would therefore seem not just logical but necessary.

Yet, despite more and more people[11] experiencing first hand and/or being affected by power's mechanisms and effects, and being able to articulate, in nuanced and incisive ways, the unjust social and structural arrangements that constitute it, humankind remains staunchly resistant to extending this knowledge and understanding to include other animals.

Indeed, numeric, textual, and visual accounts of the mechanisms and techniques of power employed against animals (Arcari, 2022) and the scale of their effects, presented in mainstream news, advocacy campaigns, documentaries, and films, have had no measurable impact on animals' escalating persecutions even in the 15–20+ years since the internet expanded the number and reach of platforms for their dissemination. Academic and literary works that articulate the logical, moral, and ethical inconsistency of *not* conceiving nonhuman animals as equally subject to mechanisms of power and techniques of oppression have also had little meaningful impact on how animals are understood, even within closely aligned critical disciplines.

Amidst increasing evidence of the connections between worsening situations for nonhuman animals and worsening human and planetary health, and the threats both pose to the stability of social, political, and economic structures, and to the basic conditions of life, scholars and advocates continue to revise (upwards) the figures that chart the scale of crises for nonhuman animals and the world's ecologies (Arcari, 2019; Jahren, 2020; Twine, 2024). However, these threats to life and ways of being are distributed unequally, with those who benefit from, and fervently uphold, normalised hierarchies having secured the means to shield themselves (literally) from their effects and advantageously conceal and/or downplay the dangers. Indeed, more time and money are being spent on efforts to realise technological imaginaries of space colonisation than on mitigating and repairing the damages wrought by humans on this planet. Drawing on Cultural Studies scholar Sarah Sharma (2017), literary journalist Mark O'Connell views this narrative of planetary escape as an exercise in patriarchal power "at the expense of a politics of care" (2020, p. 104) – a desperate impulse to assure a future for "terminal stage capitalism" (115) and also human supremacy. Starting afresh with purer models of a 'free' society controlled by sovereign individuals.

It stands to reason, then, that under the weight of a long and profitable history of colonialism, rationalism, human supremacy, and capitalism, augmented by their related ideologies of growth and progress, attempts to challenge and dismantle the reified orders on which they are built are met with forceful resistance. Those whose lives have always mattered more deploy their power to ensure that the majority of people trust in the shadows cast by them on the cave walls, believing any other purported reality to be madness – unthinkable. They depend on the duplication of these representations to defend their positions against the censure of socially dislocated Others.[12]

However, the 10% of the world's population who own over 75% of household wealth (Chancel et al., 2022), and the additional percentage who orbit this wealth to benefit politically, economically, and socially, are not able to suppress the exchange of thoughts, knowledge, experience, and ideas that supports dissent and subversive creativity. Throughout history, this creativity has found ready outlets in satire, comedy, theatre, art, storytelling, literature, activism, and critical analysis. Lacunae may persist, but as a collective zeitgeist gathers pace towards the depiction of more just futures, critical questions cannot just increasingly be asked but *heard* regarding whose lives matter in our shared futures; who counts as an 'individual' whose subjectivity deserves recognition (Higgins & Leps, 2022); and who, therefore, is part of the 'social' and included among the orders and hierarchies to be dismantled (Hetherington, 1997).

Chapter Outlines

The imaginaries this book calls forth and hopes to inspire are intended to generate serious reflection on these questions, and instigate the purposeful crafting of a stockpile of responses to that hackneyed question, showing exactly what *could* happen to all the other animals. The heterotopias gathered here reflect one or more of the three affordances I came to recognise in Foucault's concept:

1. Allowing forms of vegan, non-/post-human, or post-animal (Stanescu & Twine, 2012) *heterotopias* to be recognised, created, documented, learned from, and ideally replicated.
2. Generating spaces for *radical representations and imaginaries* of other ways of understanding and relating to other animals.
3. Creating openings for embodied experiences of disorder in relation to other animals, with the potential for the transformative dissolution or *shattering of associated hierarchies* and their mechanisms of power.

The collection comprises 14 contributions in total, organised into four sections of three to four chapters each. The sections are not clearly defined or discrete, and many chapters could easily belong in two or more sections. Section 1 explores physical sites of relational reconfiguration as examples of potential present futures being enacted in public and semi-public spaces. The next three sections offer different kinds of pathways to liberatory animal futures. The first set of pathways presents anti-anthropocentric and non-dualistic ways of thinking about and orienting ourselves bodily in relation to other animals. Another means of making the possibility of non-anthropocentric worlds thinkable is through disruptive/disordered storytelling, which is the focus of the next section. In the final section, a third possibility for liberatory disruption is conceived in moments and situations that generate a profound awareness of the human–animal duality, inviting a transformative experience of re-ordering in the relation of self to Other.

Section 1: Relational Reconfigurations in Present Futures

The journey through disorder begins in Chapter 1 with Elizabeth Tavella, who examines tensions arising in three types of heterotopic space where "multispecies relations are intentionally recalibrated through multidimensional and decolonial acts of resistance" (p. 26) – specifically, (micro)sanctuaries, food/gardening movements, and foraging initiatives. It is the inevitable entanglement of these sites with prevailing orders that constitutes these tensions. Thus, drawing on Brambilla's theorisation of bordering as a practice and the borderscape as a "space for liberating political imagination" (2015, p. 18), Tavella avoids the binary tendency to idealise heterotopias and explores the messiness and opportunities surrounding the emergence of new orders. Persistent power dynamics, dualisms, and capitalist property arrangements, as well as incomplete notions of justice are freedom, are examined through a deft interweaving of vignettes spanning Italy, the UK, the US and beyond that depict real sites striving for collective liberation. Ultimately, Tavella views borderland tensions as generative and capable of reverberating heterotopic relational ecologies more widely, both spatially and temporally.

The next chapter, by Maria Martelli (Chapter 2), invites readers to zoom in on the efforts of one sanctuary-giver in Romania to provide care and sanctuary for 'feral' pigeons – birds whose commensal existence, Martelli argues, demands that we re-think binaries. What does liberation mean for animals who appear to 'thrive' amongst humans? Martelli uses this question to highlight limits/exclusions in conceptions of nonhuman oppression, and the lack of care and also violence that can result for those who are thus invisibilised. Responding to Westerlaken's (2021) call to create multispecies worlds, Martelli explores what care-ful as opposed to care-less urban design might entail for pigeons, emphasising that this involves centering non-human interactions and entanglements while being mindful to not reinforce paternalistic models of care. The ancient technology of the dovecote or pigeon tower is used to show what such 'thinking and feeling beyond anthroparchy' might mean. Once a tool to facilitate their exploitation, it is repurposed as a place of shelter, careful care, education, and also reparation, to become a "transitional symbol from violence to multispecies world-making" (p. 53).

In this section's third chapter, Marie Leth-Espensen (Chapter 3) echoes questions raised by O'Connell and others regarding the configuration of care under capitalism. However, like Martelli, she also notes the potential for care work to be reparative. Focusing on two farmed animal sanctuaries in Denmark, Leth-Espensen examines how the work of care is constituted in these 'troubled domestic domains' shaped by past and ongoing processes of domestication. Similar to Martelli's pigeons, the ways in which resident animals find themselves in need of sanctuary reveal dimensions of nonhuman oppression that are largely ignored – violences that are less overt but equally demonstrative of other animals' disposability. However, once under sanctuary, human-centric and paternalistic care frameworks and/or the imposition of dominant orders, knowledge, and legislation often compromise the type of care these animals experience. This is especially foregrounded in prohibitions surrounding practices of mourning (human and nonhuman) and transformative care. Yet, despite these dilemmas, and taking inspiration from Donaldson and Kymlicka's (2015) vision for intentional interspecies communities, Leth-Espensen emphasises the value of sanctuaries as radical sites for reclaiming 'care' and working through the complexities of interspecies justice.

The section concludes with Chapter 4 by Paula Arcari, which zooms out again to explore the liberatory promise of seven UK-based (micro)sanctuaries, including three vegan animal sanctuaries, two non-ridden equine communities, and two rescue-chicken communities. Focusing on their material arrangements, constitutive practices, and inter- and intra-species relations, Arcari examines the potential for these heterotopias to prefigure non-anthropo-capitalo-centric futures. Arcari's analysis respectively foregrounds different human understandings of, and ways of being/not being with, nonhumans; encounters/confrontations with dominant understandings and the opening of 'lines of fragility' (echoing the borderscapes notion from Tavelli's chapter); and the risk that 'palatable disruptions' may reinforce normalised orders. Notwithstanding limitations in scope, scale, and/or political intent, these heterotopic spaces offer opportunities to 'glitch' common sense. On that basis, Arcari conceives them as essential and complementary components of a united advocacy movement, offering *pre*-liberatory prefigurations of, and embodied, transformative pathways towards, radically alternate futures.

Section 2: Conceptual and Political Re-ordering

Co-authored by Angie Pepper and Richard Healey, the first chapter in this section (Chapter 5) guides readers through a critical reflection on the structural injustice underpinning humans' relationships with domesticated animals. The aim is to challenge popular visions and frameworks for just interspecies communities that are modelled on these relationships. Pepper and Healey's enquiry is vividly contextualised by the authors' observations of a pair of Eurasian magpies. These liminal animals, belonging to no-one and choosing the kinds of interactions they have with humans, raise important and potentially disruptive questions regarding how our relations with domesticated animals come to be justified as benign and even benevolent, eliding the "extensive exercises of human power and control' (p. 105) to which these animals are subject. This power, Pepper and Healy explain, is illegitimate because it compromises self-determination and is based on an asymmetric relationship of dependence that is deliberately created and therefore morally objectionable. Addressing anticipated objections, Pepper and Healey argue that narratives of just domestication are flawed and that transitioning towards liminality is the more moral pathway to realising just relationships with other animals.

The second of this section's chapters (Chapter 6) by Michelle Westerlaken similarly unsettles normalised human-centric perspectives on our relations with domesticated animals, further problematising their instrumentality with a specific focus on dogs. Westerlaken revisits previous playful design projects involving two canine companions, reassessing their affordances through the lens of Akama and Light's (2018) notion of sensitising and attuning as applied to complex interactions. This creates a receptivity to how different histories and personalities shape the rhythm of interactions, and to acknowledging the multi- (and more-than-human) sensory capacities and experiences that other subjectivities bring to an encounter. Westerlaken beautifully describes how this enhanced sensitivity unfolds around a project involving smells and soundscapes. Spaces generated through playful encounters are hence conceived as heterotopic forums that invite sensitivity and attunement to the proposals of Others, thereby modelling the kinds of interspecies negotiations and participatory politics on which multispecies futures can be modelled. Rather than asking *if* and *how* other animals can participate in this future, we should rather learn to recognise "their already existing participation and negotiations" (p. 119).

Charlotte A. Kunkel and Scott Hurley conclude this section (Chapter 7), with a proposal for mutual interdependence that "creates a responsibility of 'being with' and accountable for all beings" (p. 132). Their proposal rejects citizenship and rights-based approaches as models for equitable futures, being part of ideologies that reinforce rather than challenge the status quo. Instead, informed by Lori Gruen's entangled empathy and Thich Nhat Hanh's notion of 'interbeing,' Kunkel and Hurley envision intimately relational spaces of freedom that promote interconnectedness and 'bodily compassion.' Addressing common understandings of Buddhism as primarily aimed at personal liberation, the authors draw on Black liberation theology and the Black Buddhism Movement to articulate an ethic of responsibility, as opposed to care, that encompasses all dimensions of oppression. An ethic of responsibility, they argue, demands not only the deconstruction of existing structures but also the creation of new ones that are regenerative and based on mutual interdependence. Kunkel and Hurley provide examples of such 'radical equitable heterotopias' drawn from feminist science fiction and also real life, noting in each case the embodiment and practice of mutual interdependence and responsibility required for the realisation of transformational futures.

Section 3: Subversion Through Radical Storytelling and Restorying

The volume's third section begins in Chapter 8 with Daniel Vandersommers' forensic disassembly of a nineteenth-century US aquarium in which he exposes how the abstraction of this highly ordered microcosm erases the far-flung extractions and exploitations required for its creation. Vandersommers dissects these externalised material processes to effectively restory the idealised aquarium, disclosing everything required, then and now, to reinforce and sustain the 'aquarial gaze.' Drawing on archives to reconstruct the historical process of aquarium-building Vandersommers takes the reader on an intriguing journey as he reveals how water, plants, nonhuman animals, rocks, and sediments were acquired and contained. This heterotopic restorying also discloses the associated supply channels and networks generated by these aquarial demands, and the human and nonhuman hierarchies that constitute them. Far from being in the past, these practices of extraction, accumulation, containment, and domestication are only intensified today. Vandersommers notes that "opening aquaria – imagining otherwise . . . means unbuilding the ideal" (p. 163), and it is in this sense that his restorying constitutes a heterotopia, one that seeks to 'shatter the contrivance' of the aquarium and allow "its mystical deceits" (p. 161) to spill out.

With Chapter 9, the focus of Section 3 shifts from restorying to radical story-telling as Amina Grunewald explores the zooheterotopian anarcho-geography presented by Jane Doe in her novel-fable *Anarchist Farm*. Recognising the potential of the novel-fable, as a genre, to mobilise counter-narratives that disturb prevailing orders, Grunewald argues that Doe's depiction of an anarchic multispecies *and* multivocal community serves as an antispeciesist intervention. Grunewald notes how Doe leverages readers' familiarity with the anthropomorphic conventions of the fable and with Orwell's *Animal Farm*, on which her story is based, to draw them into a more radical vision of animal liberation that underscores the structural power imbalance between humans and other animals. Against a discussion of other animal-centred stories, Grunewald unpacks how nonhumans are variously represented and, informed by Springer's (2016) 'Anarchist Geographies' and Miller's (2015) 'Zooheterotopia,' notes the value of nonhuman protagonists to act as "lenses to witness, teach, and deplore . . . facets of a *conditio humana*" (p. 172). The fictional worlds in which these nonhumans are imagined, Grunewald suggests, offer zooheterotopian spaces for the negotiation of alternate futures of "creaturely posthumanist co-existence" (p. 179).

Staying with radical storytelling, Sutirtho Roy focuses in Chapter 10 on animated films and specifically what he terms Gothic enviro-toons characterised by eco-critical narratives, animated characters, and a prevailing sense of mortal unease and/or terror. Highlighting a persistent anthropocentrism in popular animated enviro-toons that depict happy endings for their nonhuman characters, Roy draws on theorisations of anthropomorphism to show how three Gothic enviro-toons – *Watership Down*, *The Plague Dogs*, and *Padak* – subvert this convention. By eliciting viewer's empathy with their animated nonhuman protagonists, coupled with graphic depictions of the violences and terrors to which they are subjected at the hands of humans, the Gothic heterotopias depicted in these films challenge dominant and normalised conceptions of other animals. Acknowledging scholarship that questions the value of moral shocks for (animal) advocacy, Roy assesses each film in turn, analysing their Gothic elements in the context of the anthropocentric gaze they are seeking to disrupt. While jeremiadic storytelling offers a powerful medium for raising awareness, Roy suggests that hope may also be necessary to "unlock their potential as vessels of change" (p. 195).

Featuring elements of both radical storytelling and restorying, the final chapter in this section by Katya Krylova (Chapter 11) focuses on the subversive capacities of comedy and specifically satire. Ironic representations of cute nonhuman animals are prolific on social media. However, the majority of this content, Krylova highlights, leaves dominant constructs of human superiority unchallenged. Drawing on theorisations of superior, radical, and self-directed humour, Krylova uses Brooke Barker's @sadanimalfacts Instagram comics, along with other examples, to illustrate how posthumanist humour can defamiliarise the mundane, "make animal suffering seem extraordinary" (p. 204) and thereby unsettle the social orders that deem it ordinary. Krylova extends the possibilities afforded by these comedic heterotopias with a speculative utopia centred on an animated comedy series featuring diverse groups of animals in a guest-discussion format. The imagined series is produced by a network aiming to shift human conceptions of other animals by creating fictional narratives that prioritise the animal gaze. Conceived as a broad countercultural intervention, Krylova envisions a transformative pathway towards non-conventional futures through the affirmation of "multiple nonhuman perspectives as a new standard of content production" (p. 211).

Section 4: Personal Shifts and Transformation

The final section in this collection invites readers to consider the kinds of changes that appear to be at first sensed more than thought – a stirring of the body and/or emotions that becomes an invitation to make sense of the situation. Though more difficult to articulate, words being often inadequate, the effects of these experiences, and hence the significance of the heterotopic spaces in which they exist as possibilities, can be profound and life altering.

The section opens (Chapter 12) with Alexandra Isfahani-Hammond's exploration of altered states of consciousness that characterise fictional cross-species alliances, as depicted in Andre Alexis' novel *Fifteen Dogs* and a classic Sufi text *Layla and Majnun*. Isfahani-Hammond describes this supposed 'madness' as reverberating through history, and delicately threads through this analysis her reflections on the confusion/dissolution of identities she experienced as she ministered to her dog friend during his final months. The fictional heterotopic spaces where boundaries between human and nonhuman become blurred are thereby anchored to a present heterotopia as Isfahani-Hammond similarly resists the pathologisation of levels of nonhuman care considered to transgress normalised orders. Rather than internalising such 'transgressions,' Isfahani-Hammond encourages remaining in a state of 'bewilderment' where rational faculties are suspended. Far from indicating madness, bewilderment can reveal the artifice of separation under human exceptionalism, allowing for an attunement to a 'post-anthropocentric consciousness.' Isfahani-Hammond argues that this altered state, of madness/bewilderment, offers "a healing pathway to obliterating species dialectics, compelling humans to discard their arms" (p. 223).

In the second chapter of this section (Chapter 13), Sue Hall Pyke evocatively recounts her efforts to suspend/bewilder her normalised conceptions of, and approaches to, free-living snakes with whom she co-exists in a rural Australian heterotopia. Conceiving her efforts as finding ways to 'unhamper' their lives, Pyke draws on theorisations of personhood, 'liberties,' and oppression to describe lives characterised by more than the absence of harm. Pyke rejects Western-centric formulations of post-anthropocentric justice, turning to Tsing's (2011) notion of 'passionate immersion' defined as "a love unbounded by species, driven by curiosity, focused on reciprocity without demands" (p. 238). Drawing on First Nations cosmologies, theorisations of posthuman cosmopolitanism, and situated knowledges, this passionate curiosity, Pyke suggests, means reanimalising understandings of human wellbeing. More practically, it means minimising the risk she presents to the snakes and recognising that she, and other humans, exist at the margins of the snakes'· 'relational validities.' In aiming for a just future of unhampered hospitality, Pyke offers responsive hospitality as a way she can support the snakes' "ability to live towards a self-actualisation that is not mine to know" (p. 243).

Bringing this section and the volume to a close, the final chapter (Chapter 14) by Jennifer Rebecca Schauer demonstrates the capacity of a new pedagogy of sharing multispecies sentience to disrupt students' normal ways of thinking about and relating to other animals. Schauer draws on feminist scholarship, critical (animal) pedagogies, and religious teachings to support the grounding of this non-hierarchical, antispeciesist pedagogy in love and compassion. Acknowledging aligned pedagogies, Schauer explains that this new pedagogy is distinguished by an emphasis on "the intuitive, embodied wisdom of each individual student" (p. 252), thereby challenging the primacy of external knowledge. After describing the framework in more detail, including what it entails in practice – mindfulness practices, field observations, readings, reflections, and speculative imaginaries – Schauer shares the transformative outcomes from two groups of students led through an Animals and Society course using the new pedagogy. Students' transformative experiences within these pedagogical heterotopia are the result, Schauer argues, of the grounding in love and compassion, which fosters an embodied and deeply felt connection with nature and other animals that can help guide students towards the co-creation of liberatory futures.

This collection addresses only a fraction of the animal heterotopias that exist, can be imagined, and are needed to generate meaningful and resolute responses to all the anthropo-capito-centric questions of 'What would happen if . . . ?' Our hope is that it inspires thinking about the myriad spaces that can be conceived as heterotopia, and the radical liberatory imaginaries they can set in motion.

Notes

1 Research on fungi suggests human conceptions of sentience may be limited and subject to revision. This would have repercussions for associated legislation and demonstrates the precarious nature of protections based on changeable human-centric categories (Money, 2021; Reber & Baluška, 2021).

2 As Siobhan O'Sullivan clearly articulates in her influential work *Animals, Equality and Democracy* (2011), human-determined categories are slippery and changeable, and can work alongside and above legislation to exclude certain animals from protection.

3 Robinson is said to have been influenced by the work of moral and political philosopher Mary Wollstonecraft, whose observations regarding interconnected structures of domination based on gender, class, and species are in turn credited with anticipating contemporary ecofeminism (Seeber, 2014).

4 We do not feel a sufficient tension with plants to necessitate a human/plant binary, in addition to the human/nature binary.

5 Not wishing to draw critical attention to works that I otherwise hold in high regard, suffice to say here that my reading has encompassed leading and new/less well-known authors across many subgenres of science fiction, including those characterised as hard and soft science, utopian/dystopian, apocalyptic, cli-fi, solarpunk, feminist and queer, and black speculative fiction (see Booker & Thomas, 2009).

6 Thanks to Alexandra Isfahani-Hammond for directing me to this book.
7 Some short stories in recent solarpunk collections also warrant mention. See, for example, *Multispecies Cities: Solarpunk Urban Futures* (Rupprecht et al., 2021) and *Solarpunk Creatures* (Rupprecht, 2024). However, humanity and the human POV still tend to play a central if not leading role in these imaginaries.
8 Substitute here 'pets,' 'zoo animals,' 'farm animals,' 'racehorses,' other categories of animals, and also specific species and breeds as relevant.
9 Drawing on Lupinacci (2015, p. 180), human supremacy is understood not only as the domination of the human over all other life, but the supremacy of a particular idealised human constituted by "dominant Western industrial culture," that is, white, male, civilised, heterosexual, etc.
10 Note these are typically interlocking and compounding forms of power.
11 The global population has more than doubled in the 50 plus years since I was born, and the spatial and temporal extent of power's reach has also grown.
12 Other animals are included in the global 'social.'

References

Akama, Y., & Light, A. (2018). Practices of Readiness: Punctuation, Poise and the Contingencies of Participatory Design. *Proceedings of the 15th Participatory Design Conference, ACM Press, 13*, 1–11.

Allen, A. (2015). Emancipation Without Utopia: Subjection, Modernity, and the Normative Claims of Feminist Critical Theory. *Hypatia, 30*(3), 513–529.

Anfinson, A. (2020). Hong Kong's Paper Cities: Heterotopia and the Semiotic Landscape of Civil Disobedience. In D. Malinowski & S. Tufi (Eds.), *Reterritorializing Linguistic Landscapes* (pp. 137–159). Bloomsbury Academic.

Annist, A. (2013). Heterotopia and Hegemony: Power and Culture in Setomaa. *Journal of Baltic Studies, 44*(2), 249–269.

Apffel-Marglin, F. (2020). Co-Creating Commons with Earth Others: Decolonizing the Mastery of Nature. *Mester, 49*, 65–88.

Arcari, P. (2019). *Making Sense of 'Food' Animals: A Critical Exploration of the Persistence of 'Meat'*. Palgrave Macmillan.

Arcari, P. (2022). (Animal) Oppression: Responding to Questions of Efficacy and (Il)Legitimacy in Animal Advocacy with a New Collective Action/Master Frame. *Animal Studies Journal, 11*(2), 68–108.

Arcari, P. (2023). Slow Violence Against Animals: Unseen Spectacles in Racing and at Zoos. *Geoforum, 144*(1), 103820. https://doi.org/10.1016/j.geoforum.2023.103820.

Arcari, P. (2024). (More than) Food, Farms, and Freedom: Turning Exclusions in 'Pro-Vegan' Documentaries into Productive Interventions for Animal Advocacy. In C. Parkinson & L. Herring (Eds.), *Animal Activism On and Off Screen* (pp. 111–136). Sydney University Press.

Bailey, M. M., & Shabazz, R. (2014). Editorial: Gender and Sexual Geographies of Blackness: Anti-Black Heterotopias (Part 1). *Gender, Place & Culture, 21*(3), 316–321.

Barad, K. (2007). *Meeting the Universe Halfway: Quantum Physics and the Entanglement of Matter and Meaning*. Duke University Press.

Beckett, A. E., Bagguley, P., & Campbell, T. (2017). Foucault, Social Movements and Heterotopic Horizons: Rupturing the Order of Things. *Social Movement Studies, 16*(2), 169–181.

Blattner, C. (2019). The Recognition of Animal Sentience by the Law. *Journal of Animal Ethics, 9*(2), 121–136.

Blattner, C., Coulter, K., & Kymlicka, W. (Eds.). (2020). *Animal Labour: A New Frontier of Interspecies Justice?* Oxford University Press.

Booker, M. K., & Thomas, A.-M. (2009). *The Science Fiction Handbook*. John Wiley & Sons Ltd.

Botting, E. H. (2020). Editor's Introduction: Reading Wollstonecraft's a Vindication of the Rights of Woman, 1792–2014. In M. Wollstonecraft (Ed.), *A Vindication of the Rights of Woman* (pp. 1–18). Yale University Press.

Brambilla, C. (2015). Exploring the Critical Potential of the Borderscape Concept. *Geopolitics*, *20*, 14–34.

Chancel, L., Piketty, T., Saez, E., & Zucman, G. (2021). World Inequality Report 2022. *World Inequality Lab*. https://wir2022.wid.world/

Cooke, S. (2017). Imagined Utopias: Animals Rights and the Moral Imagination. *Journal of Political Philosophy*, *25*(4), e1–e18.

Daly, J. (2020). A Social Practice Perspective on Meat Reduction in Australian Households: Rethinking Intervention Strategies. *Geographical Research*, *58*(1), 240–251.

Dehaene, M., & De Cauter, L. (2008). *Heterotopia and the City*. Routledge.

Donaldson, S., & Kymlicka, W. (2011). *Zoopolis*. Oxford University Press.

Donaldson, S., & Kymlicka, W. (2015). Farmed Animal Sanctuaries: The Heart of the Movement? *Politics and Animals*, *1*(1), 50–74.

Dunayer, J. (2013). The Rights of Sentient Beings Moving Beyond Old and New Speciesism. In R. Corbey & A. Lanjouw (Eds.), *The Politics of Species: Reshaping Our Relationships with Other Animals* (pp. 27–39). Cambridge University Press.

Dunne, A., & Raby, F. (2013). *Speculative Everything: Design, Fiction, and Social Dreaming*. MIT Press.

Edwards, G. A. S., & Bulkeley, H. (2017). Heterotopia and the Urban Politics of Climate Change Experimentation. *Environment and Planning D: Society and Space*, *36*(2), 350–369.

Foucault, M. (1967). Of Other Spaces: Utopias and Heterotopias. *Architecture/Mouvement/ Continuite*, October, 1–9. https://web.mit.edu/allanmc/www/foucault1.pdf

Foucault, M. (1982). The Subject and Power. *Critical Inquiry*, *8*(4), 777–795.

Foucault, M. (1989 [1966]). *Order of Things: An Archaeology of the Human Sciences*. Routledge.

Foucault, M. (1994). *Ethics, Subjectivity and Truth*. The New Press.

Freeman, C. P. (2020). *The Human Animal Earthling Identity: Shared Values Unifying Human Rights, Animal Rights, and Environmental Movements*. University of Georgia Press.

Gaard, G. (2004). Toward a Queer Ecofeminism. In R. Steiin (Ed.), *New Perspectives on Environmental Justice: Gender, Sexuality, and Activism* (pp. 21–44). Rutgers University Press.

Gibson-Graham, J. K. (2008). Diverse Economies: Performative Practices for 'Other Worlds'. *Progress in Human Geography*, *32*(5), 613–632.

Halewood, M. (2017). Situated Speculation as a Constraint on Thought. In A. Wilkie, M. Savransky, & M. Rosengarten (Eds.), *Speculative Research: The Lure of Possible Futures* (pp. 63–124). Routledge.

Haraway, D. (2004 [1992]). The Promises of Monsters: A Regenerative Politics for Inappropriate/d Others. In D. Haraway (Ed.), *The Haraway Reader* (pp. 63–124). Routledge.

HARNEC. (Ed.). (2015). *Animals in the Anthropocene: Critical Perspectives on Non-Human Futures*. Sydney University Press.

Heise, U. K. (2013). Globality, Difference, and the International Turn in Ecocriticism. *PMLA*, *128*(3), 636–643.

Hetherington, K. (1997). *The Badlands of Modernity: Heterotopia and Social Ordering*. Routledge.

Higgins, L., & Leps, M.-C. (2022). *Heterotopic World Fiction: Thinking Beyond Biopolitics with Woolf, Foucault, Ondaatje*. Academic Studies Press.

Hodge, P., McGregor, A., Springer, S., Veron, O., & White, R. (Eds.). (2022). *Vegan Geographies: Spaces Beyond Violence, Ethics Beyond Speciesism*. Lantern Books.

Hook, D. (2007). *Foucault, Psychology and the Analytics of Power*. Palgrave Macmillan.

Jahren, H. (2020). *The Story of More*. Fleet.

Johnson, P. (2013). The Geographies of Heterotopia. *Geography Compass*, *7*(11), 790–803.

Knight, K. T. (2017). Placeless Places: Resolving the Paradox of Foucault's Heterotopia. *Textual Practice*, *31*(1), 141–158.

Ko, S. (2017). Addressing Racism Requires Addressing the Situation of Animals. In A. Ko & S. Ko (Eds.), *Aphro-Ism: Essays on Pop Culture, Feminism, and Black Veganism from Two Sisters* (pp. 44–48). Lantern Books.

Lent, J. (2021). *The Web of Meaning: Integrating Science and Traditional Wisdom to Find Our Place in the Universe*. New Society Publishers.

Littler, J. (2018). *Against Meritocracy: Culture, Power and Myths of Mobility*. Routledge.

Lupinacci, J. (2015). Recognizing Human-Supremacy: Interrupt, Inspire and Expose. In A. J. Nocella II, R. J. White, & E. Cudworth (Eds.), *Anarchism and Animal Liberation: Essays on Complementary Elements of Total Liberation* (pp. 179–193). McFarland and Company, Inc., Publishers.

MacKinnon, C. A. (2007 [1999]). *Are Women Human? And Other International Dialogues*. Harvard University Press, Belknap Press.

Meijer, E. (2023). Global Injustice and Animals: Towards a Multispecies Social Connection Model. *International Relations*, *37*(3), 497–513.

Miller, J. (2012). In Vitro Meat: Power, Authenticity and Vegetarianism. *Journal for Critical Animal Studies*, *10*(4), 41–63.

Miller, J. (2015). Zooheterotopias. In M. Palladino & J. Miller (Eds.), *The Globalization of Space: Foucault and Heterotopia* (pp. 149–164). Pickering & Chatto.

Mills, B. (2017). *Animals on Television: The Cultural Making of the Non-Human*. Palgrave Macmillan.

Money, N. P. (2021). Hyphal and Mycelial Consciousness: The Concept of the Fungal Mind. *Fungal Biology*, *125*(4), 257–259.

Morehouse, K. N., Maddox, K., & Banaji, M. R. (2023). All Human Social Groups Are Human, but Some Are More Human than Others: A Comprehensive Investigation of the Implicit Association of "Human" to US Racial/Ethnic Groups. *Proceedings of the National Academy of Sciences*, *120*(22), e2300995120.

Mummery, J., & Rodan, D. (2022). *Imagining New Human-Animal Futures in Australia*. Peter Lang.

Narayanan, Y. (2023). For Multispecies Liberatory Futures: Three Principles Toward "Progress" in Anti-Anthropocentric Environmental Geography. *Progress in Environmental Geography*, *2*(3), 179–190.

O'Connell, M. (2020). *Notes from an Apocalypse: A Personal Journey to the End of the World and Back*. Granta.

Oliver, C., Turnbull, J., & Richardson, M. (2023). Claiming Veganism and Vegan Geographies. *The Geographical Journal*, *190*(1), e12546.

O'Sullivan, S. (2011). *Animals, Equality and Democracy*. Palgrave Macmillan.

Palladino, M., & Miller, J. (Eds.). (2015). *The Globalization of Space: Foucault and Heterotopia*. Routledge.

Palmer, C. (2001). 'Taming the Wild Profusion of Existing Things'? A Study of Foucault, Power, and Human/Animal Relationships. *Environmental Communication: A Journal of Nature and Culture*, *23*, 339–358.

Plumwood, V. (1993). *Feminism and the Mastery of Nature*. Routledge.

Radner, H., & Fox, A. (2018). *Raymond Bellour: Cinema and the Moving Image*. Edinburgh University Press.

Reber, A. S., & Baluška, F. (2021). Cognition in Some Surprising Places. *Biochemical and Biophysical Research Communications*, *564*, July, 150–157.

Saldanha, A. (2008). Heterotopia and Structuralism. *Environment and Planning A, 40,* 2080–2096.

Sexton, A. E. (2016). Alternative Proteins and the (Non)Stuff of "Meat". *Gastronomica, 16*(3), 66–78.

Shantz, J. (2012). Spaces of Learning: The Anarchist Free Skool. In R. H. Haworth (Ed.), *Anarchist Pedagogies: Collective Actions, Theories and Critical Reflections on Education* (pp. 124–144). PM Press.

Sharma, S. (2017). Exit and the Extensions of Man. *Transmediale.* https://archive.transmediale.de/content/exit-and-the-extensions-of-man

Shove, E., & Pantzar, M. (2007). Fossilisation. *Enthnologia Europaea, 35*(1–2), 59–62.

Soja, E. (1996). *Third Space: Journeys to Los Angeles and Other Real and Imagined Spaces.* Blackwell.

Springer, S. (2013). Anarchism and Geography: A Brief Genealogy of Anarchist Geographies. *Geography Compass, 7*(1), 46–60.

Stanescu, J., & Twine, R. (2012). Post-Animal Studies: The Future(s) of Critical Animal Studies. *Journal for Critical Animal Studies, 10*(4), 4–19.

Tsing, A. L. (2011). Arts of Inclusion, or, How to Love a Mushroom. *Australian Humanities Review, 50,* 5–22.

Twine, R. (2024). *The Climate Crisis and Other Animals.* Sydney University Press.

Wadiwel, D. (2015). *The War Against Animals.* Brill.

Westerlaken, M. (2021). What Is the Opposite of Speciesism? On Relational Care Ethics and Illustrating Multi-Species-Isms. *International Journal of Sociology and Social Policy, 41*(3/4), 522–540. https://doi.org/10.1108/IJSSP-09-2019-0176

Wise, S. M. (2013). Nonhuman Rights to Personhood. *Pace Environmental Law Review, 30*(3), 1278.

Wolfe, C. (2003). *Animal Rites: American Culture, the Discourse of Species, and Posthumanist Theory.* University of Chicago Press.

Yampell, C. (2008). When Science Blurs the Boundaries: The Commodification of the Animal in Young Adult Science Fiction. *Science Fiction Studies, 35*(2), 207–222.

Young, I. M. (1990). *Justice and the Politics of Difference.* Princeton University Press.

Section 1

Relational Reconfigurations in Present Futures

DOI: 10.4324/9781003366706-2

Sites of Vegan Placemaking

A Celebration of Multispecies Alliances at the Borderlands

Elizabeth Tavella

In the video artwork *New World Order*, Hayden Fowler depicts a surreal forest of leafless trees animated by a group of heritage-bred chickens preening on a branch, scratching the scabby ground, or slipping under the tree roots into underground burrows. Their unusual plumage is markedly different from their domesticated kin, increasing the eeriness of the already post-apocalyptic atmosphere, and thus even further destabilizing the expectations of viewers. The monochromatic, hazy backdrop is meant to evoke a post-human environment in stark contrast with the bright hues and deceitful idyllic scenes of conservation estates, where 'nature' is kept in stasis with proofed fences by self-elected human protectors. Situated at the convergence of dystopia and utopia, Fowler builds a narrative of nonhuman survival immersed in a landscape where a new ecology takes shape, where a desire for resilient living gives rise to imagined alternative paradigms (Fowler, 2015, p. 247).

Immersed in a similar – but greener – atmosphere, I experienced the same feeling of alienation mixed with wonder as I was walking through a forest in the Dominican Republic, near the Laguna de Oviedo. Unexpectedly, I ran into a group of cows lying down or grazing in the shadow of trees. While the only trace left of humans was the plastic and garbage surrounding the patch of grass, which echoed the post-human ambience of Fowler's video, my physical presence enhanced the possibility of mutual familiarization and collaborative participation in the making of our shared world. I quietly sat down at a distance, hoping that my presence wouldn't disrupt their tranquility and, most importantly, that they wouldn't perceive me as a threat. I had heard with great awe of the semi-feral cows at VINE Sanctuary living in the forest at the outskirts of the pasture, largely avoiding contact with humans; yet, this encounter, in a non-surveilled space, felt even more extra-ordinary.[1] Three cows paused their grazing to look at me (Figure 1.1). Apparently, they lived on their own terms: no ear tags, no marks of ownership. And the younger cows were bonded to their feeding parent. As our gazes met, I envisioned them as a free community that responded uniquely to their own needs and interests, without having to subdue to any forced spatial boundary. For a moment, I forgot that just that same morning I had seen several small trucks on the roads, each transporting three or four cows, most likely to slaughter. As if existing out of time, this

DOI: 10.4324/9781003366706-3
This chapter has been made available under a CC-BY-NC-ND 4.0 license.

Figure 1.1 *Reciprocally Seen*. Photograph by the author (2019).

unexpected encounter generated speculative *ruminations*, leading to shared dreams of embodied hope and imaginative leaps into vegan placemaking.

In solidarity with the group of cows who helped shape this vision of liberation, in this chapter I celebrate concrete conceptualizations of community building rooted in vegan practices of care and kinship that translate speculative visions of multispecies justice into practice through embodied, relational, and sustained engagement with and within the place they inhabit. More specifically, I look at generative sites of collective liberation, from (micro)sanctuaries to gardening and foraging initiatives, where multispecies relations are intentionally recalibrated through multidimensional and decolonial acts of resistance. I argue that the selected case studies, by not settling for incremental change, offer creative and effective models for ways to bridge the fragmentation between eco-social movements. To radically tackle systemic oppression, it is in fact crucial not only to overcome intersectional approaches that have yet to reckon with anthroparchy and internalized speciesism but also to go beyond a single-issue lens and welfarist practices undermining the pursuit of collective liberation – especially within circles devoted to animal rights.[2]

Thus, while shedding light on the enduring complexities, apparent at each of the study sites, of operating within a neoliberal and capitalist framework as well as the underlying tension between a private property mindset and land sovereignty, including the entanglement of domesticated animals in extractive land practices, I propose to examine these liberatory sites through the methodological potential of the borderscape notion.[3] In other words, instead of conceiving the border delimiting the outside limit of these sites as a barrier, by embracing its permeability, the plurivocal and trans-species resistance taking place within these enclosed spaces turns into an opening toward what Catherine Oliver defines as "a politics of the possible" (p. 29). After all, farmed animals have been re-negotiating borders and teaching us about their porous qualities – both physical and epistemic – every time they transgress them by escaping from their forced confinement. To fully grasp this shift in framework, it is also vital to acknowledge that beyond being sites of hope and triumph, these are also sites of continued pain and trauma. This entails avoiding the replication of flawed binaries, such as a strict inside/outside, or romanticizing on the concept of 'safe haven,' both failing to recognize the political struggle and process of cultural rewiring happening in these liminal spaces. Sitting at the edge of these pockets of intentional solidarity, far from being marginal, provides a privileged perspective to witness the emergence of new orders based on multispecies alliances that reclaim spaces historically taken over by normative hegemonic forces.

The Farmed Animal Sanctuary: Upholding a Multispecies Body Politics

A first example of a space where alliances across species lines have the potential to flourish is the sanctuary for farmed animals. Sanctuary work has long been praised as a practice of contestation that explores a radically different kind of socio-spatial relationship with farmed animal species. Besides the liberatory opportunities sanctuaries provide by generating a space of potential autonomy, self-actualization, and social transformation, they also must deal with limitations inherent to our current conceptualization of what a sanctuary is. For instance, these sites often struggle with the reinscription of the same modes of interaction they subvert, such as the restriction of animal freedom to the spaces of the sanctuary.[4] In fact, expanding on the concept of territoriality, the practice of sanctuary becomes particularly troubled within a settler colonial context. As a consequence of these power dynamics, the act of guaranteeing refuge for species forcibly imported by settlers, such as cows, inevitably means upholding a single-issue cultural battle that has yet to come to grips with the privilege of the settler position and its complicity in maintaining hegemonic power structures.[5] Moreover, the replication of a barn model surrounded by fences represents an additional clash between settler ideas about private property and Indigenous communal relationships with the land and nonhuman relatives. In fact, fences helped institutionalize the collective recognition of private property, thus making them a neocolonial marker of land appropriation.[6]

The term 'sanctuary' itself also deters a full epistemic recalibration of nonhuman animals due to the "coexistence of the ideas of conservation and captivity in the semantics of the word" (Fusari, 2017, p. 150). It is true, for instance, that the term 'sanctuary' is often times applied to conservation parks and protected areas for so-called wildlife; yet they still rely on selective breeding and other control strategies that erase individual needs and agency in the name of population management. As a result, the risks of (internalized) paternalism remain widely spread, a mentality that is deeply tied to saviorist narratives and happy rescue stories of weak, voiceless sanctuary-seeking agents in need of humans to survive. Sanctuaries also often present as didactic farms, due to the centering and prioritization of human comfort in the spatial design (i.e., clear paths for humans to walk on, approaching residents through fences, and subjecting them to the public gaze in ways that replicate the experience of individuals held captive for human entertainment). Because of the blurred definition of sanctuary space, the term has also been appropriated for marketing purposes by zoos. In this case, the permeability of the border has been exploited by hegemonic forces, leading to an insidious rhetorical homogeneity that precludes drawing a clear distinction between sites where nonhuman animals reside to be protected and sites where they are kept for anthropocentric gain.

To challenge these problematic limitations and to push back against the dominant ideologies pressing at the borders, it is essential to confront mainstream apolitical positions that, by calling for "a gentler capitalism" (Dauvergne & LeBaron, 2014, p. 3), remain entangled in violent structures of power.[7] A concrete example of a subversive sanctuary model that intentionally draws on transformative ideas working against colonial, neoliberal, and capitalist premises is Agripunk, a self-proclaimed multispecies collective in opposition to all forms of discrimination.[8] Nestled in Val d'Ambra, a valley located in the Italian region of Tuscany, Agripunk was founded in 2015 on the premises of an intensive turkey farm owned by the Amadori Group, a company specialized in the systemic breeding and killing of chickens and turkeys for human consumption. After two years of documenting sanitary and economic irregularities, the founders of Agripunk managed to shut down the farm and its permissions were revoked, depriving the owner of the possibility to start a new operation in a different location. Prior to being owned by Amadori, the property had been besieged by the SS and Nazi soldiers, as documented by the current human residents Desirée Manzato and David Panchetti, who see it as their duty to keep the land free by holding space for those who wish to begin a new life.[9]

As proof of this purpose, the sheds and auxiliary facilities have gradually been converted into shelters where all those bodies across species that have been historically discriminated against can flourish as accomplices in their fight for autonomy and self-determination. By intentionally opposing the reification of bodies not conforming to supremacist standards, emergent subjectivities are offered a home to heal. The commitment to creating a safe(r) space for individuals of all species is tangible through visual cues as soon as you walk through the gates of Agripunk: the outside walls of the sheds are covered with murals by local artists, the kitchen door is strewn with stickers about veganism and inclusive hospitality, the communal restroom displays a wooden gender-free sign. Spotting one of the animal residents resting, without any expectation of labor, under a graffiti that calls for the abolition of oppressive classifications, is a recurrent scene that perfectly embodies the subversive optics of this space (Figures 1.2 and 1.3).

The choice to preserve the row of seven sheds and to continue to operate within the dominant western paradigm of the barn model, in this circumstance, represents a powerful act of cultural memory. In fact, recasting the architectural form without erasing the signs of its history of violence is a testament to resilience and survival that reminds us of Hartmut Kiewert's pictorial worlds of the *Animal Utopia* series. 'The valley that resists' is the slogan that echoes across the plain and that is commonly used to describe the spirit of the place. The name 'Agripunk' itself epitomizes the philosophy of rebellion at the core of the project and its constant disruption of dominant agri-cultural practices affecting precarious lives across species. The very existence of this site debunks the idealized notion of rural life so deeply tied to the Tuscan countryside, an imaginary that Italy has strategically fabricated to export products of violence worldwide through neocolonialist and

Figure 1.2 Sanctuary sheds featuring liberatory murals offer shade for nonhuman residents.

Figure 1.3 The mural was made by Italian band Pensieri Oltre and reads: Beyond Definitions, Classifications, Differences (Photograph by the author, March 2023).

nationalist marketing strategies masked under the rhetoric of the Slow Food movement or the myth of peacefully grazing animals in pastures.[10]

In response to this strenuous process of liberation, the whole ecological web reclaims space to recover: the land can heal from the devastating effects of intensive farming, while wildlife can find refuge, including from the hunters who populate the area. Unfortunately, while the land remains in the hands of the son of the farmer who ran the turkey farm operations, the risk of eviction hangs ominously over the Agripunk community. To overcome this lingering threat, the crowdfunding campaign *#agripunkbenecomune* (Agripunk Common Good) was created both to raise public awareness of their site-specific regenerative project and to promote community fundraising. The world they exemplify breaches through the borders of the sanctuary grounds every time they physically carry their message at events across Italy. Yet, the multispecies liberatory model they propose struggles to take root elsewhere. In order for it to transcend the borders of normativity imposed by dominant systems, which include national borders, it is essential not only to collectively strengthen transnational coalitions that are inclusive of non-anglophone struggles but also to ensure that social movement groups refrain from acting in isolation from each other. At the same time, it is also necessary to consider alternative strategies that question the western barn model as the unique spatial arrangement for engaging in relationships with more-than-human survivors of exploitation. The microsanctuary movement was founded as a response to these needs.

The Microsanctuary: Reframing Interspecies Co-habitation

Over the past decade, the microsanctuary movement has risen in the attempt to allow individuals who don't have access to land to still rescue more-than-humans in need with the resources available to them, no matter how limited they may be, whether that is a rented apartment or a suburban home with a garden. Operating at the borderscape of the culturally constructed opposition between urban and rural landscapes, this model of interspecies relationships is explicitly and intentionally intersectional, in that it positions speciesism in conversation with colonialist, capitalist, racist, and patriarchal systems.

In discussing the use of the term 'micropolitics' in Deleuze and Guattari, Jane Bennett refers to it as "a realm of activities that have public effect – that help to shape the tenor of collective life – but which do not fit into the traditional paradigms of political action" (2004, p. 51). What makes micropolitical tactics particularly effective, then, especially in the area of grassroots organizing, is the preference for direct action strategies and anti-institutional predispositions. In the context of microsanctuaries, besides creating more physical space for more-than-human animals in need of safety and lifetime care, their transformative potential lies in the broadening of the semantic borders of the term 'sanctuary.' According to one of the founders of the movement, Alastor Van Kleeck, this shift takes place by embracing a more holistic framework that centers sustainability as the core principle of all its operations (2022). This means making the most out of the resources available in the present moment in terms of one's abilities, finances, and spatial arrangements. As a result, the potential of crafting multispecies alliances is not limited to large farm properties but can effectively operate in a variety of places without the huge capital required to run a traditional sanctuary. The smaller scale dimension of this project, which can be limited to the cohabitation of a human caregiver and a more-than-human individual in need of shelter, creates opportunities not only to provide extensive individualized care, especially for disabled individuals, but also to build familial bonds with species normatively not considered companions. While care is provided to a wide range of species, from rats and rabbits to fishes, the individuals who most commonly find refuge in microsanctuaries are farmed animals, especially chickens. For a chicken to live in an urban setting without serving human interests means challenging not only exploitative husbandry practices but also the history of commodification of so-called backyard chickens. Obviously, it is crucial to avoid secluding the individuals who reside at microsanctuaries into the category of 'pet' due to the ties of this taxonomic label with the concept of property, which is inseparable from its colonial and racial implications.[11]

It is true that the anthropized context of our geographies inevitably restricts the freedom of the rescued individuals, especially their freedom of movement. Yet, chickens quickly adapt to living indoors by modifying their behaviors, for instance, as Heather Rosenfeld suggests, by dustbathing on blankets or on indoor dustbathing areas or by perching on furniture, which contributes to creating the best conditions for their subjectivities to thrive (2023, p. 176). Regardless of their capacity to adapt to human-centered built environments, it is important to emphasize that the humans committed to this process of liberation must also deal with structural barriers and intergenerational trauma caused by the same systems of power oppressing our nonhuman kin. An epistemic shift mindful of the interlocking, geographically embedded forms of social marginalization affecting individuals across species, is thus the key to valuable multispecies alliances rooted in interdependence and mutual care. Among the movement's greatest strengths is its potential to challenge speciesist zoning laws that segregate farmed animals outside of urban borders, which echo xenophobic immigration policies implemented at national borders. It is precisely at this borderscape that (il)legality is questioned and new cohabitation norms take shape (Chang, 2022, pp. 232–237). To further subvert cultural expectations, rescued individuals participate with their human accomplices at events and protests, thus becoming active members of their own liberation and forging imaginaries of multispecies communities existing beyond pastoral settings (Figures 1.4 and 1.5).

These moments of interspecies solidarity in urban settings trigger the well-known "out of place" cultural concern theorized by Philo and Wilbert (2000). In fact, more-than-human animals challenge their preassigned role in the eyes of society even just with their bodily presence in unexpected locations. Of course, the likelihood of also endangering human and more-than-human safety through these subversive acts is not excluded. The risk of eviction, just to mention one potential threat, remains a possibility, thus reminding us of the ways in which structural barriers, such as the maintenance of a system of private property alongside lack of access to housing, obstruct multispecies liberation. To counter these risks, the movement relies on a network of shared knowledge and strong systems of care, ranging from mental and emotional support to tips on animal care. Additionally, to tackle structural inequalities, the Microsanctuary Resource Center offers grants directed specifically to people of the global majority in support of their rescue work.[12] To further dismantle the human–animal binary and the aforementioned structural injustices kept unchallenged by the isolation of social justice movements, let us imagine multispecies microsanctuaries that welcome also human individuals in need of shelter, counting on solid networks of resistance defying national borders, a strategy that increases the chance for transformations activated at the local level to spread globally. Yet, in order to fully recognize and dismantle the overlapping vectors of oppression and privilege limiting

Figure 1.4 Corey & Human Allies holding signs with messages rooted in a multispecies intersectional framework at the Midwest Animal Rights March (Photograph by Kelsey Atkinson, 2018).

access to land and fueling both animal exploitation and systemic discrimination of human-disenfranchised communities, the problem must be viewed alongside issues of food justice.

The Farm: Cultivating Transformative Change

While veganism and food sovereignty share the common goals of eradicating systemic oppression and achieving environmental justice for all, food movements and the politics of animal liberation tend to remain disconnected when an intersectional antispeciesist lens is not adopted. For instance, radical visions of food sovereignty assume that nonhuman animals will remain in place as commodities and sources of value. An example of this disconnect can be found in the approach employed by La Via Campesina, an international, multicultural movement giving a global voice to the 'peasants' who feed the world. As delineated in their 2021 brochure, the term 'peasants' is meant to include "people who till the land to produce food, the fishers, the pastoralists, the farmworkers, the landless, the migrant workers, the Indigenous people rural workers" (La Via Campesina, 2021). Raising nonhuman animals for food, hunting, pastoralism, and fishing are all practices that

Figure 1.5 Helen & Hana Low posing for a picture at the Cleveland Pride March (Photograph by Kelsey Atkinson, 2018).

disregard the sociopolitical issues that accompany the fight for animal liberation. Any claim of living in 'harmony with nature' while relying on these practices is inherently based on anthropocentric interests, especially in Eurocentric contexts. Similarly, local food projects based on community support and mutual aid operating outside or on the fringe of the capitalist food system tend to leave unquestioned the normalization of beekeeping and chicken farming masked under the label of 'sustainability.'[13]

Equally problematic are initiatives grounded in eco-vegan principles that fail to integrate social justice concerns into their initiatives. As an example, the contemporary urban garden movement tends to promote environmental sustainability and economic development that often increase displacement pressures on local residents, a process that Kenneth Gould and Tammy L. Lewis define as "green gentrification" (2017, p. 23). Likewise, the roots of permaculture, in both vegan and non-vegan settings, are largely appropriated by mainstream western science without honoring Indigenous peoples who are the primary keepers of this knowledge.[14] Both practices are in fact based on land ownership in public/private spaces and are thus inherently intertwined with (neo)colonial ideologies, even when serving in the best interests of more-than-human animals.

The same failure to enter into solidarity relationships with Indigenous sovereignty struggles can be observed within vegan initiatives, such as the absence of a decolonization commitment in the action plan of the Vegan Land Movement, a crowdfunding campaign started in the UK and aimed at buying farmlands at auctions to 'rewild' them. In its Constitution, the founders appoint Earth as "the designated beneficiary of the VLM. All use of land held by VLM CIC (Community Interest Company) must benefit local ecosystems and its inhabitants" (Vegan Land Movement, 2024), where the word 'inhabitants' refers only to nonhuman animals, thus excluding the possibility of the co-production of landscapes through a relational framework. Additionally, access rights to humans are limited and 'taken on a case-by-case basis' as a strategy used by those in power to leave ecosystems undisturbed. While this approach may benefit the life of the nonhuman species who live within these spaces, it continues to reproduce a counterproductive nature-culture dualism and to perpetuate the notion of the human population as a homogeneous group, equally responsible for environmental damage. At the same time, despite the will to free these spaces from human presence, the exertion of governance power limited to a group of human individuals actually reinforces it, especially with the added layer of land dispossession.

A possible real-world solution to this problem is proposed by law professor Karen Bradshaw, who presents the proposal of "an interspecies system of property" (2020, p. 3) advocating "for folding animals into our existing system of property law, giving them the option to own land just as humans do" (2020, p. 1). In addition to reinforcing once again ownership ideals that impede Indigenous communities from reclaiming stolen land, this approach, clearly grounded in white, western environmentalism, is also rooted in pervasive welfarist tactics, which the same author admits do not promote systemic change:

> [I]t [the proposal] does not serve to free chimpanzees from cages or free cattle destined for slaughter. . . . Under my conception, wildlife has the land it needs, but people can still eat a burger or swat a mosquito.
>
> (2020, pp. 4–5)

A land initiative that incorporates consistent anti-oppression principles is Liberation Farm, a site of intersectional struggle founded in 2020 in the Catskill Mountains of New York. Here, cultural values and identities are radically being shaped, (re)negotiated, and contested. The founders, Nadia Muyeeb and Omowale Adewale, tend to the land by growing organic vegetables, mainly beets, while also running community events ranging from skill-building workshops to agri-therapy and outdoor camping. This project's potential for radical change lies in the choice to make Liberation Farm an "unapologetically Black space" (Liberation Farm, 2024) where the violent erasure of the history and culture of the Black diaspora is recovered and, in turn, where independent self-determination can thrive in an environment removed from white surveillance. By making this clear statement of purpose, the aim is to create an intentional space for Black healing capable of promoting a significant cultural transformation.

This approach, crafted specifically to decentralize white supremacy, also invites a deep reframing of the politics of safety, which involves coming to terms with the weaponized ideals of safety tied to a biopolitical tendency of the social group in power to deploy mechanisms of social control. On an institutional level, such mechanisms might include redlining practices, police militarization, or even Diversity, Equity, and Inclusion (DEI) initiatives that display tolerance on the façade and use a colorblind rhetoric, particularly within white vegan circles.[15] They might also include biosafety measures in 'livestock facilities,' which are implemented to designate activists attempting to disrupt operations as domestic terrorists. In discussing the need for Black-women-only spaces, Patricia Hill Collins writes, "safe spaces rely on exclusionary practices, but their overall purpose most certainly aims for a more inclusionary, just society" (2000, p. 110), thus confirming that the generative tension sustained at the borderlands inevitably reverberates throughout the social fabric.

Another pioneering trait of Liberation Farm is the intentional unearthing of the link between operating farmland and eating plant-based foods through a decolonial framework, which helps "establish cultural boundaries for Black people to thrive and share communion" (Adewale, 2022, pp. 16–17). Mindful also of the need to shift from capitalist-colonialist food systems to harvesting practices based on an intimate knowledge of landscapes, the very existence of Liberation Farm promotes a new geographic, ecological imagination that acknowledges the repercussions of local actions on a global sphere. Despite the dedication to community building and the hard work of tending to the land, Liberation Farm was recently forced to move due to a sudden eviction notice, once again revealing the contemporary perpetuation of land dispossession as well as the constant risk of infiltration by oppressive forces pressing at the border of heterotopic spaces.[16] As a consequence, Liberation Farm has relocated to a much larger plot of land in South Kortright, New York. While this moment of transition, charged with grief for the loss of the land, is unfolding in real time as I write, the future holds vibrant hope for the propagation of the revolutionary principles that Liberation Farm embodies. Similar land reclamation practices, instilled with principles of multispecies justice, can be observed in the actions of Mariposas Rebeldes, a collective that adopts a place-based approach to radically transform agro-industrial food systems.

The Forest: Foraging Our Way to Collective Liberation

In Rita Wong's poem *Recognition/Identification Test*, two columns of words are placed in specular opposition: a list of names of herbaceous plants, flowers, and trees is juxtaposed to a list of familiar corporate brands. Magnolia/macDonald's, cedar/sony, pansy/nestle (2007, p. 32). Besides criticizing the foundations of modern society and its dependence on corporate power and consumerist ideology, the poem reproduces the nature-culture dichotomy on the page as a warning of the widespread estrangement from the natural world. This distance has been historically reinforced through colonial displacement and extractive systems, which contributed to the loss of ancestral knowledge tied to the land, including Indigenous and Black food practices. In particular, foraging has a long history of oppression and criminalization enacted through racist anti-foraging laws used by white settlers to force the assimilation of disenfranchised communities and to eradicate free food supplies. As food lawyer Baylen J. Linnekin explains, after the Civil War, Southern plantation owners forcefully and systematically restrained foraging rights to chain newly freed individuals to plantation work; similarly, foraging practices of Indigenous peoples were made illegal in treaties signed by Indigenous tribes and the federal government outlining, among other things, access to gathering and, thus, to the land (2018, pp. 1011–1013). These legal dynamics and colonial power relations are reinforced today through the normalization of property laws that allow private landowners to bar foragers. As a testimony of the continued enactment of these legal instruments of control, in her documentary, *Foragers*, Jumana Manna narrates the criminalization of herb-picking cultures in occupied Palestine, which also chronicles the will of Palestinians to maintain their foraging traditions as an act of resistance to Israeli legislation and its attempts to further alienate them from their land (Manna, 2020).

In direct contrast to settler-colonial logics, the Queer, Trans Indigenous, and Latinx collective Mariposas Rebeldes has a committed mission to restore relationships to the land, imagine alternatives to capitalism, and promote food autonomy through cooperative agriculture.[17] The group initially gathered in Israel Tordoya's backyard garden in Muscogee (Creek) land, also known as Atlanta (USA), which they managed collectively by practicing a non-hierarchical organization. As part of their efforts to re-establish Indigenous foodways, they embraced a process of botanical decolonization by adopting a resilient agricultural approach that long preceded European settlements and involved the cultivation of crops such as corn, squash, and beans – also known as the 'Three Sisters' – that rely on each other for their survival. As the land modeled a system of growth based on mutually beneficial cooperation, the collective grew into a refuge for historically subjugated individuals to thrive. The overall operation of Mariposas Rebeldes was however temporarily uprooted in the middle of the COVID-19 pandemic, when Tordoya's landlord evicted them from their home, suddenly producing a condition of displacement-induced housing insecurity during a global health crisis. Yet, the collective faced this transition by continuing to organize through itinerant gatherings, even creating their own line of corn-and-bean tempeh, a high-quality vegan protein that they named MaripoSnax (Gómez-Upegui, 2021; Reign, 2021).

Following a successful fundraising campaign, Mariposas Rebeldes was recently able to acquire a forested 1/3-acre lot in the South Atlanta forest, a corridor rich in fruit and nut trees, perennial herbs, and flowers. As part of the process of land reclamation, the purchased land was renamed Bosque Itzpapalotl (Obsidian Butterfly) in honor of the patron deity of the collective, whose transformative iconography animates the spirit of the group. The location was previously used as a tire shop, now demolished, and would usually be considered in a 'state of abandonment.' However, as Gilles Clément reminds us, the term 'abandonment' itself embodies a utilitarian anthropocentrism, which discards anything that is not linked to human activity and "of use" (2011, p. 278). The area, in fact, brims with more-than-human vibrancy and the plants that are normatively considered 'weeds' or 'invasive,' such as privet and ivy, are here welcomed as migrant plants who enhance the land. This space is thus an opportunity for mutual regeneration and exchange based on principles of ethical foraging that inherently reject the violent restraints of domestication. Ultimately, the plan is to expand their urban gardening project into a food forest and community kitchen that would support the community with high-quality, nutritious food, specifically nuts, and other nutrient-dense high-calorie foods, even in the face of market shortages and climate emergency.

By attending to the unpredictability of foraging, the Mariposas are molding an example of a symbiotic and decommodified way of living that "stray[s] into the peripheries of capitalist production" (Tsing, 2015, p. 24). Yet, the intention is to go beyond ethical veganism and instead, to build political solidarity that supports systemic change; in the words of Abundia Alvarado, the co-founder of the collective, "we're trying to be more creative. We're interested in the abolition of the gender binary, capitalist farming, extraction and private property" (Alvarado & Fischer, 2023). As a result of conceiving social justice within a relational frame, liberation is inherently achieved for all those communities across species that experience exploitation and oppression caused by these interconnected systems. In line with this spirit, the Mariposas are currently involved in the collective struggle to defend the Weelaunee forest to stop the construction of a police training compound, referred to as 'Cop City,' as well as of the largest movie soundstage in the world commissioned by the massive film company Blackhall Studios. The consequences of these development projects are devastating, from drastically increasing gentrification and police force to wiping away the liveliness of the forest. But even more troubling is the perpetuation of monopoly capitalism combined with a militarization mindset, which caused the murder of forest defender Tortuguita (Manuel Paez Terán) during a raid of the Stop Cop City encampment on January 18, 2023. While the repressive containment strategies adopted by institutionalized powers are not only relentless but also fatally violent, emerging out of the borderlands is a revolutionary initiative that is paving the way for queer Indigenous futures rooted in relational ecologies of belonging.

Conclusions

The case studies examined in this chapter offer concrete visions of multispecies alliances that honor the intimate relations between liberation movements that are often insufficiently recognized and even actively negated: social and environmental justice struggles are necessarily also multispecies struggles, and vice versa.[18] In the profusion of hopeless, apocalyptic narratives, these projects set the course for futures rooted in social justice that are inherently ecological, and multidimensional. While dynamics of privilege operate differently within each site, and the histories of oppression they carry differ substantially, the change they propose is intentionally structural and systemic. Because of their disruptive nature, these initiatives all struggle in various ways for existence while resisting the constant threat posed by systems of domination violently pressing at the borders to uphold the normative status quo. But what they represent is actually more than resistance; it is imaginative power, community resilience, and radical relationality. At the same time, they illustrate the cultural and social benefits of achieving collective healing through a decolonial praxis that is mindful of how privilege affects power dynamics based on somatic, social, and ethical markers. While these liberatory sites inevitably rely on border systems to create safer territories, the recalibration of the epistemic and cultural compass extends beyond physical borders. Like scattered flames of resistance that cannot stay confined, these models have the potential to spread and be replicated, of course without neglecting to make adjustments according to the historical and socioeconomic specificities of place.

While I have hereby centered the entanglements of liberation struggles on land, unique challenges arise when the questions raised in this chapter are extended to oceans and sea dwellers. In fact, to enact systemic change, liberatory sites must intentionally dismantle multiple and mutually constituted ideologies.[19] The Whale Sanctuary soon operating on Mi'kmaq territory, otherwise known as Nova Scotia, Canada, serves as an ideal case study to examine the layers of complexity posed by bodies of water.[20] In fact, this project – the first of its kind – aims specifically at ending exploitation for whales and dolphins and becoming a model for future sanctuaries around the world wanting to offer shelter to cetaceans. While the aims are certainly laudable, there is still a risk of maturing single-issue solutions, especially if settler ideology is not addressed at the planning stage. So, for a start, in the journey to bring about reconciliation with the cetaceans who have been institutionalized for human entertainment, it is essential to avoid perpetuating displacement to make space for retired individuals. In many ways, the lives of the fishes living in the waters selected for the sanctuary will be disrupted and should thus be given high consideration. Moreover, the installation of a permanent netting system surrounding the 40 hectares of water space also unsettles pre-existing relational ecologies, which extend far beyond the prescribed confines, given the inherent permeability of water. Equally disrupted are the lives of the Indigenous communities living in these regions whose interests and needs should be prioritized on the long term, making sure they also benefit from reparation efforts. While on the project's website they claim to have been helped, guided, and encouraged by First Nation Chiefs and political leaders, it is troubling that no member of the team is Indigenous, thus suggesting a lack of Mi'kmaq presence both on a leadership level and throughout the decision-making process.

These are just some of the dangers I foresee. Conversely, the visions of trans-species justice celebrated in this chapter shake the foundations of systems of oppression, an approach that requires an active, consistent, and challenging practice of unlearning. Perhaps, it is this urge to face inconvenient and uncomfortable truths that keeps these initiatives from getting mainstream traction. Yet, the radical imagination they bring to life is shaping futuristic visions of liberation in real time. The emergence of new orders grounded in transformative worldbuilding provides positive guidance for replacing barriers and hierarchies with relational solidarity and, consequently, for building long-lasting alliances across species, beyond borders.

Acknowledgments

I am grateful to the following people whose deep knowledge has shaped my thinking and contributed to a more subtle understanding of interwoven patterns of relations: Abundia Alvarado for the conversations on the sacrality of Abundance; Margaret Robinson for the valuable insights on whales and Mi'kmaq culture. I also thank Kelsey Atkinson for permitting the reproduction of their photographs.

Notes

1 To learn more about the forest cows at VINE sanctuary, see Shen (2022).
2 Notable scholars and activists who apply the concept of intersectionality to speciesism as well as a multispecies approach to social change include among others Julia Feliz Brueck, Claire Jean Kim, Aph Ko and Syl Ko, A. Breeze Harper, Christopher-Sebastian McJetters, and Billy-Ray Belcourt.
3 For an analysis of the borderscape as a method for a geographical opposition to capitalism and of its potential to reveal the dynamic social and spatial relationships taking place in and across borders see, Brambilla (2015, 2019).
4 Recent critical perspectives on animal sanctuaries include: Abrell (2016, 2017, 2021); Blattner et al. (2020); Emmerman (2014); Donaldson and Kymlicka (2015); Jones (2014); Pachirat (2018); Scotton (2017).
5 On colonial structures within animal sanctuaries, see Boswell (2017); Gillespie (2021); Reign (2021).
6 An overview of the colonial conception of land ownership and the consequent spreading of the western notion of property through fencing systems can be found in Cronon (2003).
7 On the rise of vegan capitalism, see also Sexton et al. (2022).
8 On their website, Agripunk is defined as an "antispeciesist, antifascist, antiracist, anti-colonialist, antisexist, and ecotransfeminist" space (translation is mine). https://www.agripunk.com/home.
9 To learn about the stories of resistance and self-determination of the individuals who now live at Agripunk, see Manzato and Panchetti (2022). An English translation of the book is forthcoming.
10 According to the Slow Food's position paper on animal farming *Beyond Welfare: We Owe Animals Respect* (Goracci, 2022), "the Slow Meat campaign encourages support for meat from sustainable farming systems that respects animals" (48). The rhetoric of sustainability is imbued with unchallenged speciesism that relies on systemic violence under the guise of "animal-friendly husbandry" (24). It is thus not surprising that John Sanbonmatsu would include Slow Food among the movements that espouse "a food politics centered around the 'sustainable' enslavement and killing of nonhuman beings" alongside aquaculture, and locavorism (In Maurizi, 2013, n.p.).
11 On this topic, see Joshua Bennett (2020), particularly pp. 140–168.
12 https://microsanctuary.org/moc/
13 On the omission of more-than-human interests from sustainability discourses see Vinnari and Vinnari (2022), and Bergmann (2021).
14 The Black Permaculture Network is actively working to decolonize permaculture and reclaim their lands through radical care practices. For more details see the website of co-founder Pandora Thomas (2017).
15 On the structural barriers caused by a predominance of white spaces and the need for Black-only spaces, see Anderson (2022) and Blackwell (2018).
16 As Son Vivienne affirms, safe(r) spaces are not the reproduction of a binary opposition between inside and outside. Rather, they require "navigating a continually shifting

boundary between fear and courage, discomfort and security, trauma, and resilience" (2019, p. 219).
17 https://mariposasrebeldes.org/
18 On the exclusion of the struggles of more-than-human animals for instance from climate change movements, still too often ingrained with anthropocentric beliefs, see Weisberg and Salzani (2023).
19 To learn more about the liberation struggles of our marine kin, see Daniel Vandersommer's chapter in this volume (pp. 151–167).
20 https://whalesanctuaryproject.org/

References

Abrell, E. (2016). Lively Sanctuaries: A Shabbat of Animal Sacer. In I. Braverman (Ed.), *Animals, Biopolitics, Law: Lively Legalities* (pp. 135–154). Routledge.

Abrell, E. (2017). Interrogating Captive Freedom: The Possibilities and Limits of Animal Sanctuaries. *Animal Studies Journal, 6*(2), 1–8. https://ro.uow.edu.au/asj/vol6/iss2/2

Abrell, E. (2021). *Saving Animals: Multispecies Ecologies of Rescue and Care*. University of Minnesota Press.

Adewale, O. (2022). *An Introduction to Veganism & Agricultural Globalism*. Lantern Books.

Alvarado, A., & Fischer, D. (2023). Stopping Cop City and Reconnecting with Abundance. *New Politics*, January 14. https://newpol.org/stopping-cop-city-and-reconnecting-with-abundance/

Anderson, E. (2022). *Black in White Space: The Enduring Impact of Color in Everyday Life*. University of Chicago Press.

Bennett, J. (2004). Postmodern Approaches to Political Theory. In G. F. Gaus & C. Kukathas (Eds.), *Handbook of Political Theory* (pp. 46–56). SAGE Publications.

Bennett, J. (2020). *Being Property Once Myself: Blackness and the End of Man*. Harvard University Press.

Bergnann, I. M. (2021). The Intersection of Animals and Global Sustainability – a Critical Studies Terrain for Better Policies? *Proceedings, 73*(1), 12. https://doi.org/10.3390/IECA2020-08895

Blackwell, K. (2018). Why People of Color Need Spaces Without White People. *The Arrow: A Journal of Wakeful Society, Culture & Politics*, August 9, 1–15. https://arrow-journal.org/why-people-of-color-need-spaces-without-white-people/

Blattner, C., Donaldson, S., & Wilcox, R. (2020). Animal Agency in Community: A Political Multispecies Ethnography of VINE Sanctuary. *Politics and Animals, 6*, 1–22.

Boswell, A. (2017). Settler Sanctuaries and the Stoat-Free State. *Animal Studies Journal, 6*(2), 109–136.

Bradshaw, K. (2020). *Wildlife as Property Owners: A New Conception of Animal Rights*. University of Chicago.

Brambilla, C. (2015). Exploring the Critical Potential of the Borderscape Concept. *Geopolitics, 20*, 14–34.

Brambilla, C. (2019). From Border as a Method of Capital to Borderscape as a Method for a Geographical Opposition to Capitalism. *Bollettino della Società Geografica Italiana Roma, 13*(8), 393–402. https://doi.org/10.13128/bsgi.v8i3.409

Chang, D. (2022). Infiltrate the Cityscapes: Resisting Speciesist Segregation with Farmed Animals. In P. Hodge, A. McGregor, S. Springer, O. Véron, & R. J. White (Eds.), *Vegan Geographies: Spaces Beyond Violence, Ethics Beyond Speciesism* (pp. 217–238). Lantern Media.

Clément, G. (2011). In Praise of Vagabonds. *Qui Parle, 19*(2), 275–297.

Cronon, W. (2003). *Changes in the Land: Indian, Colonists, and the Ecology of New England*. Hill & Wang.

Dauvergne, P., & LeBaron, G. (2014). *Protest Inc.: The Corporatization of Activism*. Polity.

Donaldson, S., & Kymlicka, K. (2015). Farmed Animal Sanctuaries: The Heart of the Movement? A Socio-Political Perspective. *Politics and Animals, 1*(1), 50–74.

Emmerman, K. S. (2014). Sanctuary, Not Remedy: The Problem of Captivity and the Need for Moral Repair. In L. Gruen (Ed.), *The Ethics of Captivity* (pp. 213–230). Oxford Academic. https://doi.org/10.1093/acprof:oso/9780199977994.003.0014

Fowler, H. (2013). *New World Order* [Video artwork]. MCA, Sydney, Australia. http://haydenfowler.net/projects/new-world-order.html

Fowler, H. (2015). New World Order – Nature in the Anthropocene. In Human Animal Research Network Editorial Collective (Ed.), *Animals in the Anthropocene: Critical Perspectives on Non-Human Futures* (pp. 243–248). Sydney University Press.

Fusari, S. (2017). What Is an Animal Sanctuary? Evidence from Applied Linguistics. *Animal Studies Journal, 6*(2), 137–160. https://ro.uow.edu.au/asj/vol6/iss2/8

Gillespie Kathryn, A. (2021). An Unthinkable Politics for Multispecies Flourishing Within and Beyond Colonial-Capitalist Ruins. *Annals of the American Association of Geographers, 112*(4), 1108–1122.

Gómez-Upegui, S. (2021). Mariposas Rebeldes: Growing Food Sovereignty for Queer Latine and Indigenous People in Atlanta. *Atmos.Earth*, June 29. https://atmos.earth/mariposas-rebeldes-growing-food-sovereignty-for-queer-latine-and-indigenous-people-in-atlanta/

Goracci, J., Milano, S., Pantzer, Y., Ponzio, R., Venezia, P., & Zuliani, A. (Eds.). (2022). *Beyond Welfare: We Owe Animals Respect. Slow Food's Position Paper on Animal Farming.* www.slowfood.com/wp-content/uploads/2022/12/EN_2022_AW_SF_position_paper.pdf

Gould, K., & Lewis, T. L. (2017). *Green Gentrification: Urban Sustainability and the Struggle for Environmental Justice.* Routledge.

Hill Collins, P. (2000). *Black Feminist Thought: Knowledge, Consciousness and the Politics of Empowerment.* Routledge.

Jones, M. (2014). Captivity in the Context of a Sanctuary for Formerly Farmed Animals. In L. Gruen (Ed.), *The Ethics of Captivity* (pp. 90–101). Oxford University Press.

La Via Campesina. (2021). *English Brochure.* https://viacampesina.org/en/wp-content/uploads/sites/2/2021/12/LVC-EN-Brochure-2021-03F.pdf

Liberation Farm. (2024). https://liberationfarm.org/

Liberti, S., & Parenti, E. (Directors). (2018). *Soyalism* [Documentary]. Elliot Films.

Linnekin, B. J. (2018). Food Law Gone Wild: The Law of Foraging. *Fordham Urban Law Journal, 45*(4), 995–1050. https://ir.lawnet.fordham.edu/ulj/vol45/iss4/3

Manna, J. (2020). Where Nature Ends and Settlements Begin. *E-Flux Journal, 113*, 1–12. www.e-flux.com/journal/113/360006/where-nature-ends-and-settlements-begin/

Manna, J. (Director). (2022). *Foragers* [Documentary]. Production Jumana Manna.

Manzato, D., & Panchetti, D. (2022). *Agripunk: L'unico Allevamento Accettabile È Quello Chiuso.* Self-published.

Maurizi, M. (2013). Animal Liberation and Critical Theory. Interview with John Sanbonmatsu. *Asinus Novus*, January 13. https://asinusnovus.net/2013/01/13/animal-liberation-and-critical-theory-interview-with-john-sanbonmatsu/

Oliver, C. (2021). *Veganism, Archives, and Animals: Geographies of a Multispecies World.* Routledge.

Pachirat, T. (2018). Sanctuary. In L. Gruen (Ed.), *Critical Terms for Animal Studies* (pp. 337–355). University of Chicago Press.

Philo, C., & Wilbert, C. (2000). *Animal Spaces, Beastly Places. New Geographies of Human-Animal Relations.* Routledge.

Kay, J. (2021). Vegan-Washing Genocide: Animal Advocacy on Stolen Land and Re-Imagining Animal Liberation as Anti-Colonial Praxis. In S. Springer, J. Mateer, M. Locret-Collet, & M. Acker (Eds.), *Undoing Human Supremacy: Anarchist Political Ecology in the Face of Anthroparchy* (pp. 89–118). Rowman & Littlefield.

Reign, E. (2021). How a Queer and Trans Latinx Gardening Collective is Working to Reverse Food Insecurity in Atlanta. *Vogue*, March 8. www.vogue.com/article/queer-trans-latinx-gardening-collective-atlanta

Rodríguez, S. (2018). Animal Agriculture: An Injustice to Humans and Nonhuman Alike. In Saryta Rodríguez (Ed.), *Food Justice: A Primer* (pp. 103–128). Sanctuary Publishers.

Rosenfeld, H. (2023). Sanctuaries as Multispecies Safe Spaces. In E. Cudworth, R. E. McKie, & D. Turgoose (Eds.), *Feminist Animal Studies: Theories, Practices, Politics* (pp. 165–182). Routledge.

Scotton, G. (2017). Duties to Socialise with Domesticated Animals: Farmed Animal Sanctuaries as Frontiers of Friendship. *Animal Studies Journal, 6*(2), 86–108. https://ro.uow.edu.au/asj/vol6/iss2/6

Sexton, A. E., Garnett, T., & Lorimer, J. (2022). Vegan Food Geographies and the Rise of Big Veganism. *Progress in Human Geography, 46*(2), 605–628.

Shen, R. (2022). *Tending Sanctuary: Multispecies Entanglements at VINE Sanctuary.* https://issuu.com/rebeccashen/docs/pw2022_final_shen_rebecca2

Thomas, P. (2017). *Black Permaculture Network* [webpage]. Pandora Thomas. https://www.pandorathomas.com/black-permaculture-network

Tsing, A. L. (2015). *The Mushroom at the End of the World. On the Possibility of Life in Capitalist Ruins.* Princeton University Press.

Van Kleeck, A. (2022). *Microsanctuaries as a Form of Vegan Activism* [presentation]. Caregiving is Activism – The Radical Companionship Project, February 10. www.radical-companionship.com/presentations/

Vegan Land Movement (2024). Vegan Land Movement CIC. Organisational Structure. d https://veganlandmovement.com/cic/

Vinnari, E., & Vinnari, M. (2021). Making the Invisible Visible: Including Animals in Sustainability (and) Accounting. *Critical Perspectives on Accounting, 82*, 1–8. https://doi.org/10.1016/j.cpa.2021.102324.

Vivienne, S. (2023). *Queering Safe Spaces: Being Brave Beyond Binaries.* Lexington Books.

Weisberg, Z., & Salzani, C. (2023). No Climate Justice Without Justice for Animals. *Critical Ecologies* [Special issue] *dePICTions, 3.* https://parisinstitute.org/no-climate-justice-without-justice-for-animals/

Wong, R. (2007). *Forage.* Nightwood Editions.

Chapter 2

Careful Care Towards Animal Liberation for Feral Pigeons and Beyond

Maria Martelli

An Image of Coexistence

A bit of dust rises in the flutter of wings each time we enter the enclosure for disabled pigeons. Most of them can't fly, but some, recovering, do try to take to the air a little, especially encouraged by human presence. Silvia Moldovan, bird caregiver and artist, sits on the hay-covered ground, waiting for the one pigeon that always sits in her hair. She's showing me different pigeons, sometimes explaining how they've been since I last visited. I usually have trouble telling them apart, even if they are so different when you look close. Feral pigeons show a lesser range of color variations than domesticated ones, but compared to other free-living bird species, they're quite distinct.[1] Their blues, reds, and browns come in four patterns and mix together. When you start looking, you notice the differences in body type, eye color, beaks, and behavior. Silvia loves showing me the feisty ones who bite right back at you if you get close. She's been caring for injured birds for more than 10 years and at this point, things are starting to take shape. In 2022 she managed, with her partner Alin Spiridon and the help of friends, to move the shelter to a larger plot of land: the final destination of what will be a multispecies sanctuary and artistic educational space called Sepale (Figure 2.1).

Whenever I visit the sanctuary, Silvia tells me what else she's planning: the next construction for housing pigeons, a place to host human visitors, the tree 'fence' or the things she likes to keep as they are – weeds of many kinds, resistant to drought and to the few rabbits who hop around. She says things like "This weed is bad for humans but good for birds" with such joy in her voice that one would think she's just stumbled over a gold mine, not a spiky, poisonous plant. Then again, Silvia likes plants in themselves too, letting them take root as they see fit. Whenever I visit, I am reminded that non-anthropocentric places exist, even in this world so broken.

Looking for Animal Liberation in Places of Care and Non-exploitation

I started researching animal sanctuaries a few years ago, hoping to understand what sorts of places these are, what relationships they create, and how they contribute to animal liberation. I consider them to be heterotopias in the sense of "effectively

DOI: 10.4324/9781003366706-4

Figure 2.1 Pigeons flying above the building site at Sepale sanctuary, outside the city of Timișoara, Romania (October 2022). Photograph by the author.

enacted utopias" and "elsewhere" spaces (Foucault, 1984), away from the normalization of animal exploitation within society. Searching for such spaces in Romania, I met Silvia and, slowly after, I started focusing on (mostly feral) pigeons, the species called Columba livia. At her sanctuary, Sepale, there are hundreds of them, some in a constant coming and going, and some staying forever. Reviled birds in the Western urban space, pigeons defy the nature/culture boundary by occupying a ledge or a rooftop deemed to be for humans only. They boggle citizens' minds because they don't belong to "nature" – they don't stay in the "wild" nor in the urban parks but rather occupy the middle of the plaza, city squares, and sidewalks. Pigeons demand of us to re-think binaries just by taking space (Escobar, 2014). They put us in a position of challenging human exceptionalism, and thus are great not only to think about, or with, but also for.

This chapter aims to explore the concept of care and argue that it is an essential part of animal liberation, seen as an ongoing process. I will focus on the way care enables multispecies design within the urban space, where multispecies does not mean simply that many species reside there but rather, as Michelle Westerlarken writes, "that multiple species . . . are involved with the making of worlds, without assigning any particular order, direction, or hierarchy" (2021, p. 8). I hope that by looking at instances of intentional multispecies coexistence as an enactment towards animal liberation, we can see what works and what doesn't, mend what we can, and keep going. I will be looking at care as ethic, knowledge, practice, and work, and at how it can be part of designing multispecies spaces and prefiguring small heterotopias (Foucault, 1984), sites of contestation of anthropocentrism.

To better exemplify this, I will present the particular situation of pigeons, also called synanthropic, commensal, stray, urban, house, or street pigeons. I don't believe any of these categorizations are able to fully capture who/how pigeons are, however, in this chapter, I will use the terminology of "feral pigeon." It is important to note that the "feral" designation can wield violent power over the classified, being sometimes equated with "killable" and "ungrievable" (Probyn-Rapsey, 2016). However, I make use of it for its capacity to mark a particular set of relationships that we have with them not just as synanthropes but as abandoned ex-captives, escapees, and liminal animals (Donaldson & Kymlicka, 2011). Second, "feral" captures their ability to escape the wild/domestic binary, as categorial "outlaws" – beings whose very presence poses an ontological and epistemic challenge to the anthropocentric order. Third, it is the term Johnston and Janiga (1995) advise as most accurate in their monograph, and the term usually used in scientific publications.

The ideas explored later in the chapter are born from these past few years of thinking about animal issues from an intersectional, total liberation perspective. Some of them come out of 'grounded theory,' built on data from fieldwork/interviews with sanctuary caregivers/workers in Romania. Others arise from conversations with scholar-activists and friends.[2] This knowledge is situated, coming from a human in a country at the material and discursive periphery of Europe, usually implicitly understood to mean Western Europe.

Liberation for Feral Pigeons?

Calls for animal liberation have long been made by those who witnessed, wrote about, and fought against the exploitation, oppression, marginalization, and killing of non-human animals. To capture the different degrees of domination humans exert over other animals, ecofeminist sociologist Erika Cudworth proposes the term "anthroparchy," which refers to "a social system . . . in which the incredible diversity of non-human species are homogenized as 'animals,' identified as part of 'nature' and dominated through formations of social organization which privilege the human" (2011, p. 107). Within a multispecies liberated world, we can hope there will be no such social system, and that whatever violence may still inevitably occur, will not be part of a bigger structure from which certain groups constantly benefit. As such, we can think of the current state of the world as something to depart from when going towards a multispecies worlding of liberation, with "a determination to understand and combat all forms of inequality and oppression" (Pellow, 2014, p. 5). Total liberation allows us to truly envision other worlds, in which no one is exploited for another to profit.

It is easy to grasp how farmed animals need liberation, being bred to be disabled (Taylor, 2017) and then killed; nor is it hard to understand why "pets" who are bought, sold, and disposed of if unwanted (Shir-Vertesh, 2012) need freedom from human control. But what of feral pigeons? Like many other "feral" animals, pigeons have successfully adapted to live next to human settlements (Johnston & Janiga, 1995). What can liberation mean for beings who live and – so to say – 'thrive'[3] among humans?

From the rise of an agricultural society to the present neoliberal capitalism, pigeons have lived among humans. Their domestication happened in various places and ways.[4] When done by humans, its purpose was multiple – aside from being used for food and fertilizer, pigeons were also deeply embedded in cultural, symbolic, and religious social life (Johnston & Janiga, 1995).

Nowadays, feral pigeons can be found almost everywhere in the world because humans have brought domestic pigeons along (Johnston & Janiga, 1995, p. 14) in their global explorations and colonizations. More recently, they have been extensively used in scientific experimentation, preferred for their tameness and availability (Johnston & Janiga, 1995, p. 272), they have been bred for sports and entertainment (such as pigeon racing and fancying) and for maintaining an ethnic identity[5] (Jerolmack, 2013). In tandem with relationships of exploitation, pigeons in urban spaces have begun to be classified as "pests" (Jerolmack, 2008).

In his research on how these birds are seen in New York, in the United States, sociologist Colin Jerlomack notices how they become socially constructed as a "nuisance" through public discourse (2008). Being called "rats with wings" and treated as an economic inconvenience, as well as a (socially constructed) health hazard, creates a narrative that supports practices of control, from criminalizing pigeon feeding to more violent efforts such as shooting, gassing, or poisoning (Jerolmack, 2008, p. 72). The change in the treatment of pigeons throughout history is a prime example of how non-human animals are (de)valued based on their usefulness to humans. Their subjectivity and agency are recognized only when beneficial to human society (e.g., pigeon war hero Cher Ami) or when they cause trouble, leading to them being categorized as invasive (Timeto, 2020).

So why think of and, whenever possible, with, pigeons? First, pigeons have been around for centuries, and the pigeon-human relationship plays a specific, yet unacknowledged role in both human and pigeon lives. Second, in thinking about liberation for feral animals such as pigeons, we are confronted with the limitations of what we tend to think of as oppression in the animal rights movements. Feral pigeons are not exploited for their 'meat' or feathers (though many 'domestic' pigeons still are), and neither are they restrained (in cages) or dominated. However, when looking closely, we can see that they are marginalized, which sometimes leads to invisibilization and the lack of care and voice associated with it; and at other times to outright violence – when they are called invasive, undesirable 'pests' and killed. Moreover, their status as feral is permeable and flexible. Many domesticated pigeons become feral through abandonment or escaping, and have a tough time learning how to survive. Feral and domesticated pigeons belong to the same species, so liberation for feral pigeons must include liberation for all pigeons and thus the cessation of breeding for any reason. Such a move asks, more generally, for animal liberation, which in turn asks for total liberation.

Departing From Carelessness, Resisting the Hostile City

Today's Western cities face several problems, being developed in a capitalist system that prioritizes profit over humans and the rest of nature. They are "made for the use and consumption . . . of one well-defined group of humans, that is those who 'produce'" (Marchesini, 2016, p. 80). The few green spaces are usually designed for humans, and free-living animals are rarely taken into consideration (Borsellino, 2016), while commodified animals are largely invizibilized (Arcari et al., 2021).

Urban exceptionalism sees cities as having "somehow risen above the physical constraints of 'nature'" (Houston et al., 2018, p. 192). "Nature" becomes a resource, a raw material for the urban to grow, and its non-human inhabitants end up being ignored or driven away. As cities were cleaned of 'farmed' animals through legislations that pushed animal slaughter outside of its bounds, many synanthropes also became undesirable and were driven out, killed, and/or transformed into pets – the case of dogs, for example (Cole & Stewart, 2014; Nor, 2022). Ones that remain and adapt to "city life" are often seen as intruders, invasive, or "pests" (Jerolmack, 2008).

Not only were cities usually *not* designed for other animals but rather they were planned against them through violent architectures (Arcari et al., 2022). Spikes, wires, and nets are just some examples of anti-pigeon architecture.[6] Architect Selena Savic names these "hostile" or "unpleasant designs," a term that describes the use of architecture to control and exclude the unwanted, including the infamous example of anti-homeless benches (Teo, 2018). Unpleasant designs are discriminatory, "being pleasant to the ones that deploy it, and being unpleasant to the behavior or group of people it is used against" (Savic as cited in Teo, 2018).[7]

To counter the idea of the urban as an anthropocentric space, geographer Jennifer Wolch advises trying to "understand urbanization from the perspective of its meaning for animal life" (2002, p. 735). One tool for the multispecies city is Michelle Westerlaken's (2021) call to create multispecies worlds, a positive concept that aims to acknowledge the presence of multiple agents and open up space for them. I want to propose a similar spirit of walking/crawling/flying[8] towards coexistence, by practicing care in the urban space. That means trying out multiple solutions inspired by more-than-human interactions and entanglements, going forward, backward, and sideways to think/feel beyond anthroparchy. It means to enact, embody, and put care within urban design, so the city's environment itself becomes conducive to liberation.

Starting With Care – Nurturing the Roots of Liberation

What does a society that strives towards animal liberation look like? What does it value, push towards or pull away from? I will start with the concept of care, for my research in animal sanctuaries has shown care as one of the first steps towards multispecies coexistence. Care is not easy to do or understand, and it doesn't bring us to liberation *directly*. It happens in close contact, but it can also happen at a distance. In my understanding, care is many things: ethics, practice, knowledge, and work.

In animal ethics, there is a feminist tradition of care (Adams & Donovan, 2007) containing political analysis and understanding of systemic aspects of animal abuse. Attention is key, both towards individual suffering and well-being and towards the structural causes of violence against animals. In terms of ethics, care theory considers context rather than abstraction; it focuses on relations rather than individual reason as a source of morality; it is situated and strives to make connections known rather than demand impartiality; and it tries to reframe moral problems rather than focus on situations of conflict and choice (Gruen, 2015, C2). Importantly, it refuses the apparent separation of reason and emotion and other binaries (Gruen, 2015, C2).

In practicing care, we can start from Lori Gruen's entangled empathy, "a type of caring perception focused on attending to another's experience of wellbeing" (2015, C1). When we recognize another animal's need and act upon it, we are practicing care ethics and doing care work. Care, thus, is based on certain knowledges, which we can gain through both practice and research. "Care" needs to be understood as including both what care stands on (ethics/knowledge) *and* what it consists of (practice/work).

Feminist theory has shown how care is relegated to "women's work" and hence invisibilized and pushed out of the productive sphere (Ivancheva & Keating, 2020). However, care work is not just domestic labor but includes "the reproduction of the species, the reproduction of the workforce, the reproduction of ties and resources necessary for care and affection"[9] (Balzano, 2021, p. 114). Today women still do most of these forms of reproductive labor worldwide (Balzano, 2021, p. 115), which is one reason why feminist movements have demanded that reproductive work be paid, distributed more equally among genders, or be communalized (state or community creches, for example).

Understanding care as work shows us the politicized, unequally distributed aspect of it. Women have been deeply involved in animal movements from the beginning (Gaarder, 2011). They are generally more sympathetic to animal issues and veganism, and do more of the care and invisible work of animal activism, although men are, problematically, the more visible representatives (Gaarder, 2011).[10] Similarly, in animal rescue and sanctuaries, women often make up the bulk of the organization. An Australian study of animal rescuers found that 91.8% were women (Taylor et al., 2022) and in an unpublished study of 13 sanctuaries in Spain, 10 reported that the majority of members and volunteers were women (Franco-Barrera & Fernández-Mateo, 2021).[11]

Eco/cyborg/feminist Angela Balzano explains that even a fair division of care work (as it currently exists) among genders is not enough to reach a vision of utopia in which labor doesn't eat up all of our time, if we don't degrow and reorganize society for nurturing posthuman kin (Balzano, 2021, p. 119). So, if we consider production to serve reproduction (and not the other way round, as it currently is under capitalism), if we imagine our labor to sustain a multispecies community rather than be exploited for profit, then we can wonder: What would it look like if slaughterhouses were replaced by vegan permaculture gardens and sanctuaries? What ecosystems could emerge, if, like in Donaldson and Kymlicka's (2015) proposal, we created intentional multispecies communities instead of sanctuaries simply as spaces of refuge? What possibilities would open up for total liberation if exploitation for profit and growth ceased? If we would focus our energies on reproductive labor to nurture the more-than-human? And how would cities change if we recognized them as multispecies places where mutual care can flourish? To partially answer these questions, I will be looking at the possibilities and perils of consciously implementing housing for pigeons.

The Complexities of Reframing Dovecotes With Ecofeminist Care

Housing for pigeons, whether described as a dovecote, pigeon tower, columbarium, pigeonnier, or another term, has been around for a long time. It can almost be considered ancient technology, designed for housing these birds from the beginning of their domestication with the main purpose of using them for food and/or fertilizer. With the Industrial Revolution, chickens replaced pigeons for 'meat' and chemical fertilizers displaced natural sources, resulting in a reduction in pigeon-keeping for these purposes (Amir, 2020) and hence their housing also fell into disuse. Pigeons began occupying nooks and crannies in 'our' buildings, which offer similar functionality as the spaces in which wild Rock Doves used to nest.

After pigeons became 'useless' to humans and then reviled, the initial desire to have more of them turned into the desire to have as few as possible. Their classification as pests was accompanied by violent ways to remove them (Jerlomack, 2008). However, killing pigeons turned out to be an inefficient method of managing their populations and, often, new populations would occupy the freed-up ecological niche (Johnston & Janiga, 1995). More 'humane' and efficient methods have been researched, including the use of birth control, ultrasound, and structures known as contraceptive dovecotes (Haag-Wackernagel, 1993; Dobeic et al., 2011).

The contraceptive dovecote provides shelter to pigeons who might otherwise live in overcrowded attics and derelict buildings, ending up killed when renovating (Figure 2.2). It functions by having volunteers/workers (known to the pigeons) replace their eggs with artificial ones.[12] The idea of the contraceptive dovecote has gained international support among researchers, veterinarians, animal activists, and city councilors. It provides healthy nesting conditions for pigeons while also managing their populations – seemingly a win-win solution. While animal activists see the primary purpose of the contraceptive pigeon house to be pigeon welfare, others, for example, city councilors, see it as a way to minimize the annoyance, dirt, and economic damage pigeons cause in the urban space. Reducing pigeon populations addresses the "pigeon problem" while also minimizing intra-species disease transmission and infant mortality. Moreover, signage around dovecotes can

Figure 2.2 Baby pigeons rescued from a renovated attic. Photograph by Silvia Moldovan.

be used to promote proper pigeon nutrition to the passing public,[13] suggesting alternatives to the processed waste foods they usually rely on (Figure 2.2).

As Silvia Moldovan sees it, modern dovecotes are a sign of humans taking responsibility for having previously used domesticated pigeons to their advantage (Ivana, 2023), including selecting them for being prolific breeders. Because city pigeons lack nest hygiene behavior as a result of domestication (Weyrather, 2021), by cleaning them, it can be said that humans engage in an act of repair manifested as care for the health of current pigeon populations.[14] The dovecote as a current response to our long-standing relationships with pigeons becomes a site of transformation – a place that holds the past, present, and future together; a complex, troubled, yet familiar architecture for both humans and pigeons.

When pigeons are made to be out-of-place, by urban regulations that ban public feeding (Martelli, 2021), technologies that deter nesting and roosting, and policies that support relocation, dovecotes are a way to legitimize their presence and also care for them, constituting a sort of heterotopia. However, gaining support for these structures can be challenging, as councils aren't eager to invest money in architecture for non-human beings, even if it might be more cost-effective than cleaning and restoring damaged historical buildings, as has been shown in Germany (People for Pigeons, 2008).

In Timişoara,[15] where the public feeding of animals was banned in 2021, Silvia Moldovan has been proposing pigeon towers for years with little advancement (Figure 2.3). The interdiction of feeding only exacerbated the need for them, as pigeons are used to their daily routines and can recognize the people who feed them. While not perfect, and with a tainted history, dovecotes are potential multi-species architectures that require cooperation. Pigeons will only flock to a dovecote if it is located in places they already frequent. Ultimately, the dovecote is a space whose meaning we can change together – humans and pigeons.

Currently, modern dovecotes are a response to the hatred pigeons receive, providing legitimacy to pigeon presences and nudging humans to see them as part of the city, as co-residents. In an ecofeminist ethic of care, dovecotes can be seen as situated reactions to the violence and marginalization feral pigeons face. Ideally, they provide protection, minimize human-pigeon conflict, and support adequate nutrition. Moreover, the dovecote could be a way to make up for past harm. Even if it continues to employ control over pigeon reproduction, the contraceptive pigeon house may be considered a way of doing more-than-human transitional justice (Celermajer & O'Brien, 2021).[16] Yet how do we reconcile reproductive control with feminist care? Truth is, we can't. This care we can practice is haunted by

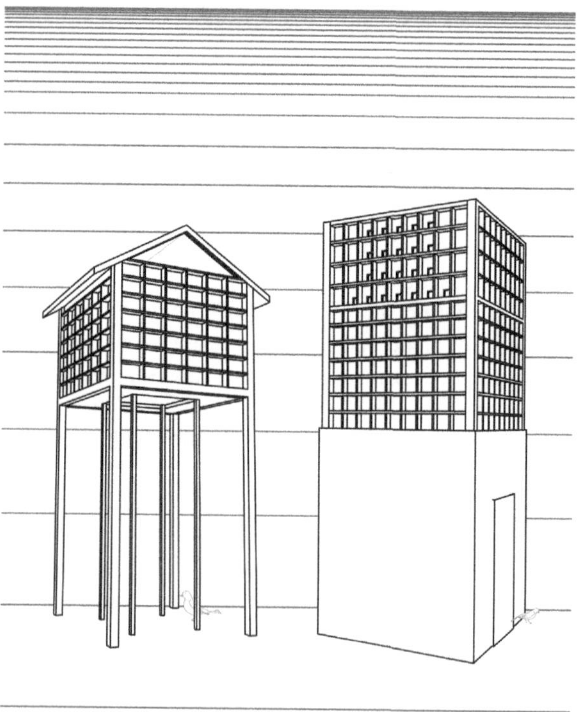

Figure 2.3 Pigeon tower models designed by Silvia Moldovan (2013–2020).

history and shaped by the society in which it's given. It's imperfect, it's still not *there*.[17] But it's what we can already do, and it's *doing* that will teach us where to go next. Recognizing our practices on an axis of time/space is essential for us to be able to push towards total liberation.

Reflecting on our past relationship of exploitation, our present relationship of marginalization and invisibilization, and on future possibilities for coexistence, the dovecote can work as a transitional symbol from violence to multispecies world-making. Thereby, it is important to consider its historical use and understand that more-than-human architecture is neither innocent nor new. Even in Timişoara, right in Operei's Plaza, where it is said that pigeons provide the most disturbance, there is a water fountain which was, anecdotally, made for them.[18] However, because of its faulty design, the water in it is not cleaned and is thus a source of infection. This, and other histories, tend to be forgotten. We need to be reminded that both coexistence and terror have already happened.

One way to go towards a multispecies city, with attention to its pigeon population, is to take care of the basics: shelter, water, and food. According to Silvia, that might mean having a dovecote, flowing water, and adequate feeding stations in every other neighborhood. We can reach a modicum of general population health through multispecies design, always learning, changing, and listening to other animals, trying to grasp how the ecosystem works. Banishing pigeons through various means, including sudden cessation of feeding, given human involvement in this dependent relationship, is at least "morally problematic" (Palmer, 2003) if not violent and outright unjust. A right-to-the-city argument for non-human animals has already been made, requiring access to resources as a simple first step (Shingne, 2020). In the quest for total liberation, solidarity between housing, right-to-the-city, and antispeciesist movements could be further explored – the city is not for profit, nor is it human only.

Careful Care – For Pigeons and Beyond

While some aspects of urban architecture have proven beneficial for pigeons, others are deadly. Walking around Timişoara's city center, particularly at the Punctele Cardinale roundabout, it is rare not to notice fallen feathers or, occasionally, inanimate pigeon bodies on the main road. Pigeons flock there to drink from the iconic fountain situated in the middle of the roundabout and surrounded by a patch of grass. It's like an oasis in the middle of a perilous mechanical storm. Pigeons who have trouble flying or reacting quickly are often hit by cars. Car accidents, string foot,[19] and the impact of pollution on their health are just some of the ways in which urban living disables these birds (Jiguet et al., 2019). To the best of my knowledge, there is no structural, city-coordinated response *for* the well-being of pigeons. The one official health concern regarding pigeons is the zoonoses they arguably spread to humans. This recurring argument is used to classify them as dangerous pests and to assure their swift removal, even though the likelihood of them passing diseases to humans is extremely small (Haag-Wackernagel, 2004).

Certainly, if zoonoses were considered to be a serious threat, animal agriculture would be targeted (Blattner, 2020), not pigeon colonies.

Given our history of domestication and exploitation, given that it is urban design that causes them harm (Martelli, 2022), and given that humans are already in deeply entangled relationships with particular pigeon populations, I argue humans have a duty of care to city pigeons (Figure 2.4). This caregiving could be a basis for pigeon liberation from the (infra)structural violence and marginalization that they face, which results in early deaths, mutilated feet, and sickly populations.

Yet how do we stay clear of paternalistic care and of further domesticating ferals? This question must stay with us, complicating our thoughts and practices. For example, pigeon towers/dovecotes have been proposed by animal activists, including Silvia, to safely house pigeons evicted from rooftops and lofts. However, they come with contraceptive designs to control their populations. This population control is supposedly also beneficial for pigeons, minimizing the spread of disease within colonies. Pigeons can breed all year round and tend to do so when resources are plenty (Johnston & Janiga, 1995). How to not confuse caregiving with projections of our own desires and interests? How do we avoid "conflating empathy with murder, conflating killing with caring" (Sanbonmatsu, 2018, 46:05) as farming and breeding practices often do? And how much care work can we actually do?

I say it *matters what kind of care we care with*, thinking of Westerlaken's "it matters what designs design designs" (Westerlaken, 2019, 10:03), which in turn

Figure 2.4 Pigeon being helped to drink water by Silvia Moldovan. Photograph by the author.

paraphrases Donna Haraway's "It matters what thoughts think thoughts. It matters what knowledges know knowledges. It matters what relations relate relations" (Haraway, 2016, p. 35). Relations of care happen in various contexts and include 'bad care' relations in which utilities extracted from non-human animals are primary (Ouellette-Dubé, 2022). Instead, Maude Ouellette-Dubé argues that good care is "when individuals in a caring relation can flourish" towards their own ends, rather than human ends (2022, p. 39).

My proposition is to ground our antispeciesist efforts in contextual knowledge-feeling, rather than abstract moral categories. This means we consider each situation and population in its specificity, answering to the present with an understanding of the past and a dream of the future. For example, this means acting now for pigeons' wellbeing and, more precisely, analyzing whether a specific site is proper for a pigeon tower by looking at multiple factors. Silvia underscores that pigeons must already be present there, otherwise, it won't function, but neighboring humans must be also welcoming to pigeons, otherwise, it might harbor conflict to their detriment.

Moreover, I am proposing we stay away from "careless care." In the case of pigeon towers, *careless care* means prioritizing the benefits they bring to the human population rather than the pigeons, including the extreme dwindling of their numbers. *Careless care* in relations with other animals, just like 'bad care,' is anthropocentric and instrumental, valuing animal welfare only as long as it benefits humans. It doesn't engage in entangled empathy but aims for efficiency; it doesn't value the process/relationship but the result/endpoint. *Careless care* is care work done without care ethics. It uses knowledges of care directed towards survival (of those receiving care), optimizing outcomes favorable to humans rather than individual or collective multispecies flourishing.

As humans, we can strive for a multispecies, feminist, *careful care* that responds and attends to the call of all animals, as both individuals and collectives. We can practice a care that is wary of anthropocentric projections and engages in anthropomorphism critically, seeing how non-human animals are both similar to and different from ourselves. An ongoing, *careful care* is care that learns while doing. Many sanctuary caregivers are already practicing it, looking at how other animals themselves care for each other,[20] recognizing their agency, and disavowing human domination. One point of *careful care* is the practice itself, the togetherness, in an entanglement beyond species that isn't assimilation or blurring of power relations, but mutual closeness. Instead of having disabling designs, we can hope to have caring ones. Instead of careless places, we can strive for cities in which care thrives, and we, too, as human and non-human animals, thrive with it.

From Anthroparchal to Caring Multispecies Relations

In the winter days of 2022, Silvia was working on constructing the sanctuary and improving her designs for pigeon towers.[21] At that time, she was calculating the costs of one model to be built as an art installation. In her practice, art works as a subversive force bringing non-human animals face to face with human ones, asking us to look, to hear, and to pay attention.[22] And art is a good enough excuse for some humans to accept something deemed as "useless" as a pigeon tower. To design caring places and multispecies cities, we must go beyond what is "useful." Anthroparchal relations are part of the roots of the current failing socio-economical system that is destroying the Earth and endangering its inhabitants. Placing care at the roots means creating another world altogether.

When we understand total liberation as freedom from domination and oppression, we can see that it also requires taking responsibility for past and current wrongdoings (Martelli, 2022). This can mean changing our narratives, our practices, and our material infrastructures – or at least, how they are used. Saying "care is at the roots" is a way of reminding us of this important, contextualized knowledge, ethic, practice, and work. It's putting ourselves in spaces of care, which are both careful towards us and conducive to care towards others. Through care we sustain each other, we get to know each other's desires, and we find out ways to meet them. Care works better when it is collective, beyond the individual or the human. When it's not in the service of capital but rather of multispecies community. When it's recognized in all its dimensions. When it's embedded into the space we live in and when it's seen in its messiness. So, if we look at pigeon towers with careful care we can see how they protect and aid pigeons, as well as how they might further control and impose human preference. Recognizing this duplicity keeps us aware that the road to liberation is long and that it must yet be imagined.

"Do you think you know what the future will look like?" is the last question we asked the audience in a short speculative film about pigeons created by Silvia and myself.[23] Overimposed on images of baby pigeons, the question is meant to decenter the anthropocentric view of time, reminding us that birds are evolutionary descendants of dinosaurs. After daring to imagine ahead into the climate crisis we are shaken back into our current seats. We only need to stand up and step ahead with careful care, to truly be "together" with those that are now, here, alive. Care is at the roots of total liberation, in the sense that it makes a liberated multispecies world possible, it brings it closer and nurtures it into being.

Notes

1 The plumage of pigeons has been intensely studied in science, from Charles Darwin to the present day (Johnston & Janiga, 1995). The matter of "reverting" to their original colors is not at all straightforward, as ferals tend to adapt to new environments rather than simply become "wild" again (Gering et al., 2019).

2 Such as Aron Nor and Mina Mimosa from the *just wondering . . .* collective, people from the *Queer Vegan Community* in Romania, chats with Silvia Moldovan, and ideas risen from/with the non-human animals I've had the luck to live with (Nuna, Aki, Mushi, Aria). Also, I want to thank the two peer-reviewers, Susan Pyke and Jennifer Schauer, as well as Paula Arcari, for the helpful comments on this piece. I underline this to make transparent that (multispecies) imagination is always drawn from a collective pool, that no work is individual, and that all ideas are born from other ideas – and more so, all are crafted from the mundane acts of living together, day by day, feeding and caring for each other, entangled in glocal networks of unequal exchanges.

3 This is often said of feral pigeons due to their population growth. However, it is questionable whether 'thriving' is the proper description. Pigeons in the urban space tend to be ill, disabled by their environment, and live short lives. While their numbers might be 'high' – and this is another debate about when/why certain species are deemed to be too many – their quality of life is frequently not. On violence against ferals, see Probyn-Rapsey (2016).

4 Johnston and Janiga, in their monograph, explain that pigeons were domesticated sometime between 5,000 to 10,000 years ago (1995, p. 6) from the wild Rock Pigeon, in a combination of capture, breeding, and synanthropy.

5 "If you grew up with pigeons, if it is in your culture, then it is in your blood," a migrant interviewee of Jerolmack states (2013, p. 110), explaining how pigeon-keeping is essential to keeping his Turkish identity in a foreign country.

6 Violent technologies and designs deeply affect animals (including humans) as more of them are being developed across wide market sectors from the military to pest control and hunting (Scotton, 2019).

7 You can find more examples of anti-pigeon architecture on Slavic's blog entry "Unpleasant for pigeons," Unpleasant, retrieved https://unpleasant.pravi.me/unpleasant-for-pigeons/.

8 A reference to multiple, multispecies ways of movement, from the title of the film "we fly, we crawl, we swim – a short film about climate justice", just wondering . . . 2021. Retrieved www.justwondering.io/we-fly-we-crawl-we-swim/

9 Translation from Italian by the author.

10 This is often noted by vegan ecofeminists (see the Vegan Ecofeminist Network). I believe that one reason for the over-representation of women and queers in the animal movement is that patriarchal societies withdraw care and emotion from (cis)men, both of which are associated with women *and* animal advocacy. When (cis)men *are* involved, it tends to be in leadership (Wrenn, 2022), and/or louder or more dangerous forms of activism such as direct action and sabotage, as Pellow (2014) noted in his research on radical movements.

11 One sanctuary respondent indicated that their member/volunteer base is equally composed of cis women and non-binary or trans people. Ties between queerness and the animal movement have been surfacing more frequently with the work of queer sanctuaries such as VINE in the United States, books such as "Queer and Trans Voices: Achieving Liberation Through Consistent Anti-Oppression" edited by Julia Feliz Brueck & Zane McNeill; and non-hierarchical activist groups such as The Queer Vegan Community (@ cvqro) in Romania.

12 The French model advises shaking the eggs to prevent the birth of new pigeons while the Augsburg model replaces the eggs with artificial ones (People for Pigeons, 2008). Within the Sepale sanctuary, Silvia gives the eggs to be eaten by corvids, fostering an ecological system in which what is taken from one non-human species is given to another, and one in which not all food comes from human labor or economic exchange within capitalism.

13 A panel can be attached to the dovecote with information for the general public on what is appropriate for pigeons to eat.

14 Depending on the local council and organizations involved, this work can be either voluntary or paid.

15 Timișoara is the main economic, social, and cultural center in western Romania. It is also where Silvia Moldovan, whom I research with, founded her sanctuary, and where, in 2021, the illegalization of public feeding of animals was proposed (and passed).

16 Transitional justice aims to account for past wrongdoings and proposes conditions to attain peace, aiming to change the "relationships that constitute the social order" (Celermajer & O'Brien, 2021, p. 130)

17 *There* is a signpost for liberation. *There* is the utopia we cannot reach as a destination, only as a process.

18 Mentioned in the news report "Oficial! În Timişoara s-a interzis hrănirea porum-
    beilor şi se permite plimbarea câinilor şi pisicilor cu STPT" by Roxana Deaconescu,
    25 May 2021, retrieved www.tion.ro/stirile-judetului-timis/oficial-in-timisoara-s-a-
    interzis-hranirea-porumbeilor-si-se-permite-plimbarea-cainilor-si-pisicilor-cu-
    autobuzul-1440108/.

19 String foot, also known as toe mutilation, is often found in urban pigeons, particularly in
    densely populated and polluted areas, as a study in Paris has shown (Jiguet et al., 2019).

20 There are countless examples of friendships and mentorships developed within sanctu-
    aries between non-human animals nurturing others to health or helping them socialize
    and learn the daily routines. Silvia's sanctuary practice is based on non-humans being
    interdependent with each other rather than with her, as much as possible. Many exam-
    ples can be found on social media, for example Nima Sanctuary (www.facebook.com/
    photo.php?fbid=536530258588886&set=pb.100066957487271.-2207520000&type=3)
    and Spirit Animals Sanctuary (most recently a dog befriending a pig – www.facebook.
    com/watch/?v=1573162686786431).

21 Sketches of Silvia's designs can be found on https://silviamoldovan.art/pigeon-towers-
    2013-2014/, https://silviamoldovan.art/dovecoats/ and https://sepale.ro/columbiada/.

22 Silvia's 2023 solo show "Columbiada" featured large portraits of pigeons and advo-
    cated for pigeon towers as a signpost against habitat encroachment and as a step
    toward cohabitation. The portraits could be seen from outside the Meta Spaţiu Gallery.
    Retrieved www.facebook.com/events/208993858547191/.

23 The film "In Time, You Will See How The Sky Is Great Enough" was initially shown
    at the art exhibition Cozzzmonautica 2022, at Indecis Art Space, and later appeared
    in Specula Mag. Retrieved https://speculamag.com/the-loud-ocean-and-in-time-you-
    will-see-how-the-sky-is-great-enoughthe-loud-ocean/.

Bibliography

Adams, C. J., & Donovan, J. (Eds.). (2007). *The Feminist Care Tradition in Animal Ethics:
A Reader*. Columbia University Press.

Amir, F. (2020). *Being and Swine: The End of Nature (As We Knew It)* (C. Russell, Trans.).
Between the Lines.

Arcari, P., Probyn-Rapsey, F., & Singer, H. (2021). Where Species Don't Meet: Invisibilized
Animals, Urban Nature and City Limits. *Environment and Planning E: Nature and Space,
4*(3), 940–965. https://doi.org/10.1177/2514848620939870

Arcari, P., Probyn-Rapsey, F., & Singer, H. (2022). Violent Architecture. *Architect
(Architect Victoria)*, Edition 3, 72–76. https://www.architecture.com.au/vic-chapter/
architect-victoria-design-for-all-life-architecture-and-planning

Arruzza, C., Bhattacharya, T., & Fraser, N. (2019). *Feminism for the 99%: A Manifesto*.
Verso Books.

Balzano, A. (2021). *Per farla finita con la famiglia: Dall'aborto alle parentele postumane*.
Meltemi Editore.

Blattner, C. E. (2020). From Zoonosis to Zoopolis. Derecho Animal. *Forum of Animal Law
Studies, 11*(4), 41. https://doi.org/10.5565/rev/da.524

Borsellino, L. (2016). Animali liminali in città: Spazi, resistenza e convivenza. *Liberazioni,
27*, 43–57.

Celermajer, D., & O'Brien, A. T. (2021). Alter-Transitional Justice; Transforming Unjust
Relations with the More-than-Human. *Journal of Human Rights and the Environment,
12*(Special), 125–147.

Cole, M., & Stewart, K. (2014). *Our Children and Other Animals: The Cultural Construction of Human – Animal Relations in Childhood* (1st ed.). Routledge.

Cudworth, E. (2011). *Social Lives with Other Animals: Tales of Sex, Death and Love*. Palgrave Macmillan.

Dobeic, M., Pintari, Š., Vlahović, K., & Dovč, A. (2011). Feral Pigeon (*Columba livia*) Population Management in Ljubljana. *VETERINARSKI ARHIV, 81*(2), 285–298.

Donaldson, S., & Kymlicka, W. (2011). *Zoopolis: A Political Theory of Animal Rights.* Oxford University Press.

Donaldson, S., & Kymlicka, W. (2015). Farmed Animal Sanctuaries: The Heart of the Movement? *Politics and Animals, 1*(1), 25.

Escobar, M. P. (2014). The Power of (Dis) Placement: Pigeons and Urban Regeneration in Trafalgar Square. *Cultural Geographies, 21*(3), 363–387. https://doi.org/10.1177/1474474013500223

Foucault, M. (1984). Of Other Spaces: Utopias and Heterotopias. *Architecture/Mouvement/Continuite*, October, 2, 1–9 ("Des Espace Autres," March 1967 Translated from French by Jay Miskowiec).

Franco-Barrera, A., & Fernández-Mateo, J. (2021). *Expressions of Animal Ethics: Animal Sanctuaries, the Case of Spain* [Conference presentation]. EACAS Conference on Appraising Critical Animal Studies, June 24–25, online, Edge Hill University. https://youtu.be/UCw-AApI4oA

Gaarder, E. (2011). Where the Boys Aren't: The Predominance of Women in Animal Rights Activism. *Feminist Formations, 23*(2), 54–76. https://doi.org/10.1353/ff.2011.0019

Gering, E., Incorvaia, D., Henriksen, R., Conner, J., Getty, T., & Wright, D. (2019). Getting Back to Nature: Feralization in Animals and Plants. *Trends in Ecology & Evolution, 34*(12), 1137–1151. https://doi.org/10.1016/j.tree.2019.07.018

Gruen, L. (2015). *Entangled Empathy: An Alternative Ethic for Our Relationships with Animals*. Lantern Books.

Haag-Wackernagel, D. (1993). Street Pigeons in Basel. *Nature, 361*, 200.

Haag-Wackernagel, D. (2004). Health Hazards Posed by Feral Pigeons. *Journal of Infection, 48*(4), 307–313. https://doi.org/10.1016/j.jinf.2003.11.001

Haraway, D. (2016). *Staying with the Trouble: Making Kin in the Chthulucen*. Duke University Press.

Houston, D., Hillier, J., MacCallum, D., Steele, W., & Byrne, J. (2018). Make Kin, Not Cities! Multispecies Entanglements and 'Becoming-World' in Planning Theory. *Planning Theory, 17*(2), 190–212. https://doi.org/10.1177/1473095216688042

Ivana, M. (2023). The Contraceptive Dovecote – a Coexistence Experiment. *Asociatia Culturala Contrasens*. www.accontrasens.ro/artsens-articles/the-contraceptive-dovecote-a-coexistence-experiment/

Ivancheva, M., & Keating, K. (2020). Revisiting Precarity, with Care: Productive and Reproductive Labour in the Era of Flexible Capitalism. *Ephemera: Theory & Politics in Organization, 20*(4), 251–282.

Jerolmack, C. (2008). How Pigeons Became Rats: The Cultural-Spatial Logic of Problem Animals. *Social Problems, 55*(1), 72–94. https://doi.org/10.1525/sp.2008.55.1.72

Jerolmack, C. (2013). *The Global Pigeon*. University of Chicago Press.

Jiguet, F., Sunnen, L., Prévot, A.-C., & Princé, K. (2019). Urban Pigeons Losing Toes Due to Human Activities. *Biological Conservation, 240*, 108241. https://doi.org/10.1016/j.biocon.2019.108241

Johnston, R. F., & Janiga, M. (1995). *Feral Pigeons*. Oxford University Press.

Kimmerer, R. W. (2015). *Braiding Sweetgrass: Indigenous Wisdom, Scientific Knowledge and the Teachings of Plants*. Milkweed Editions.

Marchesini, R. (2016). Animals of the City. *Angelaki, 21*(1), 79–91. https://doi.org/10.1080/0969725X.2016.1163825

Martelli, M. (2022). Despre porumbei şi oameni. În căutarea unei responsabilităţi non-antropocentrice. *Iscoada*. https://iscoada.com/text/despre-porumbei-si-oameni-in-cautarea-unei-responsabilitati-non-antropocentrice/

Miklósi, A. (2015). *Dog Behaviour, Evolution, and Cognition*. Oxford University Press.

Nor, A. (2022). Inamicii economiei socialiste, intruşii civilizaţiei occidentale. *Gând vagabond*. https://aronnor.ro/inamicii-economiei-socialiste-intrusii-civilizatiei-occidentale

Ouellette-Dubé, M. (2022). What Are Good Multispecies Relations? An Analysis Through the Concept of Caring Relations. In E. Cudworth, R. E. McKie, & D. Turgoose (Eds.), *Feminist Animal Studies: Theories, Practices, Politics* (pp. 35–49). Routledge. https://doi.org/10.4324/9781003222620

Palmer, C. (2003). Placing Animals in Urban Environmental Ethics. *Journal of Social Philosophy, 34*(1), 64–78. https://doi.org/10.1111/1467-9833.00165

Pellow, D. N. (2014). *Total Liberation: The Power and Promise of Animal Rights and the Radical Earth Movement.* University of Minnesota Press.

People for Pigeons. (2008). *Safe-Feeding Zones for Pigeons.* https://peopleforpigeons.blogspot.com/2008/09/safe-feeding-zones-for-pigeons.html

Probyn-Rapsey, F. (2016). Five Propositions on Ferals. *Feral Feminisms, 6,* 18–21.

Sanbomatsu, J. (2018). *Lady Macbeth at the Rotisserie: 'Femivores,' Violence, and the New Maternalism in Animal Agriculture* [Conference presentation]. UPC's Seventh Annual Conscious Eating Conference: What Are the Most Compassionate Choices? March 10. www.youtube.com/watch?v=hFRT4W6KHPk

Scotton, G. (2019). Taming Technologies: Crowd Control, Animal Control and the Interspecies Politics of Mobility. *Parallax, 25*(4), 358–378. https://doi.org/10.1080/13534645.2020.1731004

Shingne, M. C. (2020). The More-than-Human Right to the City: A Multispecies Reevaluation. *Journal of Urban Affairs, 44,* 1–19. https://doi.org/10.1080/07352166.2020.1734014

Shir-Vertesh, D. (2012). "Flexible Personhood": Loving Animals as Family Members in Israel. *American Anthropologist, 114*(3), 420–432. https://doi.org/10.1111/j.1548-1433.2012.01443.x

Taylor, N., Fraser, H., & Signal, T. (2022). 'Rescued and Loved' Women, Animal Sanctuaries, and Feminism. In E. Cudworth, R. E. McKie, & D. Turgoose (Eds.), *Feminist Animal Studies: Theories, Practices, Politics* (pp. 218–234). Routledge.

Taylor, S. (2017). *Beasts of Burden: Animal and Disability Liberation.* New Press.

Teo, M. (2018). Unpleasant Design: The Rise of the Silently Hostile City. *Azure Magazine.* www.azuremagazine.com/article/unpleasant-design-hostile-architecture/

Timeto, F. (2020). *Bestiario Haraway: Per un Femminismo Multispecie.* Mimesis Edizioni.

Westerlaken, M. (2019). *It Matters What Designs Design Designs: Speculations on Multispecies Worlds* [Conference presentation]. WUD Silesia conference in Katowice, Poland, November. https://michellewesterlaken.com/2020/04/22/it-matters-what-designs-design-designs/

Westerlaken, M. (2021). What Is the Opposite of Speciesism? On Relational Care Ethics and Illustrating Multi-Species-Isms. *International Journal of Sociology and Social Policy, 41*(3/4), 522–540. https://doi.org/10.1108/IJSSP-09-2019-0176

Weyrather, A. (2021). *Basics for an Efficient, Animal Welfare-Friendly City Pigeon Management in (Large) Cities in Germany.* A manual for practical implementation. Menschen für Tierrechte – Bundesverband der Tierversuchsgegner e.V. (People for Animal Rights – Federal Association of Opponents of Animal Experiments, registered non-profit association).

Wolch, J. (2002). Anima Urbis. *Progress in Human Geography, 26*(6), 721–742. https://doi.org/10.1191/0309132502ph400oa

Wrenn, C. (2022). Building a Vegan Feminist Network in the Professionalised Digital Age of Third-Wave Animal Activism. In E. Cudworth, R. E. McKie, & D. Turgoose (Eds.), *Feminist Animal Studies: Theories, Practices, Politics* (pp. 235–248). Routledge.

Chapter 3

Unveiling Shared Histories

Crafting Sanctuary and the Work of Care in Troubled Domestic Domains

Marie Leth-Espensen

> The sanctuary provides a glimpse into a completely different world where you can experience interactions and share a life with animals you are only used to seeing in a refrigerated display or similar settings. As you witness this, you realise the potential for something beyond . . . a recognition that it could be different, and I believe this realisation inspires a unique kind of change.
>
> *Sanctuary caregiver at the Cornflower Refuge*[1]
> *Denmark, September 2020*

> I envision a future where we return nature to animals, and it would actually be humans living in reserves or confined spaces. In this scenario, animals would be free to inhabit the natural environments they truly deserve.
>
> *Sanctuary caregiver at the Little Green Cottage*
> *Denmark, October 2020*

Introduction: Interspecies Relations in the Domestic Domain

A recent body of literature across the humanities and social sciences re-examines history to reposition the role of nonhuman animals within it. What are the implications of entangled historical narratives for the multispecies present and future? This question serves as the starting point for the reimagining of interspecies relations undertaken, for example, by animal studies scholars Sue Donaldson and Will Kymlicka (2011) in their influential work *Zoopolis*, in which they develop a political theory of animal rights focused on viewing nonhuman animals as members of political communities. Their approach emphasises human relational duties towards other animals, rooted in shared pasts, deviating from the focus on animals' inherent rights independent of their value to humans, as often found in conventional animal ethics and philosophy (see, for instance, Regan, 1983). Donaldson and Kymlicka also advocate for articulating a broader vision for interspecies relations, arguing that the animal movement must envision a comprehensive future beyond simply ending exploitation and wrongs of the past. An additional critical consideration in exploring shared more-than-human histories is the necessity to frame perspectives

DOI: 10.4324/9781003366706-5

on interspecies just futures within the historical contexts of extractivism, colonial exploitation, and capitalism without succumbing to tendencies towards human exceptionalism but rather by acknowledging the agency and role of nonhuman animals in shaping our shared world (Gillespie, 2022; Noske, 1997).

Similar to this call for new images of multispecies futures anchored in human and animal past relationships, an alternative strand of literature has recently raised concern with what is perceived as a tendency to over-emphasise human control in prior scholarly work on domestication, thus resulting in the further marginalising of nonhuman animals, their histories and agency (Swanson et al., 2018). For example, the authors of the edited volume *Domestication Gone Wild* problematise the characterisation of domestication as a manifestation of complete domination (naming the often-cited overview of the topic by Clutton-Brock, 1999) and wish to adopt a more open-ended approach to domestication (Swanson et al., 2018; see also Lien, 2015). However, the call for more 'nuanced' perspectives on domestication tends to overlook specific and extremely troubling dimensions of human–animal relations, such as the extensive violence done in the name of domestication. Alternatively, critical animal scholarship addresses the historical legacy of domestication and specific institutions based on the subjugation of domesticated or tamed animals, such as the farm or the zoo, bringing new insights into shared histories while still recognising animals as historical agents (Colling, 2021; Hribal, 2007). This body of literature acknowledges the many pasts in which humans have sustained themselves through the use and consumption of animals without putting aside the question of the injustices inflicted upon other animals within these arrangements.

In this chapter, I delve into how sanctuary making for previously farmed animals can be considered an attempt to address the challenges inherent in improving the circumstances for a distinct group of domesticated animals, namely those subjected to agricultural control and management of their lives. In this context, the recent emphasis on shared histories prompts important inquiries, such as the specific obligations that fall upon human beings when considering past and present histories and relationships between humans and domesticated animals (Donaldson & Kymlicka, 2011; Gruen, 2013). As readers familiar with this topic will know, farmed animal sanctuaries (FASes) play a prominent role in animal liberation and critical animal studies literature. In fact, FASes have previously been proposed as a valuable context for exploring these complicated ethical and political issues (Abrell, 2021; Donaldson & Kymlicka, 2015; Gillespie, 2019; M. Jones, 2014; p. jones, 2010), even though they differ in many of their practices (Abrell, 2021).

Given sanctuaries' commitments to the flourishing of animals who have escaped from sites of exploitation, they come to embody an important – although not entirely uncomplicated – position in the broader landscape of animal advocacy (Chang, 2018; McMain & Torres, 2023; Pachirat, 2018).[2] However, a continually debated issue concerns whether sanctuaries should be considered a manifestation of future interspecies relationships based on reciprocal care and compassion or if the broader objective of sanctuaries commits them to make themselves redundant. Put simply, if the ultimate purpose is to eradicate the abuse and exploitation of other animals entirely, there should be no necessity for nonhuman animals to live in human-dominated environments, as is the case in sanctuaries (though human-driven rehabilitation might arguably still be needed). However, based on my observations during this study, I am increasingly convinced that there is no definitive or clear-cut answer regarding sanctuaries' role in providing concrete alternatives to the complex interspecies relations of the present. Instead, I want to emphasise the dual function of sanctuaries: deeply involved in resisting human domination while actively working to foster innovative forms of interspecies caring relationships. Consequently, sanctuaries are rooted in the acknowledgement that, as humans, we bear the responsibility to undertake the reparative efforts due to the burden of our species' historical relations and practices, at least until these practices and relational arrangements cease to dominate (Gruen, 2013).

Based on ethnographic vignettes that stem from interviews and participant observations conducted at two sanctuaries in rural Denmark, the chapter underscores critical aspects of the effort of crafting sanctuary in what I characterise as *troubled domestic domains*. These domains are understood as spaces shaped by both historical and ongoing processes of domestication that embody a unique form of speciesist boundary-making characterised by power-laden binaries such as nature/culture and human/animal. Moreover, I emphasise the significance of care modelled at sanctuaries as essential in prefiguring and envisioning multispecies just futures, given its central role in re-doing and enhancing interspecies relationships in the present. Indeed, I situate the efforts of sanctuaries against the fundamental crisis in care that characterises the many urgent and multifaceted issues that define human relationships to the nonhuman world in contemporary neoliberal and capitalist societies, including continuous environmental degradation, extinction, and general decline in animal abundance. By crisis *in* care, I refer to the long history of feminist activist and scholarly interventions that have highlighted how the issue of the unequal distribution of care is not simply one of a deficit or a lack of care but concerns how care is currently configured in global structures of inequality, exploitation and the commodification of nonhuman beings (Crary & Gruen, 2022; The Care Collective, 2020).

The two sanctuaries included in this study can be placed in the larger movement that has emerged globally during the past decades, although divergent in their caregiving practices and principles (Abrell, 2021). With the intention to rescue, care, and rehabilitate previously farmed and abused animals, *Farmed Animal Sanctuaries* – as they are often named – operate with a foundation of compassion and empathy in their caregiving efforts. Individuals provided refuge at the sanctuaries are mainly domesticated land animals, such as cows, pigs, chickens, geese, sheep, goats, and rabbits. The two Danish sanctuaries I visited are comparatively small, housing approximately 40 and 60 individual animals, respectively. Each sanctuary is managed by a single human caregiver who resides on the property, although they receive support from a community of volunteers. These sanctuaries prioritise meeting the individual and social needs of their nonhuman residents. In contrast to well-known sanctuaries in North America, such as *Farm Sanctuary* established in

1986, the two sanctuaries informing this study are less accessible to the public and place a stronger emphasis on maintaining meaningful and caring practices in their daily operations.

In the following sections, I delve into the day-to-day operations of the sanctuaries, with less emphasis on differentiating between them, aiming to understand the nature of care needed for the rehabilitation of previously abused and traumatised individuals. Within this context, I observe how FASes seek to accommodate the needs, interests, and desires of previously abused and traumatised individuals while actively working towards fostering more compassionate and responsible terms for coexistence between humans and other animals. The chapter is organised according to three analytical points related to the critical work of care at farmed animal sanctuaries. The first section proposes sanctuary making as unrecognised care work, parallel to earlier feminist critique in the realm of intra-human care. The second section draws attention to the ways in which the effort of providing life-long care for previously farmed animals juxtaposes sanctuaries against the legal standards of care as currently shaped by industrial farming. The third section centres on mourning the death of sanctuary members as an ethical and political act that underlines care as a transformative force in society. Finally, the chapter concludes by discussing how sanctuary care work challenges hegemonic images of interspecies life in the domestic domain. On this point, I contend that sanctuary caregiving provides a stimulating context to ponder the question of shared histories, pointing to some specific absent issues in mainstream animal advocacy while simultaneously adding to ongoing scholarly conversations about the ethics and politics of entanglement (Giraud, 2019). At the same time, the care performed and modelled at the sanctuaries is key to significantly improving and bringing about more just interspecies futures.

Unrecognised Care Work

I find myself standing beside one of the enclosures that separates the sanctuary from the outside world. I stay for a while, observing a donkey and a goat basking in the sun-drenched pasture. They move slowly as they eat. They don't seem to mind the crowd watching them at a distance. It's one of those rare occasions when the sanctuary opens its doors to the public. Pauline, a passionate advocate, introduces the sanctuary's residents to me and other visiting guests. She shares each of their individual stories, including their histories of trauma and how they eventually made it to the sanctuary.

Later, Pauline informs me that the goats are the most recent arrivals at the sanctuary, having joined the community about six months ago. Their previous guardians, a small family consisting of two adults and two kids, had been forced to relocate them due to numerous complaints from annoyed neighbours disturbed by the goats' bleating. As the family lived in a small town, it was up to the local authorities to decide if the family should be allowed to keep the goats. As it turned out, the goats would soon find themselves in need of a new home.

This wasn't the first time the goats had been forced to relocate, I was informed. In fact, the goats had previously lived on the property of a residential treatment centre for at-risk youth, where their presence had had calming effects on troubled teens. However, at some point, the 'care service' provided by the goats was no longer deemed useful, and it was decided that they would be "put down," to use the common euphemism normalising this practice. Only because one parent of the family intervened after learning about their situation, the two goats avoided the destiny set out for them.

Visit to the Cornflower Refuge
August 2020

This ethnographic vignette illustrates some crucial points about the uncertainty and precariousness that characterise the lives of many animals in human custody. The residential treatment centre's decision to dispose of the goats, despite years of building and sustaining relationships with them, illustrates how the deaths of nonhuman animals are often regarded as what Dinesh Wadiwel (2015) has referred to as *a non-event*. Despite the daily relationships that had formed with these individuals, the goats were discarded when they no longer served their initial purpose, with seemingly no extended effort made to preserve their lives. There is a reason I highlight the goats' story and not the many more overt cases of abuse and neglect that I learned about during the time I spent at two sanctuaries. The goats' story is not so much about ill intentions or moral corruption but rather serves as an illustration of the fundamental issue of anthropocentrism and fraught ideas of 'care' in society. Even when individual animals are part of seemingly non-exploitative relationships, which they could potentially also benefit from, humans still often fail to safeguard their interests and wellbeing. It is chillingly easy to decide on the disposal of an animal.

The caregivers at the sanctuaries I visited explained how it is not rare for them to take in individuals due to an unforeseen change in the lives of their human caretakers. In fact, and surprising to me, many of the sanctuary members did not arrive from 'primary' sites of exploitation, such as the farming industry. Instead, some have been used for recreational purposes or have been kept by hobby farmers who, too late, had learned what it requires to keep and *care* sufficiently for animals. Often, sanctuaries are asked to take in animals to prevent their immediate killing. This is not because these animals were initially destined for the meat or dairy industry but rather due to a change in circumstances that left their human caretakers seemingly unable to provide for them any longer. Furthermore, some sanctuary members have managed to escape the facility where they were initially kept and sometimes individuals have intervened after witnessing instances of abuse and mistreatment.

Recent literature on animal labour offers insightful interpretations of intra- and interspecies caregiving as a particular kind of labour. For example, following the classical insight produced in feminist activism and scholarship, labour studies scholar Kendra Coulter (2016, p. 200) observes, "animals' own forms of caregiving are rarely recognized as a kind of care work." However, even when other animals' care work is acknowledged (at least partially), the responses it provokes remain limited. A particular issue concerns the general lack of awareness of and curiosity about other animals' preferences in terms of fostering and nourishing relations of care with other animals on their own terms. In other words, much of the interest in nonhuman care work depends on what we, as humans, might gain or benefit from being in these relationships (Gruen, 2013).

As a consequence of this, standard definitions of animal sanctuaries, like the one I provided in the previous paragraph, emphasise the aspirations of the human beings involved, defining sanctuaries according to the *human* intentions of caring for, rescuing, and rehabilitating animals. Yet, as Sue Donaldson and Will Kymlicka (2015) suggest, the sanctuaries might better be shaped by the vision of what they term an interspecies *intentional* community, where individuals of various kin and kinds coexist and share a delimited place. This latter definition, focusing on the conditions for communal living, introduces an important distinction between the human-conceived idea of the sanctuary and the actual experiences and relationships unfolding within it. Still, with regard to their message to the outside world, the sanctuaries I visited emphasise their vision of creating *a safe haven* against how farmed animals are commonly conceived as inevitably destined to be eaten or killed. However, as the following sections will reveal, the reality of the sanctuaries is much more complex.

Against the Law: Hegemonic Standards of Care

According to the Danish Animal Welfare Act, animals "must be treated properly and protected in the best possible way from pain, suffering, distress, lasting injury and substantial nuisance."[3] What implications does this legally sanctioned 'care' for animals, intended for industrialised animal production, have for the caregiving efforts at the sanctuary? I turn to an example provided by the sanctuary caregiver, Caroline. In the following passage, she recalls an incident in which two police officers arrived to inspect the sanctuary and ended up reporting it for their failure to provide sufficient care for an old boar, named Walter, a long-serving resident at the sanctuary who likes to keep to himself. Caroline recalls the conversation with the police:

> When the two officers arrived, they took a close look at Walter. It was a winter morning and Walter had just got out of bed. One of the officers asked: *Is he staying in there?* [pointing towards the entrance of the old farmhouse] Based on the tone of his voice, I could tell he was sceptical about those arrangements. *Yes, he is.* [The officer exclaims:]. *But look! He is trembling and not walking properly. You need to get a veterinarian to assess him and you got two days to get it done. You probably will be charged,* he told me. A veterinarian arrives the next day and asks: *Why do you wish to keep him alive? No one will miss him once he receives the shot* [implying the use of lethal injection]. The second veterinarian who examined him told me I could not expect him to grow very old. I knew that. Still, she disagreed that Walter needed to be put down at this very instance.
>
> *Excerpt from Interview with Caroline*
> *September 2020*

Caroline tells me about her many challenging encounters with police officers, state-employed veterinary inspectors and private veterinarians. Even when she meets veterinarians who understand what the sanctuary is trying to accomplish, there is a lack of knowledge about farmed animals outside the context of industrial production and, therefore, a failure to understand the many implications of providing life-long care for this particular group of animals. Caroline recalls the numerous times she has been told that there is no available treatment, even when companion animals such as dogs would have received treatment for the same condition. Of course, veterinarians, on their part, might claim that a given treatment is not desirable based on their concern with the potential suffering inflicted on the animal at the receiving end. Given the complexity of obtaining consent when it involves other animals, the decision to provide or not provide treatment becomes intricate. Consequently, the human sanctuary members find themselves navigating challenging ethical dilemmas and taking responsibility for their decisions, especially in cases of end-of-life care (Abrell, 2021). For this reason, it is troubling that the sanctuaries experience a substantial lack of expertise and guidance offered by veterinarians. The shortage of expertise in the health of domesticated animals such as pigs and cows is even more disconcerting when considering the substantial number of farmed animals in Denmark. However, due to the fact that present-day agribusinesses kill and dispose of animals at an ever-younger age, there is little knowledge of what long-term care might involve.[4]

In addition, the current legislation hinders the sanctuaries' capacity to provide rescue and care. For example, the sanctuaries are faced with regulations that restrict the number of animals that they are allowed to keep, even for non-commercial purposes, which seems paradoxical when compared to neighbouring facilities comprising animals in the thousands, as noted by one of the sanctuary caregivers. Danish zoning law regulates the number of animals kept for non-commercial purposes, including which species and sexes can be kept. In addition, since the sanctuaries keep cows and pigs, they are required to register as food producers and comply with different administrative regulations including keeping records and applying methods of identification such as the use of ear tags. Moreover, there is the close monitoring of certain species due to the risk of spreading diseases, which also means that deceased pigs cannot be buried on site.

The problem of the current legal framework does not only apply to restrictive zoning and regulations rooted in environmental protection but also more fundamental issues are at stake when it comes to the standard interpretation of animal protection law. Indeed, the concept of animal welfare, as legally defined, can actually prevent the sanctuaries from providing lifelong care for chronically ill and/ or injured animals. As it has often been claimed, anti-cruelty and animal welfare provisions only offer minimal levels of protection for animals in industrialised farming, where they are kept in large facilities, bred to grow as fast as possible, restrained and forcefully impregnated, and subjected to a stressful death, such as pigs who are slowly suffocated in underground gas-filled chambers. Paradoxically, whereas animal welfare legislation is often interpreted loosely in terms of protecting the welfare of animals within the farming industry, the opposite seems to hold true for sanctuaries. As previously noted, the regulatory frame places sanctuary caregivers at risk of being reported for breaching specific animal welfare provisions (Leth-Espensen, 2023).

As made evident by these examples, sanctuary caregivers are confronted with a specific regime of care embedded in animal welfare law. This regime of care might acknowledge that animals are living and even sentient beings – that is, more than simple commodities or property – but also promotes the idea that a good animal life can be defined as one without 'unnecessary suffering.' As apparent in the example with Walter, the strong focus on suffering results in a situation where the attempt to provide life-long care for animals suffering from lasting injury and chronic conditions can become a potentially unlawful act. This is due to the fact that the existing legal structure mandates so-called euthanasia of individual animals with chronic conditions and injuries. Paradoxically, sanctuaries' efforts to provide life-saving treatment can run counter to the idea of 'merciful' killing that the law sanctions. As a result, private veterinarians and veterinary officers, acting in the name of the law, may instruct human caregivers to kill animals rather than attempting to establish the best possible conditions for continued life. Thus, a presumably 'humane' or 'painless' death is considered preferable to a life of chronic illness, testifying to the widespread disability injustice (Taylor, 2017) and the general societal discomfort with loss of abilities or functions. Sanctuaries are places in which such perceptions about what constitutes a good life can be challenged. Indeed, in this regard, mourning seems a powerful tool.

Mourning and Transformative Care

There are several transformative aspects of the caregiving activities provided at the sanctuary. In contrast to the exploitative practices of agribusinesses, sanctuaries believe that human moral obligations towards other animals extend beyond the duty to protect animals in the best possible way against harm, as legally defined. Equally important, the sanctuaries acknowledge other animals' ability to care for their offspring and other members of their species, and the creation of close bonds across species lines. Many times, the sanctuaries witness friendships develop between individuals of different species and the human caregivers describe how the entire sanctuary community is affected when a member passes away. Caroline has many times witnessed the sorrow in other animals when a close friend or companion passes away. In previous writing, I have suggested thinking of the sanctuaries' caregiving activities as a kind of *post-domestic care*, noting how sanctuary caregiving recuperates 'animal care' from its current constraints defined by agricultural control, management, and captivity (Leth-Espensen, 2023).

Feminist animal geographer Kathryn Gillespie (2016) writes powerfully about the politics of interspecies grief and mourning. Gillespie notes the transformative aspects of grieving against the backdrop of the widespread indifference to the death of nonhuman animals. Mourning is a way of coping with grief at personal and collective levels (Gillespie, 2016). At the sanctuary, human members find support in each other but might also find support in the broader animal movement. As part of the advocacy-oriented goals, the sanctuaries also sometimes choose to communicate the passing of a sanctuary member to a broader audience. By sharing the personal story of the individual animal and the circumstances of their death, the sanctuaries seek to create awareness and encourage more empathetic attitudes towards other animals. However, their mourning rituals, which are integral to their care practices, equally face challenges due to anthropocentric legal frameworks, here prohibiting sanctuaries from burying certain species – an otherwise common way for humans to mourn the passing of their loved ones.

In the following interview excerpt, sanctuary caregiver Caroline recalls an incident in which the Danish Veterinary and Food Agency and the Danish police had been informed that she had buried two deceased pigs on site.

> Bo was buried on the field up there. I had not made any secret of it. I had someone come and dig a big hole and place the pigs in there but my neighbours informed the police. The Veterinary and Food authorities and two police officers arrived and asked if I had buried a pig? *Yes, I have*, I told them. *Well, you have to remove it again and you got until Monday to get it fixed* [the officers told her].

Caroline reflects on this decision:

> Dogs and cats you are allowed to bury. Horses, if you get permission, but not pigs, sheep, and goats. They need to be sent off to destruction. Still, you can store a pig in a fridge and eat it. It does not need to be processed or anything – you can simply store a part of a pig right next to your tomatoes. Bizarre, don't you think?

The prohibition against burying pigs is primarily motivated by concerns about pigs as carriers of diseases and the potential for soil or groundwater contamination. While these issues are closely linked to the large-scale operations of the farming industry and its extensive use of antibiotics, they also extend to the use of concentrated anaesthetic agents used to 'euthanise' large animals under the illusion of a painless death. These agents pose risks if the body of a deceased animal is dug up and consumed by another animal. However, acknowledging that these risk-mitigation measures associated with industrial practices are not applicable to multispecies sanctuaries and households, advocates should push for sanctuaries to be allowed to perform their mourning rituals. Indeed, mourning and sharing grief provide a means to challenge and disrupt the anthropocentric and capitalistic logics that currently underpin the regulation and endorsement of the prevailing interspecies arrangement rooted in exploitation (Gillespie, 2016).

Concluding Reflection: Care in Past, Present, and Future

The examples of interspecies care discussed here underscore the indispensable role of farmed animal sanctuaries in the larger animal liberation movement. They illustrate how sanctuary caregiving responds to entrenched structures of abuse and exploitation, which lie at the heart of the broader crisis in animal care, as initially outlined in this chapter. Through the intricate work of care in the sanctuary setting, various manifestations of anthropocentrism come to the fore: from inadequate responses to the needs, desires, and interests of nonhuman animals within legislative frameworks and veterinary authorities, to the pervasive influence of the animal industry shaping current policies governing animal life (and death) and human–animal relationships. In challenging such practices and dominant paradigms of interspecies life within the domestic sphere, sanctuaries embark on radical efforts to reclaim 'animal care.' Through their overarching aspirations and caregiving endeavours anchored in a relational approach, they challenge the existing paradigm, replacing patronising and paternalistic visions of care with approaches finely attuned to the needs of those on the receiving end. As a result, sanctuaries introduce new meanings of care into the domestic domain, which has historically been shaped by the legacy of animal farming. Despite ongoing ethical dilemmas and challenges, sanctuaries serve as spaces where alternative forms of multispecies life can be envisioned. Grounded in efforts to challenge specific social structures, sanctuaries can thus play a crucial role in developing nuanced responses to the complex issues surrounding interspecies justice.

The practices of care at sanctuaries intersect with violent interspecies pasts. Thus, informed by Donaldson and Kymlicka's project of formulating a relational political vision for interspecies presents and futures and the aspiration of Critical Animal Studies in its emphasis on animal subjectivity and agency, I believe that anchoring the question of how to address ethical responsibilities in the intertwined interspecies history marked by human violence and dominance holds great potential. This approach enables us to envision a future that promotes justice across species lines. Adopting a historically informed, *relational* approach can contribute to a forward-looking examination of human relationships with other animals, improving interactions with non-human members of this shared world. Moreover, by employing an approach rooted in shared histories, guided by sanctuaries' ethical commitments, we can avoid the pitfalls of depoliticising species relationships associated with increasingly popular, though uncritical, conceptions of 'species entanglement' that fail to challenge exploitative arrangements. Instead, the work of care at sanctuaries demonstrates that insisting on animal agency can coexist with critiquing extensive human exploitation and prevailing social norms. In this sense, sanctuaries hold the potential to move beyond mainstream approaches in animal advocacy, presenting radical ideas of interspecies care for the present and future.

Notes

1 The sanctuaries and their residents have all been given pseudonyms or fictitious names.
2 Indeed, as highlighted by these and other authors, sanctuaries situated on Indigenous lands face the imperative of addressing colonial relations and forging multispecies and post-colonial alliances. Moreover, there exists a significant risk of perpetuating 'white' and 'human saviorism' when sanctuaries celebrate human individuals, thereby reinforcing pervasive images of other animals as passive beings. See also, Gillespie (2022).
3 Section 2 of The Animal Welfare Act no. 133 of 25 February 2020 (author's own translation)
4 Denmark is one of the countries in the world with the largest numbers of animals consumed per capita. In particular, Denmark has a significant pig production (Sanders, 2020).

References

Abrell, E. (2021). *Saving Animals: Multispecies Ecologies of Rescue and Care*. University of Minnesota Press.

Chang, D. (2018). *Farmed Animal Sanctuaries on Stolen Land: De-Weaponizing and Re-Subjectifying Domesticated Animals as Part of Anti-Colonial Struggles*. Paper presented at Animaladies: II, University of Wollongong. https://bpb-ap-se2.wpmucdn.com/uowblogs.com/dist/8/2931/files/2018/11/Program-abstracts-and-bios-1xoezkc.pdf

Clutton-Brock, J. (1999). *A Natural History of Domesticated Mammals*. Cambridge University Press.

Colling, S. (2021). *Animal Resistance in the Global Capitalist Era*. Michigan State University Press. https://doi.org/10.14321/j.ctv15nbzx9

Coulter, K. (2016). Beyond Human to Humane: A Multispecies Analysis of Care Work, Its Repression, and Its Potential. *Studies in Social Justice, 10*(2), 199–219. https://doi.org/10.26522/ssj.v10i2.1350

Crary, A., & Gruen, L. (2022). *Animal Crisis: A New Critical Theory*. Polity.

Donaldson, S., & Kymlicka, W. (2011). *Zoopolis: A Political Theory of Animal Rights*. Oxford University Press.

Donaldson, S., & Kymlicka, W. (2015). Farmed Animal Sanctuaries: The Heart of the Movement? *Politics and Animals, 1*(1), 50–74.

Gillespie, K. (2016). Witnessing Animal Others: Bearing Witness, Grief, and the Political Function of Emotion. *Hypatia, 31*(3), 572–588. https://doi.org/10.1111/hypa.12261

Gillespie, K. (2019). For a Politicized Multispecies Ethnography. *Politics and Animals, 5,* 17–32.

Gillespie, K. (2022). An Unthinkable Politics for Multispecies Flourishing Within and Beyond Colonial-Capitalist Ruins. *Annals of the American Association of Geographers, 112*(4), 1108–1122. https://doi.org/10.1080/24694452.2021.1956297

Giraud, E. H. (2019). *What Comes After Entanglements? Activism, Anthropocentrism, and an Ethics of Exclusion*. Duke University Press.

Gruen, L. (2013). *Entangled Empathy: An Alternative Ethic for Our Relationships with Animals*. Lantern Books.

Hribal, J. C. (2007). Animals, Agency, and Class: Writing the History of Animals from Below. *Human Ecology Review, 14*(1), 12.

Jones, M. (2014). Captivity in the Context of a Sanctuary for Formerly Farmed Animals. In L. Gruen (Ed.), *The Ethics of Captivity* (pp. 90–101). Oxford University Press.

jones, p. (2010). Roosters, Hawks and Dawgs: Toward an Inclusive, Embodied Eco/Feminist Psychology. *Feminism & Psychology, 20*(3), 365–380. https://doi.org/10.1177/0959353510368120

Leth-Espensen, M. (2023). Care in a Time of Anthropogenic Problems: Experiences of Sanctuary-Making in Rural Denmark. In K. Aavik, K. Irni, & M.-M. Joki (Eds.), *Feminist Animal and Multispecies Studies. Critical Perspectives on Food and Eating* (pp. 97–116). Brill.

Lien, M. E. (2015). *Becoming Salmon: Aquaculture and the Domestication of a Fish*. University of California Press.

McMain, E. M., & Torres, J. (2023). Saviors, Nurturers, or Magically Insane: A Braided Reading of White Women Characters in Three Ecological Narratives. *Feminist Media Studies*, *23*(4), 1643–1658. https://doi.org/10.1080/14680777.2022.2041692

Noske, B. (1997). *Beyond Boundaries: Humans and Animals*. Black Rose Books.

Pachirat, T. (2018). Sanctuary. In L. Gruen (Ed.), *Critical Terms for Animal Studies*. The University of Chicago Press.

Regan, T. (1983). *The Case for Animal Rights*. University of California Press.

Sanders, B. (2020). Global Animal Slaughter Statistics & Charts: 2020 Update. *Faunalytics*, July 19. https://faunalytics.org/global-animal-slaughter-statistics-and-charts/

Swanson, H. A., Lien, M. E., & Ween, G. B. (Eds.). (2018). *Domestication Gone Wild: Politics and Practices of Multispecies Relations*. Duke University Press Books.

Taylor, S. (2017). *Beasts of Burden: Animal and Disability Liberation*. The New Press.

The Care Collective. (2020). *The Care Manifesto: The Politics of Interdependence*. Verso.

Wadiwel, D. J. (2015). *The War Against Animals*. Brill Publications.

Chapter 4

Non-ridden Horses, Implanted Chickens, and Vegan Sanctuaries

The Liberatory Promises and Limits of Animal Heterotopias

Paula Arcari

Figure 4.1 A self-directing chicken (author's own image).

DOI: 10.4324/9781003366706-6

Introduction

This chapter is formulated around Bauman's claim that

> the overall order of things is not open to options; it is far from clear what such options could be, and even less clear how an ostensibly viable option could be made real in the unlikely case of social life being able to conceive it and gestate.
>
> (2000, p. 5)

Viable alternatives to the overall order – ones less harmful, more just, and more sustainable – are modelled every day in a multitude of manners and spheres of political, economic, social, and cultural life. Clearly, for many, the overall order is not the inescapable cage Bauman portrays it to be. There are "other ways of thinking and acting" (Manzini, 2019, p. 9) characterised by contradiction and difference (Lefebvre, 1991), disorder (Foucault, 1989), and resistance (hooks, 2015). Such counterspaces or 'heterotopias' (Foucault, 1967) include animal heterotopias that resist and negate the overall order for nonhumans, bending or breaking cage bars and redrawing speciesist maps of 'otherness.' However, Bauman may be right that the *broader* viability of options is tied to the capacity of 'social life' – with its extensively normalised arrangements of practices (Schatzki, 2013) – to not only conceive but also accommodate them. This is compounded by the negation and stifling of options by beneficiaries of the dominant order, and what Bauman sees as a progressive disempowerment of collective action (2000, p. 14). One of the primary tasks for animal advocacy, then, is not (only) to encourage individual behaviour change, the end of certain practices, or better welfare (potentially sustainable but always precarious changes) but to disseminate alternate ways of understanding and living alongside nonhumans; to provide revolutionary countermaps comprising different imaginaries and viable pathways towards them – to not just bend or break a few bars but shatter the cage entirely (Figure 4.1) (Hertzberg, 2022). As Manzini argues, "incremental changes do not prepare systemic [emancipatory] changes" (2019, p. 11). This chapter explores if, where, and how animal heterotopias fit into this liberatory plan.

Background

Discussions of 'multispecies' relations and futures have gained momentum in a variety of cross-disciplinary contexts. However, not all multispecies worldings (Haraway, 2016) presently being embodied and enacted are borne of a radical imagination. Many do not shatter or even challenge existing orders and the futures they point to for animals are ambiguous in their liberatory promise. Specifically, conceptions of multispecies justice typically fail to contend with those animals over which humans claim legal ownership. Poetic visions of vibrant common worlds full of sympathetic wonder and cellular co-mingling thus commonly exclude highly commodified and instrumentalised animals (Arcari et al., 2020).

Just futures for such animals, and pathways towards them, are unarguably harder to articulate than for their free-living kin. However, their exclusion undermines proclamations for multispecies futures that, while challenging ontological dualisms, "continue to perpetuate ethical dualisms" (Srinivasan, 2022, p. 2). To more effectively puncture the enduring global ideology of humane use, our most problematic nonhuman entanglements need to be addressed. Only through exploring different ways of being in relation with *these* animals can more rigorous and transparent formulations of just multispecies futures emerge. Heterotopic spaces that reject or "fuck" (Cudworth & Hobden, 2017, p. 141) heavily capitalocentric bonds of designation and value are thus positioned as key political and potentially prefigurative nodes in a wider project of emancipation.

To explore the emancipatory promise of these spaces, this chapter examines the material arrangements, constitutive practices, and inter- and intra-species relations at seven heterotopic sites comprising humans alongside equines, chickens, cows, pigs, sheep, dogs, and other animals. Objectives are to: 1) explore what constitutes heterotopias for nonhumans; 2) discern the boundaries/edges of these spaces, their permeability, and/or forms of resistance encountered with the 'overall order'; and 3) explore the possibilities and limitations they present and what these mean for understandings of 'liberation.' These themes offer critical fora for assessing the capacities of animal heterotopias to fuck or 'jam' the anthropo-capitalo-centric machine (Calarco, 2008; Cudworth & Hobden, 2017) and – as perhaps the most important dimension of that – to jam the "anthropocentric process of 'becoming human'" (Herbrechter, 2021, p. 214).

Anthropal 'jams' or glitches where social conditioning, underpinned by binary thinking and hierarchical orders, becomes at odds with embodied experience appear regularly in this study's data. They present opportunities for a factory reset of sorts, whereby cumulative lifetimes of 'human' programming are confounded and even dismantled, sometimes in an instant, creating space for an immanently responsive ahuman praxis (Iveson, 2015). As such, whether instantaneous or more protracted, these jams are conceived as the core, or heart, of the necessary human transformation towards liberatory and truly multispecies futures.

Research Design and Methodology

Fieldwork sites include three vegan animal sanctuaries, two non-ridden equine communities, and two rescue-chicken communities. Each comprises animals rescued from industrial systems, abandonment, and/or neglect, their human rescuers and now guardians, and other free-living animals who permanently or periodically share these spaces.

Site selection was purposive, beginning with a set of criteria defining what are considered, for the purposes of this chapter, to constitute animal heterotopias. Foucault defines heterotopias as spaces "whose functions are different or even the opposite of others" (2002, p. 361), embodying a "simultaneously mythic and real *contestation* of the space in which we live" (1967, p. 4, emphasis added).

Hence, heterotopias are conceived, first, as real or virtual spaces characterised by counter-practices of 'undoing' and, second, as spaces that actively contest 'normal' space through forms of effortful resistance and/or disruption. Animal heterotopias therefore do more than offer rescued animals the opportunity to live in safe environments where they are cared for and about. Caring may be provided without challenging prevailing systems and can even become an additionally formalised, legitimate, and legitimising dimension of associated industries, for instance, industry-funded adoption schemes, self-described 'sanctuaries' selling 'meat' and other animal products, and rescue groups paying farmers for 'wastage' animals. Such activities might benefit large numbers of animals. However, they also perpetuate welfarist paradigms by elevating the social credentials of associated industries and reinforcing 'pet'-keeping practices.

As well as rejecting dominant valuations of nonhumans, animal heterotopias must purposefully push back in some way against the anthropological machine by seeking <u>to overturn the epistemological *and* material terms of nonhuman commodification</u>. It is this pushing back that holds the potential for radical disruption and shattering orders and serves as the marker of something worth exploring in each case study – the glow of a liberatory futurity. The extent to which the push is deliberate or an indirect outcome of endeavouring to maintain the heterotopic space varies. Speciesist limits in the scope of push-back also emerge where some heterotopias maintain a foot (or hoof/trotter) in Foucault's wider, uncontested space, while others are more comprehensive in their offence.

In terms of challenging and/or undoing the epistemological dimensions of nonhuman commodification, the human members of the heterotopias examined here do not conceive their nonhuman co-habitants en masse as (primarily)[1] resources for human use but as individuals with distinct emotions, wills, preferences, capacities, and proclivities for attachments and bonds of community. Materially, each site also seeks to withdraw technologies of 'enhancement' or somatechnologies (Dalibert, 2011), including 'high-performance' diets engineered to enhance 'productivity' (whether measured in eggs, body mass, or speed) (Bryden et al., 2021; Mactaggart et al., 2021), antibiotics, hormones, vaccines, anti-inflammatories, and other injected or ingested bio-technologies seen as necessary to this optimisation.[2] A lot of care is typically involved in managing this undoing. This is on top of the intensive levels of physical care many rescued animals need to recover or learn to live with injuries and disabilities sustained and/or inherited as a result of standard industry practices.

These can be considered minimum epistemo-material conditions for animal heterotopias. However, it is the more active or forceful push-backs that are of interest here and each heterotopic model illustrates variations in this regard. Limited by logistical and budgetary factors, this study aimed to identify two sites for each model.

The Sites

Among the *chicken-rescuer community*, the hormone implant Suprelorin (Deslorelin) is often used to arrest the unnatural rate of egg production genetically encoded into 'layers.' Without the physical exhaustion of producing on average six eggs per week for 52 weeks compared with four to nine eggs per year for their wild progenitors (Cheng, 2010), rescued birds slowly recover their health, regrow a rich plumage, are relieved from laying-related problems including prolapse and peritonitis, and generally live longer, happier, more self-directed lives. Implants are controversial for their cost, effects, and perceived denial of animals' natural reproductive impulses – concerns addressed by Rosenfeld (2021). Nevertheless, they represent the application of a counter-technology aimed at reversing or mitigating the historically compounding effects of the many somatechnologies that have played a part in materialising avian bodies and lives to fit the anthropo-capitalo-centric paradigm with ever greater efficiency.

A recruitment notice was posted on six UK hen-rescue Facebook groups seeking participants interested in sharing their experiences of using Suprelorin. Two participants were selected – one located in the East Midlands of England and the other in Wales.

By definition, *non-ridden equine* advocates defy centuries of entrenched 'knowledge' and practice founded on equines' primary purpose and value to humans. Confronting what is often vociferous criticism from traditional quarters, these individuals and groups seek to change perceptions of non-ridden equines, often described pejoratively or affectionately (depending on the narrator) as 'field/pasture ornaments.' As well as negating the dominant use category of equines, members of non-ridden communities also address the damage inflicted on bodies not designed to carry humans (Cocozza, 2019; Holmer et al., 2007) and, as with chickens, the historical and compounding efforts to make them 'fit-for-purpose' (Jones, 2022; Librado et al., 2017).

Following a recruitment notice on the Facebook page of the Non-Ridden Equine Association seeking participants interested in discussing their alternate human-equine relations and practices, two participants were selected, both located in the West Midlands of England.

Among hundreds of listed *animal sanctuaries* and rescue centres in the UK, some cater to specific nonhuman species, categories (e.g., ex-racing greyhounds or horses), and even breeds, while others have more inclusive policies. In line with my definition of heterotopias, I am interested in sanctuaries that do more than offer care and shelter to animals ejected (sometimes temporarily) from the Animal-Industrial Complex (A-IC), but also set rules and boundaries (on humans) that subtly and more obviously 'glitch' the speciesist machine. Initial filtering was undertaken through a review of sanctuary websites to determine their aims, missions, and operating models. Self-described sanctuaries whose operating models resemble zoos, in that human visitor experiences are prioritised with animals accessible and 'on display' most days of the year, were excluded. Ones that host onsite cafes serving 'meat' and other animal products, often of the same species to which they offer 'sanctuary,' were also excluded.

Of the selected sanctuaries, one located in Scotland (S1) and another in South East England (S2) accepted invitations to participate. Visitors are restricted to three days per year for one and Sundays only during summer for the other.

Numbers are also capped. Given their similar models, an additional sanctuary was recruited that follows an anti-speciesist philosophy but engages more actively with the public. Also located in South East England, this sanctuary (S3) opens four days a week for four hours between March and November and hosts an on-site vegan café.

Site visits and interviews were conducted between February and July 2022 following the easing of UK Covid restrictions. Visits lasted between three and seven hours (over two days), with an average of four and a half hours per visit. Names of sites and their human and nonhuman residents have been changed.

As the following three sections illustrate, each of these spaces does more than offer resident nonhumans the opportunity to live in spaces where they are cared for and about. They also seek to contest, unsettle, and undo, and in so doing are potentially prefiguring non-anthropo-capitalo-centric futures, or at minimum, pathways towards them.

1. Nonhuman Heterotopias – Liberation of and Between Bodies

The nonhuman perspective on animal heterotopias arguably matters most but can only be surmised. Giving representation to nonhuman perspectives is fraught with difficulties (Hortle et al., 2021). However, all representation, including lack thereof, is political (Hediger, 2013) and a vast gap must be bridged to weaken and silence human-centred narratives, leaving space for the other (Barcz, 2017; Sands, 2019).

Animal autobiography is one approach (Arcari, 2023). Here, my focus is on human efforts to engage with nonhuman perspectives and, specifically, the alternate ways of inter- or intra-acting these efforts demand and produce (Barad, 2007).[3] For animal heterotopias are about more than simply extending human understandings to encompass nonhuman perspectives and going beyond the alleviation of suffering to consider animals' dislikes, preferences, pleasures, and joys. More significantly, they are about the prefiguration of radically *different human* understandings and ways of being/not being with nonhumans. Humans immersing themselves in others' lives without critically interrogating the purpose of this multispecies intent risks capturing and representing the other under persistently anthropocentric auspices (e.g., Schmeer, 2019; van Dooren et al., 2016). Hence, while I am concerned with how the lived experiences of animals in these heterotopias might be an improvement on their previous lives, or what is considered 'normal' for their species, following Herbrecheter (2021), Calarco (2015), and others, and aligning with my call for more critical *human* studies (Arcari, 2019), I especially want to foreground different ways of being (or not being) 'human' that precipitate, arise from, and are fostered through these arrangements. Only through our being differently human do nonhumans become more fully and equally present.

Being differently human means that, in these heterotopias, dominant meanings inscribed on nonhuman bodies are rejected so that species, breed, gender, age, size, ability, and non-humanness are no longer variables on a scale of value, describing what and how much humans can *take*, but are among many qualities that shape what humans *offer*, how, and when – generating and inviting different kinds of response. Nonhuman interests and even cultural rules are prioritised with the effect, if not conscious aim, of maximising animals' opportunities for self-determination and freedom (albeit within the limits of their human-constituted material arrangements). Exemplifying this common sensitivity to a liberatory, respectful, and non-invasive intent is the attunement of human heterotopians to the unique characters, distinct needs, sensory capacities, and variable, moment-to-moment preferences and wants of their nonhuman community members.

As Kerry (S2) explains, "we interact with each of them very differently. They interact with us quite differently to each other. They need different things from us. Our roles change as they go through their lives . . . [from] stand-in mothers . . . brother and sister . . . to carers, nurses." Key to this type of interaction is listening – a 'deep listening' that senses beyond meaning (Heddon, 2017), coupled with the embodied knowledge and intuition of shared sentience – "if you listen . . . they make it clear what they need" (Kerry) or as Sarah (S1) explains, "what we focus on is how they tell us who they are" (Figures 4.2 to 4.4).

Figure 4.2 Sebastian, the pig, relaxing in the conservatory at S1 joined by Flora the chicken (image from Sanctuary Facebook page).

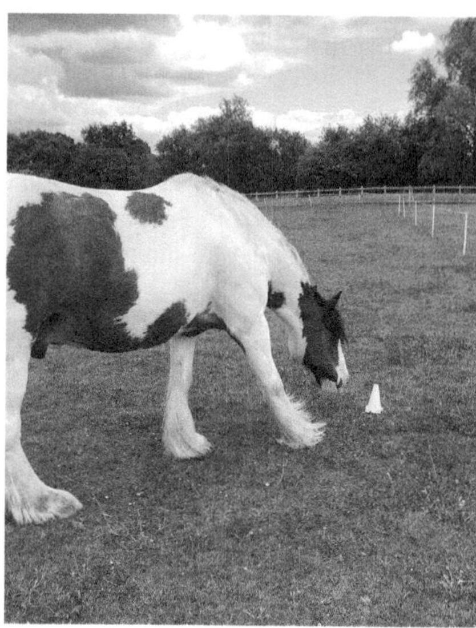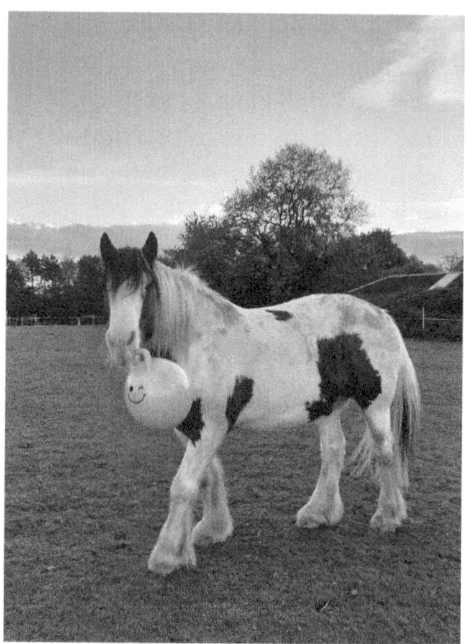

Figure 4.3 Donald, the non-ridden equine, participating in scent games and interacting with different objects.

An openness to what the other needs in order to be as emotionally and physically free as possible requires letting go of lingering traces of anthropocentric authority that insist 'you' (the 'saviour' or giver of sanctuary) are the person best placed to know and provide what they need (Meijer, 2021). Mark (S3) describes how difficult it has been, and the years it has taken, for him to learn to put aside his own attachments to each animal and "think what is right for them, not what is right for me." It may be that the sanctuary is not the best place for some. You must, he explains, take account of where a particular animal came from, and "who that animal is at the shelter." He dismisses the notion of there being blanket rules for what is best for 'a horse' or 'a cow.' Each is an individual and "[i]f you are prepared to sit and listen you will learn from them."

Kerry and Tom (S2) also advocate deep listening. Resident animals provide the cues for how and when they interact or occasionally intervene to do 'what is right for them.' For example, moving youngsters away from older animals when the latter become 'fed up,' and dehorning certain confrontational cows who would otherwise have to live isolated from the herd. When Cher, an epileptic chicken, was being bullied by the flock, chicken rescuer Claire responded by providing a second sanctuary in her home. Another called Blondie lets her know when she wants to go to her coop for the night by perching on her knee (Figure 4.5). It is about "being aware of how they are without putting restrictions on them" (Kerry).

Figure 4.4 James' chickens roaming freely into James' house, enjoying a car trip for take-away coffee and pastry, engaging with food puzzles, and hanging out in their customised home complete with mirrors, mirror balls, hammocks, laser lights, music, chime feeders, different roosting levels and areas, automated doors, and cameras.

Autonomy is a common theme. The normalised default to human 'expertise' and control is muted so that animals are permitted to be self-directing, for example weaning themselves when they choose, communicating "what they like and don't like" (James), deciding who (human and nonhuman) they want to be with and, most importantly, having their preferences respected. There is Geoffrey the sheep who had no time for other sheep and spent his life among a herd of horses who accepted him as one of their own. And an unbreakable bond between Figaro the cow and Basil the sheep who both arrived at S3 as youngsters and 'adored' each other. These bonds are respected even when it comes to rehoming. Mark will never separate a bonded pair, group, herd, or flock – they are rehomed together or not at

Figure 4.5 Blondie actively seeks Claire's company.

all. Kerry also keeps families together as much as possible at S2, where one Hereford group comprises great-grandmother, grandmother, mother, son, and auntie. Their bonds are clear to see, with individuals choosing to stay physically close to and touching their family members.

Bonds extend to humans, with Fiona the chicken (S3) seeking out a particular visitor who she 'loved' and would sit with while she was there, and Larry the turkey who developed a strong attachment to a woman he 'adored.' He would "spot her through the crowd" and stay with her overnight in the holiday lodge. Allowing space for animals' free and uncoerced self-expression leads humans into different states of being as fellow animals, where human supremacy is momentarily quietened, dialled down, or 'glitched.' These moments can seem minor but are deeply touching for the fully embodied, emotional, affective, and multisensory connectedness (or intra-action) they represent.

Once a group of pigs had settled into their new life at the sanctuary, Sarah noticed when she leaned close, they made what she describes as a monkey-like noise. Contrary to being a threat, which was her initial interpretation, Sarah learned this meant they considered her a valued family member. In turn, she says, "I let them know that they're part of the family by looking after them. So, it's a two-way thing." Claire describes some of her rescued chickens as being very aloof and not at all interested in her, while others want to sit and spend time with her. James recounts the development of a unique relationship between his father and one of his rescues, Gloria, when his father was digging in the garden:

He lifted the spade, he turned in over and loosened [the earth] and he stopped and Gloria came in, picked out the worms. She would step aside and then Dad would carry on. Only a couple of times Dad had to show her and it would be that rhythm. He could dig all day and she wouldn't get in the way. She would stand, wait, and then go in.

James says he used to wonder if his chickens were only interested in him for food but noticed how one of his 'girls' will come up to him and shake her wings at him. He says, "I ignored it a couple of times, she did it again, picked her up, she wanted to be picked up." Another, when she was unwell and being nursed by James in his home over several days would fall asleep on his lap. I suggest to James these chickens gain something from their relationship with him (besides food and health care) that is not available from their kin or other species – a positive experience of a particular nature and quality that they seek out. Just as our emotional and affective worlds can be enhanced in different ways through relations with nonhumans, it stands to reason that so too can those of nonhumans. However, until those relations are un- (or minimally) conditioned genetically, socially, and materially by humans, they cannot be described as wholly consensual or the result of free choice (i.e., liberatory). Furthermore, as we know, positive human interactions are the exception and nonhuman preferences for no interaction with humans must be equally respected.

Duchess the cow arrived at Mark's sanctuary at eight years old after having multiple calves taken from her. Mark says she has clearly "seen enough of humans" and "is never going to come round." Mark's site is also home to a horse born to an abandoned mother who he describes as "the nearest thing to feral I have ever met in a domestic setting. "She is like, 'I have seen enough of you lot. . . . There's not a good one amongst you.'" A group of sheep who have been living at Sarah's sanctuary for five years still do not trust her or her partner. They had a bad start, she says, and will never change their minds, "and that's fine, we'll look after them."

There is no impulse to change these animals. Decisions regarding human interactions are left to them, with time and space understood from their perspective, or as belonging equally to them, not quantities to be externally controlled. As Kerry illustrates:

These [young male calves] are going to take a long time [to settle] because they were in the shoot ready to be pushed into the knocking box. I would think with that fear it's going to take them a long time. But they have got time. They have got as long as it takes. If it's 10 years, if it's a minute before they die, it doesn't matter. That is up to them.

In contrast to every moment preceding their present lives, the normal human–animal hierarchy is flipped and these animals are permitted to decide what they want and when. Particularly in sanctuaries, many resident nonhumans, including pigs, sheep, chickens, turkeys, ducks, geese, peacocks, goats, donkeys, cows, dogs, and cats, are free to wander and mingle in various multispecies configurations, sometimes grooming each other, sleeping together, dropping in on amicable human neighbours, and entering human dwellings whenever they want. As Sarah explains, "everything is on their terms and we're rearranging this place in a way that suits them, not us." The heterotopic space is conceived very clearly as one of possibilities rather than limitation. Notwithstanding unavoidable scheduling of feeding and health care, nonhumans dictate the terms and timings of their interactions with others.

Some sanctuary models are able to accommodate a more comprehensive notion of animals' agency than others, and animals' individual natures also have to be considered. For instance, for the wellbeing of nonhuman as much as human animals, it is not possible for Mark (S3) to let their large groupings of pigs, sheep, cows, horses, dogs, and other animals roam freely in a setting designed to be publicly accessible. Nor can Kerry (S2) allow Delilah the pig to continue sleeping on her bed, along with the occasional lamb, given her size and belligerent character which also means she can no longer safely mingle with the other animals. At S1 however, goats, pigs, sheep, donkeys, the occasional cow, chickens, turkeys, dogs, and cats muddle along together perfectly well. Taking a "non-interventionist" approach, Sarah and her partner "just let them get on with it . . . they have to sort stuff out."

Circumstances that challenge normalised orders can unsettle human expectations of how things ought to be and may generate or demand an associated adjustment. This adjustment can give rise to what I have described as a glitch in the anthroparchic mindset, or a jam in the anthropocentric machine. Whether instantaneous or a slow burn, this process speaks to the boundaries/edges of heterotopias, forms of resistance encountered, and possibilities to liberate the (non) mind that are the focus of the next section.

2. Encounters, Confrontations, and Lines of Fragility

During interactions with vets, people associated with animal-based industries, members of the public, family, and friends, heterotopic practices and perspectives encounter the wider society of which they are part. These encounters can be more or less formalised but are characterised by human members of these heterotopias holding and defending both nonhuman *and* non-/less-anthropocentric human space – maintaining its integrity in the face of pressures to acquiesce to anthropocentric norms. These situations are illuminating as power's dimensions are articulated in its response to resistance, contestation, and dissent (Foucault, 1982).

Each study site (by design and/or dictate) makes a concerted point of challenging human supremacy, learned entitlement, and the dominant ordering of things by imposing rules on visitors (particularly children), refusing normalised demands and expectations, and taking animals' macro-agency seriously (Meijer, 2021). This includes prohibiting certain activities (such as 'petting,' riding, and bringing animal-based food products on site); removing anyone who behaves inappropriately towards animals; cancelling activities according to animals' expressed needs and (dis)comfort; allowing animals to set the terms of engagement, including no engagement; flipping the sponsorship script by allowing animals to choose their sponsor; insistently referring to animals by name instead of number during inspections; demanding 'spent' or 'wastage' animals receive levels of care and treatment traditionally reserved for 'pets' or their more useful counterparts; and resisting everyday social norms that denigrate these animals and question the allocation of time, money, and emotional energy on them.

By holding the heterotopic space for animals at moments of encounter, the epistemic boundary between humans and nonhumans, and the discursive, emotional, and sensory work required to maintain and challenge it, become more evident. However, drawing on Deleuze, Todd and Hynes argue that bodily encounters can "only ever be sensed," lending them a more singular capacity to bewilder thought, to unsettle, unravel, or 'glitch' the unity of a hegemonic *common* sense (2017, p. 735). Hence, they conceive the emergent qualities of experience as crucial to creating openings of non-sense where more experimental and inclusive modes of being with animals become thinkable. While broadly agreeing, I caution against underestimating the hold that power, 'knowledge,' and dominant discursive and emotional associations can exert even during encounters, shaping the reconfiguration of anthropocentric human–animal relations towards, for instance, more 'ethical' practices, higher welfare standards, and utilitarian expressions of care and respect (Arcari, 2019).

The transformative potential of a transgressive encounter involving dis-/dys-ordered bodies hinges on the discomfort it creates when an "affective registering" (Wilson, 2020, p. 2) challenges all that is 'known' up to that point. Discomfort is generated by the sense of having arrived at an ontological crossroads between change or continuity (ibid.). Among the available responses are those that seek continuity, as indicated earlier – "the conservative impulse of the commonsense image of thought, which seeks to assimilate that which threatens to confound recognition" (Todd & Hynes, 2017, p. 736). This can manifest as resistance and even doubling down on the normalised order. Other responses might support an instantly transformative experience or contribute to a more protracted process of dissolution and reconfiguration. In these instances, transgressive encounters create and/or extend 'lines of fragility' understood as virtual fractures that open up "a space of concrete freedom, i.e., of possible transformation" (Foucault, 1988, p. 36).

Each heterotopia in my study offers examples of resistance and change and both are important for understanding the promissory potential of these spaces. However, with limited space and an emphasis on foregrounding pathways towards liberatory animal futures, I focus here on change, both transformational and more protracted, and the contexts in which it unfolds.[4]

Kerry tells the story of two sisters (vegans) who visited the sanctuary with their non-vegan father. They asked permission for him to visit the cows. As Kerry explains, "Well, he took up with Derrick [the cow]. . . . He just stayed with Derrick, sat with him and was stroking him. This man was off having a conversation with Derrick the whole time." Some weeks later, one of the daughters let Kerry know her father became vegan that day and there was "no sign of him changing." Tom adds they "have a lot of crying" on open days. It is, he continues "almost inexplicable . . . the person suddenly gets it and bursts into tears."

It may take longer for resolutely anthropocentric containers to spring a leak (Gillespie, 2018b). A couple once visited Mark's sanctuary, non-vegans both, and sponsored a newly arrived young cow called Bella. A few years later, Bella took ill and Mark let them know she was to be euthanised. One partner requested to be present. Mark describes her breakdown as unlike anything he has seen before or since. His attempts to console her were to no avail as she repeatedly wailed, "[Y]ou don't understand." Some months later, she returned to the sanctuary and explained to Mark that all the time they were sponsoring Bella, she and her partner reared and killed cows. However, the death of Bella with whom she had formed a deep connection became the discomforting catalyst for a profound transformation. She swore to Mark that the only thing they now farm is grass and so it would remain.

For James the chicken rescuer and the two non-ridden equine communities, the transformative impulse of uncomfortable encounters has been more limited. Cringing at the memory, James remembers sitting in the garden with his chickens, not long after their rescue, eating chicken drumsticks. He vividly recalls his realisation that, "Oh, it's you lot on a plate here" and how the resulting discomfort led him to stop eating chickens, although other animals remain edible. For Chloe and Stephanie, expectations of riding their rescued equines were challenged only when they discovered these horses' bodies were too damaged from previous use. The transgression into their lives of equines that were, by normative standards, unfit-for-purpose placed them at a crossroads between their desire to ride and their affection for these horses. For both, their affection and the poor future prospects for these horses outweighed their desire. Chloe describes the day she sold her riding gear as "very symbolic" and she "grieved terribly."

Letting go of the deeply rooted notion that horses are for riding relieved Chloe and Stephanie from the discomfort they felt when this anthropocentric norm coincided with the embodied connections they felt that ran counter to this 'common' sense. However, neither are anti-riding as long as it is "part of a relationship" (Stephanie). Rather, they believe every part of the equestrian industry "should be catering for non-ridden. It should be seen alongside riding" (Chloe). Much like meat alternatives, it becomes another option to be added to standard practices.

It might be argued that visiting sanctuaries and rescuing animals ejected from their respective industries indicates these are people who are already, even unconsciously, becoming differently human by being open to the possibility of animals being other/more than their socially constructed categories. However, one further example demonstrates how encounters can not only widen existing lines of fragility but also perhaps instigate the originating fracture, as Gillespie (2018a) also describes.

Throughout S2's first inspection by the UK Government's Department for Environment, Food and Rural Affairs (DEFRA), Kerry recounts both inspectors remained "very officious" and "grim." The following year's inspection was conducted by one of these same men who was at the time DEFRA's chief cattle inspector. At the end, Kerry continues:

he says, 'I must tell you, the last time I came turned out to be the most important day of my life.' He looked up and said, 'I went vegan. Not only that, I've converted the whole family to vegan . . . and, I'm actively campaigning within DEFRA to have vegan food across all DEFRA offices a couple of days a week.'

As a result of his transformation *and* his transgression, DEFRA no longer permits this man to conduct inspections at sanctuaries due to potential bias.

Further stories are related of vets becoming increasingly willing to treat older 'farm' animals who they would not generally encounter, and animals whose injuries, disabilities, or ailments would not normally be considered worth treating because of their category and/or deemed useless-ness. Initial discomforts shift and dissipate, as routinised categories of value are disturbed, thoughts momentarily bewildered, and understandings of nonhumans, and of being human, reconfigured. These changes are frequently the result of the steadfastness of people like those interviewed here, and their determination to maintain the integrity of their heterotopic space against anthropocentric norms.

What presents as resolute resistance may thus be part of an evolving process of transformation whose constitutive diffractions and accumulations are difficult to discern and impossible to map or predict. One (more) encounter may be the difference between resistance and transformation. Of course, discomfort cuts both ways. Those who derive comfort in, and endeavour to defend, heterotopic spaces and practices spend significantly more time than the 'majority' population in states of discomfort as they contend with the epistemic smoothness (Wadiwel, 2015) and everyday normalcy of nonhuman oppression. Over time, changing patterns of comfort and discomfort, as reflected in everyday practices and the landscape of a society's heterotopias and hypertopias (Arcari, 2023), are indicative of broader shifts in power (Eaves et al., 2023). Steering these patterns more decisively towards discomfort with oppression and discrimination in all its forms involves destabilising, upending, and glitching dominant understanding of 'the human.'

Encounters with the 'other,' especially those that are "between sense" (Iveson, 2015), absent verbal exchange or accompanying narrative – a "being-with prior to any recognition . . . prior to the constitution of the 'human' or the 'nonhuman'" (ibid., p. 96) – have the capacity to cut through decades, centuries even, of encoded learning (Wilson, 2020). They offer a vital component in the 'how to' kit for disturbing and undoing the anthropo-capitalo-centrism at the root of the world's most pressing crises. Having touched on the 'how,' it is vital to ask: what alternate ways of relating with nonhuman others are there? Are animal heterotopias prefiguring better kinds of future posthuman 'normalcy' – the kinds that Cudworth and Hobden imagine can "fuck the capitalocene"?

3. Prefiguring Liberatory Arrangements: Manifesting the Spirit of Liberation

Yates defines prefiguration as "the attempted construction of alternative or utopian social relations in the present" (2014, p. 1). The term is primarily associated with social movements (Asara & Kallis, 2022; Maeckelbergh, 2011), and a related body of literature that Cornish et al. contend marginalises "everyday spaces commonly considered outside the public sphere" (2016, p. 121). Preferring to view everyday space as equally political, I draw on Kinna's looser and more inclusive conception of prefiguration as "the continuous exercise of testing the imaginary landscapes against . . . the subterranean flows of daily life" (2017, p. 202). The notion of *liberatory* prefiguration is used to emphasise that the kinds of spaces being radically imagined are spaces of freedom (Foucault, 1988), explicitly excluding prefigurative practices (including 'movements') that increase inequalities, and intensify relations of domination and oppression (Fians, 2022; Yates & De Moor, 2022).

Contesting and inverting hegemonic practices, defamiliarising the every day (Yates, 2022), and extending a heterotopology of counter-sites are vital dimensions of liberatory prefiguration. However, wary of the notion that all such efforts should be equally valued and supported, I suggest it is important to define measures against which new imaginaries can be held to some account – principles that help articulate, in progressively more refined ways, what we do and do not want them to include, thereby hedging off less desirable developments. As Hage argues:

> an 'anti-politics' concerned with the overthrow of existing orders need[s] to be supplemented with an equally vibrant and passionately 'alter-political' thought capable of capturing the possibilities and *laying the grounds* for new modes of existence.
>
> (2015, p. 9, emphasis added)

For example, drawing from the animal heterotopias examined here, particular expressions of human-nonhuman relations may be conceived as transgressive – running counter to the dominant order. They may provide an individual or several nonhumans the opportunity to enjoy lives of relative autonomy and freedom, and represent a qualitative shift in understandings of the human *vis a vis* nonhuman animal. However, these relations might still be part and even supportive of the A-IC, may not carry across to other nonhuman species/ categories (or even other animals within the same category), and may still reflect anthropo-capitalo-centric conceptions of liberation. Here, lines of fragility may in geologic terms become an intrusion – a crack that over time solidifies/normalises under modified arrangements (Schatzki, 2013) that co-opt the liberatory impulse towards anthropo-capitalo-centric ends.

Consider chicken rescue groups, including one Claire works with, who pay industrial producers more for each 'spent' bird than they receive from supermarket chains or the slaughterhouse; industry-backed 'retirement' and 'rescue' schemes for ex-racehorses and racing greyhounds, and as part of zoo programs, that bolster these industries' welfare credentials; bonds between humans and particular nonhumans that fail to extend to others of the same species; and, material arrangements that constitute lives of *relative* freedom but do not provide for completely autonomous self-determination. These efforts are palliative in that they mitigate the effects of underlying power structures while leaving the structures themselves unchanged (Meijer, 2021). Worse, such 'palatable disruptions' (Clay et al., 2020) can further legitimate these structures by normalising the notion that compensatory restitution – compensating nonhumans (and redeeming humans) for previous harms – is both possible and sufficient (Emmerman, 2014).

If humans continue to determine nonhumans' reproductive lives, limit their space, choices, and interactions with others, and make them dependent on humans for food and care – to what extent can they be said to be liberated? As Kerry admits, "We can't give them freedom. They're not able to take food and they wouldn't survive." They are, like the animals at VINE sanctuary, only "as free as possible" (Gillespie, 2018a, p. 137).

I do not exclude the potential for nonhumans to choose to interact with humans and gain something unique from these interactions that is not possible with con-specifics or other species. However, questions to foreground as a measure of the cost of these relations, and therefore the extent of their liberatory force, include: is the relationship predicated on the bodies of others brought into existence, assessed, measured, exploited, abused, spent, wasted, and killed by humans? Is it contingent on enabling systems of oppression and death? If yes, then we are not yet prefiguring the material arrangements of these animals' liberation but responding to the failings of overarching arrangements. Unfortunately, this is currently the case for all commodified animals (including those rescued from the A-IC). 'Lucky' individuals remain, by definition, exceptions and are in situations of dependency on humans (for their safety, care, wellbeing, nourishment, and existence) as constituted by their commodification. As long as humans are required to assume the roles of rescuer, protector, and carer, the risk of perpetuating hierarchical relationships remains (Meijer, 2021).

Despite their limitations, animal heterotopias, as spaces that contest dominant understandings and uses of commodified animals, are perhaps "the best option available from an array of unsatisfying options" (Emmerman, 2014, p. 227). They provide alternatives mostly for animals who have escaped, been abandoned, or ejected from the A-IC. A small number are born into these lives and here, I suggest, are glimmers of pathways towards more liberatory arrangements. Where, for example, second- and third-generation ex-battery chickens roost in trees, return to more natural rates of egg production, and live close to autonomous lives, and where these and many other animals experience nothing of the cruelty and denigration that marks the lives and bodies of most commodified animals. From this point, it is possible to conceive the function of these spaces shifting to become less about providing places of sanctuary defined by wider landscapes of harm to places of liberatory prefiguration where new social relations are co-evolving. Such evolutions, however, must contend with the possibilities and consequences for native ecosystems of 'liberating' non-native animals including chickens, sheep, goats, pigs, and cows (Manchester & Bullock, 2001). This complicates Mark's assertion that domesticated animals could evolve/revert to living wild and Kerry's view that "if we could just not interfere, but support, they would sort themselves out." As Sarah admits, "It's a proper quagmire when you start looking into it."

That said, for individual animals, the importance of these heterotopic spaces, whatever their scope, scale or political intent, is unquestionable. They offer counter-spaces that refute the normalised order and its generative hypertopic arrangements where commodified and other animals affected by human practices have the opportunity to shed (as far as possible) the physical and emotional weight of their socially constituted use categories. They are also spaces that model different ways of being human, and the glitches in 'common' sense that can take place within, and radiate from, these spaces are arguably their most powerful force for liberatory prefiguration. As Herbrechter contends, "Jamming th[e] anthropocentric process of 'becoming human' is . . . what lies before *humanity* today" (2021, p. 214).

From this progressively liberated human (non-)mind or reconfigured sense-making, more liberatory nonhuman arrangements will hopefully evolve. As it stands, while existing heterotopic arrangements may be prefigurative of something better, the extent to which this 'better' is also liberatory is unclear. Without critically questioning, and continuing to question, the conditions surrounding apparent consent and co-authorship, commonly parsed (and fundamentally unchallenging) accounts of relational encounters, entanglements, intra-actions, and agency are unsatisfactory. As Spannring cautions, ". . . not every behavior constitutes agency. Instead, it is often a performance devoid of inner experience and meaning and geared towards the fulfilment [sic] of human expectations" (2019, p. 11). Thus, in any encounter underpinned by a liberatory intent, we have a responsibility to ask ourselves: to what extent are we (the human) still a 'consumer'? How much of the experience is scripted? What possibility exists for "the unknown' or the "mutually interesting engagement and shared experience"? (ibid., p. 9):

> Numerous are the anthropocentric practices that inhibit a coming into presence of and with animal Others, especially by doing something to them or making them do something . . . They are difficult to unravel, especially where there is an apparent sympathy for the nonhuman animal that hides the refusal to see him/ her as a subject.
>
> (ibid., p. 15)

Conclusion

Owen et al. argue that uncomfortable geographies, essentially Foucault's heterotopologies, are inherently political, revealing "deeply entrenched power dynamics" (2022, p. 1) and providing opportunities to challenge established norms. Animal heterotopias constitute a diverse landscape of norm-challenging, with innumerable, competing, and conflicting narratives and practices making for unpredictable outcomes. Nevertheless, radical change remains one possible outcome, and the stories that emerged from my heterotopian encounters show how alternate futures for animals might be possible.

This possibility is evinced less through the material arrangements of these heterotopias. These remain problematic to varying degrees in terms of animals' originating contexts, extent of self-determination, and the scope of species inclusion, not to mention the lack of regulation over 'sanctuaries' and other rescuers, and indeed private animal 'ownership' generally. Rather, it is the radical change in animals' everyday experiences, ways of being, and relations with humans and other animals, and the shifts in understandings of human-ness that constitute, are generated by, and radiate from these sites that are their most promissory and hopeful aspects. In this respect, these heterotopian explorations demonstrate that these situated, embodied, and lived realities of mutual change and co-evolution cannot be designed or pre-determined on the basis of species or use category. They have to be worked through, trialled, and felt by all beings involved, allowing every individual to speak and be listened to.

However, transgressive encounters in spaces constituted as being apart from the 'normal' everyday risk becoming part of a modified order whereby the experience is corralled as an expected exception with little to no bearing on the wider normal (Arcari, 2019). The presence of a broader, intersectionally cohesive anti-oppression movement, ideology, and discourse could offer vital leverage in these instances. Indeed, the potential for heterotopian glitches, cracks, and lines of fragility to keep extending outwards – to other nonhuman individuals, species, and categories (including those whose associated violences are more socially smoothed) and to other people and their practices – makes animal heterotopias and the encounters they foster an invaluable and complementary part of a united advocacy movement that supports the end of nonhuman oppression; a movement that is arguably yet to emerge (Arcari, 2022).

Of course, anthropocentrism is not the only challenge. As well as reflecting a lifetime of schooling in the human–animal binary, many feelings and judgements regarding what is good/bad, right/wrong for animals are socially and culturally conditioned. Westerners tend to bemoan the lives of street dogs in other countries, unaware of the extent to which our apparently superior values apply to only a small proportion of western dogs. The majority, bred in larger numbers than they would naturally, live lives that are differently compromised through abuse, neglect, isolation, abandonment, exploitation for breeding, conditioning, various neuroses and stereotypies, the effects of breeding, and being deemed of no worth or surplus to market requirements. At the same time, most westernised societies are less tolerant of dogs and other domesticated animals moving freely in public spaces. Is there a liberatory middle ground where less fawning, neurotic, and human-dependent creatures are able to live safely yet (semi-) autonomously as part of mutually caring communities? Such communities can only emerge from a radical shift in what it means to be human. As Kerry observes, "liberation is a state of mind." If such a shift is not imaginable, is it multispecies justice to keep bringing domesticated animals into existence if they will never be liberated? Humankind traditionally prioritises "the enormous benefit of life" more than the quality of that life (Salt, 1976).

I was hoping through this research to collate a range of existing and realisable options for liberatory nonhuman arrangements. However, such arrangements are not possible by westernised humanity in its present embodiment, however well intentioned. The only hope is they might emerge and develop over time, distinct from the reactive circumstances, needs, and relations surrounding current sanctuary/rescue (see Meijer, 2021), and based on mutual responsiveness to unique happenings in specific contexts. As Sarah and Tom have learned over 25 years, "the practicalities of it are way more difficult than the ethics of it" making me inclined to regard present animal heterotopias as exhibiting only *pre*-liberatory prefigurations. Humans are only just beginning to seriously question what it means to be human. It is hubris to think we are ready to imagine let alone manifest liberatory multispecies societies when undoing human-ness is just one of many projects, alongside undoing white-ness (Ahmed, 2007) and masculinity (Pulé & Hultman, 2021), that are central to dismantling the deeply rooted anthropocentrism, supremacism, Eurocentrism, and capitalocentrism that fuel the consumption of life in all its forms. These are the co-constitutive historical, political, economic, social, and ideological orders that need to be shattered on the way to imagining radically alternate futures.

Acknowledgements

I wholeheartedly thank the participants in this research project for welcoming me into their spaces and being so generous with their time and hospitality during a period unsettled by both severe storms, the UK's largest-ever outbreak of Avian Flu, and ongoing impacts of the Covid pandemic. Many thanks also to Jane Daly, Iris Bergmann, and Richard Twine for lending their time to reviewing this chapter during its development and providing valuable suggestions for improvement.

Notes

1 Acknowledging that understanding (certain) nonhuman animals as subjects does not preclude seeing them and/or their species also as resources.
2 These include wearable or embodied technologies in precision farming (Zhang et al., 2021) and horse racing (Marcellin, 2016), underscoring the constitution/understanding of these nonhuman bodies as imperfect, not optimumly abled, and requiring 'micro-prosthetic' technologies of simultaneous debility and capacity to habilitate them for maximal commodification (Puar, 2017). They are bodies not just inhabiting disciplinary spaces, but increasingly inhabited *by* them (Preciado, 2008). Once removed, the unsuitability of these bodies to their singularly nonhuman life is foregrounded and re-habilitation begins.
3 Barad's 'intra-action' affirms the ontological inseparability of entities. However, it is primarily a prefigurative device because the differential 'mattering' of human and nonhuman bodies, and the mechanisms that make this so, are constitutive of, and still dominate, these heterotopic spaces. Apparatuses of power are not neutralised through notions of ontological equity but are part of what is actively reconfigured through intra-actions that open up "possibilities for change in [power's] topology and dynamics" (Barad, 2007, p. 246)

4 My differentiation between transformational and protracted change is theoretical and cannot take account of all previous, potentially contributory processes, experiences, and encounters.

References

Ahmed, S. (2007). A Phenomenology of Whiteness. *Feminist Theory, 8*(2), 149–168.

Arcari, P. (2019). *Making Sense of 'Food' Animals: A Critical Exploration of the Persistence of 'Meat'.* Palgrave Macmillan.

Arcari, P. (2022). (Animal) Oppression: Responding to Questions of Efficacy and (Il)Legitimacy in Animal Advocacy with a New Collective Action/Master Frame. *Animal Studies Journal, 11*(2), 68–108.

Arcari, P. (2023). Slow Violence Against Animals: Unseen Spectacles in Racing and at Zoos. *Geoforum, 144*, 103820.

Arcari, P., Probyn-Rapsey, F., & Singer, H. (2020). Where Species Don't Meet: Invisibilized Animals, Urban Nature and City Limits. *Environment and Planning E: Nature and Space, 4*(3), 940–965.

Asara, V., & Kallis, G. (2022). The Prefigurative Politics of Social Movements and Their Processual Production of Space: The Case of the Indignados Movement. *Environment and Planning C: Politics and Space, 41*(1), 56–76.

Barad, K. (2007). *Meeting the Universe Halfway: Quantum Physics and the Entanglement of Matter and Meaning.* Duke University Press.

Barcz, A. (2017). *Animal Narratives and Culture: Vulnerable Realism.* Cambridge Scholars Publishing.

Bauman, Z. (2000). *Liquid Modernity.* Polity Press.

Bryden, W. L., Li, X., Ruhnke, I., Zhang, D., & Shini, S. (2021). Nutrition, Feeding and Laying Hen Welfare. *Animal Production Science, 61*(10), 893.

Calarco, M. (2008). *Zoographies: The Question of the Animal from Heidegger to Derrida.* Columbia University Press.

Calarco, M. (2015). *Thinking Through Animals: Identity, Difference, Indistinction.* Stanford University Press.

Cheng, H. W. (2010). Breeding of Tomorrow's Chickens to Improve Well-Being. *Poultry Science, 89*(4), 805–813.

Clay, N., Sexton, A. E., Garnett, T., & Lorimer, J. (2020). Palatable Disruption: The Politics of Plant Milk'. *Agric Human Values, 37*(4), 945–962.

Cocozza, V. S. (2019). *Core Conditioning for Horses: Yoga-Inspired Warm-Up Techniques: Increase Suppleness, Improve Bend, and Unlock Optimal Movement.* Trafalgar Square Books.

Cornish, F., Haaken, J., Moskovitz, L., & Jackson, S. (2016). Rethinking Prefigurative Politics: Introduction to the Special Thematic Section. *Journal of Social and Political Psychology, 4*(1), 114–127.

Cudworth, E., & Hobden, S. (2017). *The Emancipatory Project of Posthumanism.* Routledge.

Dalibert, L. (2011). *Posthumanism and Somatechnologies Exploring the Intimate Relations Between Humans and Technologies.* PhD Thesis, University of Twente.

Eaves, L., Gökarıksel, B., Hawkins, M., Neubert, C., & Smith, S. (2023). Political Geographies of Discomfort Feminism: Introduction to the Themed Intervention. *Gender, Place & Culture, 30*(4), 517–527.

Emmerman, K. S. (2014). Sanctuary, Not Remedy. In L. Gruen (Ed.), *The Ethics of Captivity* (pp. 213–230). Oxford Academic.

Fians, G. (2022). Prefigurative Politics. In F. Stein, J. Robbins, A. Sanchez, M. Candea, R. Stasch, S. Lazar, & H. Diemberger (Eds.), *Cambridge Encyclopedia of Anthropology.* University of Cambridge.

Foucault, M. (1967). Of Other Spaces: Utopias and Heterotopias. *Architecture/Mouvement/ Continuite*, October, 1–9. https://web.mit.edu/allanmc/www/foucault1.pdf

Foucault, M. (1982). The Subject and Power. *Critical Inquiry, 8*(4), 777–795.

Foucault, M. (1988). *Politics, Philosophy, Culture: Interviews and Other Writings, 1977–1984.* Routledge.

Foucault, M. (1989). *Order of Things: An Archaeology of the Human Sciences.* Routledge.

Foucault, M. (2002). Space, Knowledge, and Power. In J. D. Faubion (Ed.), *Power: The Essential Works of Michel Foucault 1954–1984 V.3* (pp. 349–364). Penguin Books.

Gillespie, K. (2018a). *The Cow with Ear Tag #1389.* The University of Chicago Press.

Gillespie, K. (2018b). The Loneliness and Madness of Witnessing: Reflections from a Vegan Feminist Killjoy. In L. Gruen & F. Probyn-Rapsey (Eds.), *Animaladies* (pp. 77–87). Bloomsbury.

Hage, G. (2015). *Alter-Politics: Critical Anthropology and the Radical Imagination.* Melbourne University Press.

Haraway, D. J. (2016). *Staying with the Trouble: Making Kin in the Cthulucene.* Duke University Press.

Heddon, D. (2017). The Cultivation of Entangled Listening: An Ensemble of More-than-Human Participants. In A. Harpin & H. Nicholson (Eds.), *Performance and Participation: Practices, Audiences, Politics* (pp. 19–40). Palgrave.

Hediger, R. (2013). Our Animals, Ourselves: Representing Animal Minds in Timothy and the White Bone. In M. DeMello (Ed.), *Speaking for Animals Animal Autobiographical Writing* (pp. 35–48). Routledge.

Herbrechter, S. (2021). *Before Humanity Posthumanism and Ancestrality.* Brill.

Hertzberg, L. (2022). Absolutely Personal: A Countercurrent in Moral Philosophy. In S. A. Salskov, O. Beran, & N. Hamalainen (Eds.), *Ethical Inquiries After Wittgenstein* (pp. 105–124). Springer.

Holmer, M., Wollanke, B., & Stadtbaumer, G. (2007). X-ray Alterations on Spinal Processes of 295 Warmblood Horses Without Clinical Findings. *Pferdeheilkunde, 23*(5), 507–511.

hooks, b. (2015). *Yearning: Race, Gender, and Cultural Politics.* Routledge.

Hortle, E., McKay, L. J., & Flynn, C. (2021). Animal Perspective: Breaking the Language Barrier. *Griffith Review, 71,* 205–214.

Iveson, R. (2015). *Zoogenesis: Thinking Encounter with Animals.* Pavement Books.

Jones, E. (2022). 'Will We Be Riding in 20 Years' Time?' Horse World Must 'Walk the Talk' for Sport to Survive. *Horse & Hound.* www.horseandhound.co.uk/news/will-we-be-riding-in-20-years-time-horse-world-must-walk-the-talk-for-sport-to-survive-781040.

Kinna, R. (2017). Utopianism and Prefiguration. In S. Chrostowska & J. D. Ingram (Eds.), *Political Uses of Utopia: New Marxist, Anarchist, and Radical Democratic Perspectives* (pp. 198–215). Columbia University Press.

Lefebvre, H. (1991). *The Production of Space.* Blackwell.

Librado, P., Gamba, C., Gaunitz, C., Der Sarkissian, C., Pruvost, M., Albrechtsen, A., Fages, A., Khan, N., Schubert, M., Jagannathan, V., Serres-Armero, A., Kuderna, L., Povolotskaya, I., Seguin-Orlando, A., Lepetz, S., Neuditschko, M., Thèves, C., Alquraishi, S., Alfarhan, A., . . . Orlando, L. (2017). Ancient Genomic Changes Associated with Domestication of the Horse. *Science, 356*(6336), 442–445.

Mactaggart, G., Waran, N., & Phillips, C. J. C. (2021). Identification of Thoroughbred Racehorse Welfare Issues by Industry Stakeholders. *Animals (Basel), 11*(5), 1358.

Maeckelbergh, M. (2011). Doing Is Believing: Prefiguration as Strategic Practice in the Alterglobalization Movement. *Social Movement Studies, 10*(1), 1–20.

Manchester, S. J., & Bullock, J. M. (2001). The Impacts of Non-Native Species on UK Biodiversity and the Effectiveness of Control. *Journal of Applied Ecology, 37*(5), 845–864.

Manzini, E. (2019). *Politics of the Everyday.* Bloomsbury.

Meijer, E. (2021). Sanctuary Politics and the Borders of the Demos: A Comparison of Human and Nonhuman Animal Sanctuaries. *Krisis | Journal for Contemporary Philosophy*, *41*(2), 35–48.

Owen, J., Walker, A., & Ince, A. (2022). Editorial: Uncomfortable Geographies. *Emotion, Space and Society*, *42*, 100871.

Pulé, P. M., & Hultman, M. (Eds.). (2021). *Men, Masculinities, and Earth: Contending with the (m)Anthropocene*. Palgrave Macmillan.

Salt, H. S. (1976). Logic of the larder. In: Regan, T. and Singer, P. (eds.) *Animal rights and human obligations*. PrenticeHall, Englewood Cliffs, USA, pp. 185–189.

Sands, K. L. (2019). Shared Spaces on the Street: A Multispecies Ethnography of Ex-Racing Greyhound Street Collections in South Wales, UK. *Leisure Studies*, *38*(3), 367–380.

Schatzki, T. (2013). The Edge of Change: On the Emergence, Persistence, and Dissolution of Practices. In E. Shove & N. Spurling (Eds.), *Sustainable Practices: Social Theory and Climate Change* (pp. 31–46). Routledge.

Schmeer, J. (2019). Xenodesignerly Ways of Knowing. *Journal of Design and Science*. https://jods.mitpress.mit.edu/pub/6qb7ohpt

Spannring, R. (2019). Ecological Citizenship Education and the Consumption of Animal Subjectivity. *Education Sciences*, *9*(1), 41.

Srinivasan, K. (2022). Crafting Scholarly Alliances for Multispecies Justice. *Dialogues in Human Geography*, *12*(1), 79–83.

Todd, R., & Hynes, M. (2017). Encountering the Animal: Temple Grandin, Slaughterhouses and the Possibilities of a Differential Ontology. *The Sociological Review*, *65*(4), 729–744.

van Dooren, T., Kirksey, E., & Münster, U. (2016). Multispecies Studies: Cultivating Arts of Attentiveness. *Environmental Humanities*, *8*(1), 1–23.

Wadiwel, D. (2015). *The War Against Animals*. Brill.

Wilson, H. F. (2020). Discomfort: Transformative Encounters and Social Change. *Emotion, Space and Society*, *37*, 100681.

Yates, L. (2014). Rethinking Prefiguration: Alternatives, Micropolitics and Goals in Social Movements. *Social Movement Studies*, *14*(1), 1–21.

Yates, L. (2022). How Everyday Life Matters: Everyday Politics, Everyday Consumption and Social Change. *Consumption and Society*, *1*(1), 144–169.

Yates, L., & De Moor, J. (2022). The Concept of Prefigurative Politics in Studies of Social Movements: Progress and Caveats. In L. Monticelli (Ed.), *The Future Is Now: An Introduction to Prefigurative Politics* (pp. 179–190). Bristol University Press.

Conceptual and Political Re-ordering

DOI: 10.4324/9781003366706-7

Chapter 5

The Magpies
Reflections on Liminality, Domestication, and Animal Agency

Angie Pepper and Richard Healey

Introduction

Philosophers who work on issues of animal rights increasingly argue that we can develop just relationships with domesticated nonhuman animals (e.g., Cochrane, 2012; Donaldson & Kymlicka, 2011; Nussbaum, 2022).[1] Many of these writers focus on the best of our existing relationships with domesticated animals to illustrate and anticipate what a just interspecies community may look like in the future. Importantly, this position is motivated by the idea that "the love and care that many humans direct to their animal companions is not misdirected sentiment to be despised but a powerful moral force to be harnessed and expanded" (Donaldson & Kymlicka, 2011, p. 155). Accordingly, though our relationships with domesticated animals require a radical reconstruction, we can love and live with them in ways that respect their most fundamental interests.

Contrary to this optimism, we worry that the human desire to keep domesticated animals distorts our moral thinking. It gives us a strong motivation to find a justification for our practices and serves to obfuscate structural injustices inherent in relationships between domesticated animals and their human guardians. Consequently, we think that our existing relations with domesticated animals are not a good starting point from which to think through the future of domestication. Instead, we propose that our relationships with liminal animals – those who live among us but not with us – offer a more promising paradigm for reflection on what morally decent human-nonhuman relationships might look like. What is more, we suggest that reflecting on our relationships with liminal animals reveals how the practices and processes of domestication are inherently problematic and incompatible with the demands of justice.[2]

The chapter is structured as follows. In Section 1, we begin by offering a personal reflection on our relationship with two Eurasian magpies (*Pica pica*). We then suggest, in Section 2, that these reflections reveal why liminality allows and enables the self-determination of nonhuman animals in ways that domestication cannot. Moreover, we argue that relationships between humans and domesticated animals are structurally unjust and that our exercise of power over domesticated animals is inherently tyrannical. We further claim that this gives us good moral

DOI: 10.4324/9781003366706-8

reasons to transition away from domestication towards liminality. Section 3 considers one objection to our view: in ideal circumstances, domesticated animals are better off than liminal animals because we can ensure that their basic needs are always met. In response, we suggest that freedom has non-instrumental value for sentient nonhuman animals, which means that they have an interest in being self-determining even when being so is burdensome and risky.

1. One for Sorrow, Two for Joy

In the tree outside the window of our first-floor flat is a pair of magpies. It is February and they have been industriously preparing their nest. The male breaks large twigs from the tree and throws them on the floor before swooping down and carrying them off. Sometimes he takes on more than he can handle, hopping along the pavement with his swag. The tree gives them a good vantage point from which to observe the comings and goings on our street and to spot food and nest-building materials to be scavenged. And, if we are home, they sit in the tree shouting at the window until we put nuts out on the windowsill, or so we like to think.

We have become quite fond of these magpies. They are inquisitive, smart, and funny – especially when they are chasing squirrels out of *their* tree. But we are under no illusion about magpies and their less wholesome activities. We know that magpies are predatory. We know that they will eat insects, mice, and voles; and that they are notorious for eating the eggs and young of other birds. We also know that they can be noisy, cunning, thievish, and even aggressive towards larger mammals, including humans. Indeed, a quick survey of recent news items involving magpies sees them described as "brazen,"[3] "evil,"[4] and "vicious"[5] creatures who "besiege,"[6] "invade,"[7] and "terrorize"[8] urban environments and unwitting human residents.

Nonetheless, they continue to delight us. No one is perfect after all, and their misdemeanours pale in comparison to those of the many humans with whom we share the neighbourhood. Most humans consume the eggs, milk, and flesh of other animals. And it is humans who pollute and damage the environment with cars, waste, and colonisation of wild spaces. It is humans who live unsustainable lives, destroying the planet to satisfy our desire for more needless things. Seen from that perspective, the magpies, while not moral saints are hardly deserving of the vilification and scorn so regularly heaped upon them. They certainly do not deserve to be trapped, shot, or poisoned to reduce their numbers.

Magpies are an example of what are sometimes referred to as liminal animals, which places them conceptually and geographically between the wild and the domestic (Donaldson & Kymlicka, 2011, p. 210). They are those animals who don't live with us in our homes but, rather, live among us and benefit from coexistence in human settlements. For example, liminal animals may utilise cultivated plants to find vegetables, fruits, and seeds; scavenge in our bins; or eat and drink from intentionally placed sources such as bird feeders. This means that while we (typically) do not keep them as pets or employ them as labourers, many of them are dependent on human ways of life for food, water, and shelter and are consequently not fully wild. Besides magpies, foxes, mice, sparrows, rats, pigeons, and squirrels (among many others) can all be liminal in this sense.

Though we co-exist with liminal animals, our relationship with them is very different to our relationship with domesticated animals (Palmer, 2011; Donaldson & Kymlicka, 2011). Liminal animals are not usually brought into existence to serve human ends, and they are rarely dependent on one specific human to meet their needs.[9] As such, liminal animals do not *belong* to anyone as domesticated animals do. Furthermore, whereas domesticated animals generally exist to serve human interests in food, labour, biomedical research, or companionship, liminal animals rarely have a human-assigned role to fulfil. And importantly, unlike domesticated animals, liminal animals are free to determine the extent of their relationships with us and the overall shape of their lives.

Consider the magpies. The magpies outside our window are not *our* magpies nor were they created to satisfy our desire for companionship, affection, or exercise. We do not attempt to control them, make demands of them, or exert power over them (at least not directly). The magpies are thoroughly self-determining creatures: they act to satisfy their desires and preferences, and they are free to do as they please in most spheres of activity. Their days unfold in accordance with their own preferences, choices, and actions. Most importantly, besides the snippets of their magpie life that *they* allow us to see – we do not hide behind curtains, or spy on them with binoculars – we have no idea what they do or where they go. We do not know where their nest is. We do not know where they sleep. We do not know whose company they keep. We do not know how far afield they fly. We do not know where they go at night. We do not know how they keep themselves safe and what they do to protect themselves against urban predators such as cats and foxes. So much of their life is a mystery to us as ours is to them.

We think reflecting on the liminality of the magpies is suggestive of important lessons concerning what morally decent relationships between humans and animals might look like. Since liminal animals do not exist because of us, in no way belong to us, and are not entirely dependent on particular humans, they experience and possess a significant degree of control over their own lives. They also possess the ability to determine for themselves the kinds of relationships they have with us. Some of these animals are a persistent presence in our lives, wandering nonchalantly about our communities with the same degree of confidence and indifference as the many humans going about their business. These animals may interact with us (especially for food) or choose to keep their distance; the choice is theirs. Others are rarely seen; they live in spaces and places that humans are less likely or less able to access such as rooftops and sewers, or they are more active when humans are typically asleep. Either way, the animals we rarely see will usually seek to keep it that way – ensuring that their relationship with us is only the very minimal one of living nearby.

For the most part, liminal animals do not choose to develop thick interpersonal relationships with us. We catch only glimpses of their lives and often from afar. This is true even when we foster relationships with them such as when we feed birds in our gardens or indeed put nuts out for them on the windowsill. Wanting to be fed is very different from wanting to be friends (contra Milburn, 2022). Although some liminal animals might use us to satisfy their needs and wants, they may also have a strong desire to avoid the kind of intimate relationships that we currently enjoy with domesticated animals.

What's important here is that humans generally refrain from interfering with the agential capacities of liminal creatures who are everywhere around us. We do not force them into proximity with us, we do not seek to control their every activity or surveil their every move. For the most part, we leave liminal animals to decide how they live their lives and how close they want to be to us. The physical and social distance between us and liminal animals gives them the space and opportunity to be self-determining.

To be sure, it would be misleading to categorise this current propensity to non-interference as rooted in moral respect. Though some humans do respect the rights of liminal animals, our relationships with liminal animals are more aptly characterised as relationships of mutual indifference. As a species, we simply do not care that much about (most of) the animals with whom we share our settlements. In part, this is because there are simply too many of them to care about. But it's also because, insofar as they keep out of our way and do not interfere with us, we have little reason to give them much thought.

Of course, liminal animals do not always keep out of our way, and when their interests conflict with those of humans they are likely to be labelled as a nuisance or a pest. In such circumstances, humans will stop at nothing to eliminate or expel these creatures with little consideration for what kinds of treatment are morally acceptable. In this respect, the relationship between humans and liminal animals is morally shameful. Sentient liminal animals, like all sentient creatures, including humans, have significant interests in avoiding suffering and in continued life. Resolving conflicts by resorting to violence without exploring other possibilities is thus unacceptable since unnecessary violence unjustifiably sets back the most fundamental interests of thinking and feeling beings.

Aside from the intentional harm that humans do to liminal animals when our interests conflict, there are at least two other ways in which our relationships with them are morally imperfect. First, we do unintentional harm to liminal animals when we design our settlements in ways that are at odds with their needs and interests. For example, tall buildings with reflective surfaces are a known hazard to birds (Klem, 2009, 2015) and our roadways kill millions of animals every year (Abra et al., 2021). Second, humans are notoriously preoccupied with observing the private lives of animals and an increasing number of people desire to watch liminal animals in contexts in which those animals have chosen not to be seen. If you search YouTube, you will find endless footage from cameras placed in bird nests, squirrel nests, fox dens, bat roosts, and badger setts, to name but a few. And some nosy humans have seen fit to film the private abode of magpies. If one grants that animals have significant interests in privacy (Pepper, 2020), then our encroachment into their private spaces is morally problematic.

While we undoubtedly have the capacity to harm liminal animals – a capacity which we exercise all too frequently – there nonetheless remains something promising about the mutual disinterest that is common between us and them, and our ability to live relatively peacefully in quite varied mixed-species settlements. As our reflections on the magpies outside our window indicate, we already *do* live among animals in ways that respect their basic interests. What we need moving forward are better ways of managing conflicts between humans and liminal animals and a more mindful approach to the ways in which our actions and ways of life affect them as well as us.[10]

2. What the Magpies Tell Us About Domestication

Though we do not want to overly romanticise our existing treatment of and relations with liminal animals, we think it is in these relations that we observe most clearly what morally decent human–animal relationships might look like. Furthermore, we think this exercise helps to reveal why our relations with domesticated animals can never be fully legitimate. The key difference lies in the *structure* of the relationship that obtains between humans and animals in the two contexts. Domesticated animals are structurally positioned to be subject to extensive forms of human power and control. The exercise of this power severely limits domesticated animals' ability to be self-determining, and the fact of this power renders them persistently vulnerable to human abuse. We call this the *problem of human tyranny*. By contrast, magpies and other liminal animals enjoy a significant degree of control over the content and shape of their own lives, including their relative intimacy with humans. To be sure, liminal animals remain subject to human power and are vulnerable to human abuse. But they are not so positioned *by the very nature of the relationship itself.*

There is a lot going on here and much to disagree with so let's unpack things more slowly. Domesticated animals are those animals whom we have brought into our societies to live with us and who have been bred by humans over successive generations to exhibit certain desirable traits, such as tolerance of people, sociability, docility, smaller teeth, smaller (or larger) bodies, coat colour, and so on. Importantly, liminal animals who are friendly towards humans, perhaps entering people's homes or eating from people's hands, are not domesticated. Domestication is not the same as tameness: "taming is conditioned behavioral modification of an individual; domestication is permanent genetic modification of a bred lineage that leads to, among other things, a heritable predisposition toward human association" (Driscoll et al., 2009, p. 9972). The kinds of animals that have been domesticated include (but are not limited to) dogs, cats, horses, cows, sheep, goats, ducks, alpacas, guinea pigs, camels, reindeer, hamsters, mice, rats, chickens, and canaries.

As we noted earlier, domesticated animals are usually deliberately brought into existence by humans to serve some human purpose(s): to be sold or exchanged, to be food, to produce food, to hunt other animals for food, to produce non-edible commodities such as wool, to kill pests, to be a research subject, to be some human's companion, to improve some human's welfare, to take part in competitive sports, to transport humans and their goods, to protect some human's property, and so on. To satisfy these human ends, domesticated animals have been selectively bred to make them fit for purpose. This means that domesticated animals are by their very design instruments of human ends.

The moral ills of domestication are well documented (du Toit, 2016, 2022). Many domesticated animals are neglected, abused, and abandoned. Even when domesticated animals are valued or loved, the conditions of their captivity are often sub-optimal. Think for a moment of the many birds and rodents kept in small cages, or the rabbits made to live in outdoor hutches. Think also of the many dogs who must spend long periods of time alone and who are entirely dependent on their owners for access to outside spaces and exercise. Moreover, many domesticated animals suffer debilitating and sometimes fatal physical and psychological problems, because of selective breeding for traits that humans find desirable. For example, "virtually all 'purebred' dogs suffer from some or other genetic ailment, the most common being cancer" (du Toit, 2016, p. 30). And animals who are industrially farmed for meat, such as chickens, suffer from numerous genetic problems that are a direct result of selective breeding for fast early growth rates and feed conversion (e.g., European Commission Scientific Committee on Animal Health and Animal Welfare, 2000).

While advocates of animal rights share a deep concern for these problems, many maintain that they are not necessary features of our relations with domesticated animals. As Alasdair Cochrane argues,

> the real problem of contemporary animal domestication as it is currently instituted, is the immense suffering and diminished opportunities that it involves. Fortunately, such harms are not intrinsic to domestication: we have both the ability and the duty to eradicate them.
>
> (2014, p. 172)

Thus, we can rescue practices involving domesticated animals by, for example, refraining from knowingly breeding animals who will experience compromised health and by improving the lives of those domesticated animals who come into existence. Of course, this would require a massive overhaul of our existing practices. There would, for instance, be no factory farming and no farming of animals for meat; at most, there may be small-scale farming operations for eggs and dairy (e.g., Cochrane, 2012, pp. 86–89). Our relationships with our pets would also be radically different. They would not belong to us as property but rather be included as co-members of the family and wider community (Donaldson & Kymlicka, 2011; Kymlicka, 2022).

While we agree that we can (and should) bring about significant improvements in the lives of existing domesticated animals, we believe that attempts to reimagine and reform these relations, and thus rescue the wider practices (e.g., of animal companionship) fail to address the deeper structural problem. To see why, let's imagine that changes such as those suggested earlier were implemented. It is our view that there would continue to be a deep-seated problem with the nature of our relationships with domesticated animals.

Central to this moral problem is the fact that domesticated animals are subject to extensive exercises of human power and control. At the domestic level, individual humans have the power to determine what their animals eat, where they sleep, whom they socialise with, how much exercise they get, how much space they have to move freely, how they are educated and the extent of that education, whether they reproduce, what time they get with their young, whether they are required to labour for human ends, and so on. At the institutional level, domesticated animals are regulated through a system of policies and laws which prohibits or permits them to be in certain spaces, under certain conditions, including whether they are allowed to exist in the community at all (consider breed-specific legislation that prohibits the existence of certain dogs). This regulation imposes all manner of legal duties on human guardians regarding their animals and demands that individuals exert extensive control over their animals or else be sanctioned.

The power we possess and exercise over domesticated animals is not morally benign by default; like all social and political power, it stands in need of justification. When power involves restricting what others can do, compelling others to do what they otherwise might not, and denying others access to essential goods and resources, it must be justified if it is to be legitimate. Accordingly, when those in dominant social positions wield power over their subordinates without adequate justification, they lack the moral authority to do so and thus act as tyrants.

We hold that our possession and exercise of power over domesticated animals is generally illegitimate, for two main reasons. First, we think that the extensive exercise of power over domesticated animals – concerning what they can do and where they can be at all times – necessarily sets back important interests in self-determination. Second, the deliberate establishment of a relationship in which humans have this extensive power, and domesticated animals are respectively subordinate and vulnerable, is often morally objectionable in and of itself.

Start with the former idea. Domesticated animals have their capacity for self-determination severely thwarted by the exercise of human power. Unlike the magpies who frequent the tree outside our window, domesticated animals are unable to determine for themselves the shape and content of their lives or the degree of intimacy they have with us. The practices and processes of domestication – including the training and socialisation required to make domesticated animals welcome members of the household and community – involve thoroughgoing exercises of power over animals that prevent them from doing what they want and limit their ability to live free of our control.

Next, the fact that humans possess this extensive power over domesticated animals means that those animals exist in relationships of asymmetric vulnerability and dependence with their human guardians.[11] A domesticated animal is always vulnerable to the whims and vices of their human guardians; a domesticated animal is always dependent on their human guardian for the fulfilment of their most basic needs; and humans always have discretionary control over what domesticated animals do, how the animal is treated, with whom the animal interacts, and whether the animal has access to basic goods. By contrast, humans are not dependent on their companion animals in these ways. Though some humans may rely upon their animals for assistance, companionship, or economic gain, they are never as thoroughgoingly vulnerable or dependent as the animal in the relationship. The inherent asymmetry in power that exists between humans and domesticated animals means that domesticated animals are *always* vulnerable to abuses of power, and humans *always* have the power to harm them. Though some humans will treat *their* domesticated animals well, how much control an animal has over their life and whether their life goes well is ultimately determined by their human guardian.

Now importantly, we do not think there is a moral objection to all relationships that involve dependency.[12] However, we do think there is a significant moral objection to such relationships when they are unnecessary, yet knowingly created and reproduced (at both individual and social levels) (Goodin, 1985). Insofar as it is *pro tanto* wrong to intentionally expose individuals to vulnerability and significant risks of harm, we need a good justification for creating or allowing the persistence of these asymmetric relationships. In the case of domesticated animals, we do not think such a justification is available. After all, it is surely not *necessary* that we maintain this form of cohabitation with animals. Of course, many people *want* to live with or use domesticated animals, but it is hard to see why our desire for animal companionship or labour would serve to justify exposing them to this serious risk of harm.

When taken together, these dimensions of domestication amount to the problem of human tyranny over domesticated animals: domesticated animals are subject to extensive human power, their capacity and ability for self-determination are severely thwarted, and their vulnerability is both deliberately created and unnecessary. Crucially, the problem is structural, which means that all the time we require domesticated animals to live with us – in our homes and communities – they will stand in the same kind of asymmetric relationship with us, no matter if most people treat them better than they do at present. For this reason, when thinking about the development of just interspecies relations, we suggest that we look to the example provided by our relations with liminal animals instead. Though liminal animals are subject to centralised political power – both directly (e.g., regulation on pest control, habitat protection, culling) and indirectly (e.g., planning permission, building regulations, traffic regulation) – they are not ascribed social roles by humans and so are free to shape their own lives without human demands or responsibilities.

3. Happy Dogs, Unhappy Magpies

So far, we have suggested that the relationship between humans and liminal animals has more moral promise than the relationship between humans and domesticated animals. Indeed, we have argued that the practices and process of domestication are fundamentally at odds with the rights of animals and the structural nature of the problem means that domestication cannot be made good by reform. Given this, we think that we have good moral reasons to transition away from domestication towards liminality.

Transitioning from domestication towards liminality will require a radical transformation of our current practices and attitudes towards other animals. While it is beyond the scope of this chapter to propose a full suite of policies that would achieve that end, it's worth pointing out what we think our argument commits us to. First, moving away from domestication requires bringing an end to the practice of creating or capturing animals with the intention of socialising them into our communities and using them to satisfy our ends. At the supply end, this means that animals should no longer be bred for human use such as for food, research, labour, or companionship. At the demand end, humans should not purchase animals for any purpose, including for companionship. Of course, we must care for the domesticated animals who are here already, and where it's appropriate we should enlarge the domain of freedom for those animals. And it might be that with the right support, some domesticated animals who currently live with us could live more independently of us. (Consider, for example, existing populations of street dogs who do not belong to particular humans and who are more free-living than their captive kin.) However, continuing to bring animals into existence (either deliberately or negligently) who have no choice but to stand in an asymmetric relation of power to us is morally impermissible.

Second, we must show restraint in our relations with animals whether they be domesticated or liminal. Sentient animals are wilful creatures with interests in self-determination and interests in controlling the level of intimacy they have with us. Consequently, animals are not ours to touch, hold, or watch, and we must respect the rights of all animals to live free of us if they so choose. Keeping a respectful distance from other animals, allowing them to dictate the terms of our interactions, and indeed whether such interactions occur, is central to ensuring that our relations with them are not characterised by tyranny.

At this point, we anticipate that someone may object that our view of liminality *is* too romantic. Independent life is hard. An animal needs food, water, and a place to sleep, hibernate, and raise their young. Depending on their circumstances these essential goods might be hard to come by and access may be unreliable. Moreover, life is full of hazards. An animal may be attacked, killed, eaten, trapped, injured, poisoned, diseased, orphaned, or lost, among many other bad things. Freedom comes at a significant cost.

For some, this cost is too great. On one such view, the interest of animals in self-determination is only instrumental and so does not compete against their general interest in not suffering (Cochrane, 2009). On such a view, insofar as we can reliably ensure that domesticated animals' needs are satisfied, they have a preferable form of life to those who must work to meet their own needs. For these commentators, transitioning to liminality from domestication makes animals worse off and is therefore morally prohibited. If there were a way to turn the whole world into a zoo – where animals would be safe from predators and have their needs for food, water, and healthcare met – then we should do it!

We reject this line of argument for three reasons. First, even if one thinks that animals' interest in self-determination is only instrumental, much of our interaction with domesticated and captive animals suggests that we are not well-placed to meet their welfare interests. Animals kept in zoos, farms, labs, and homes often have sub-optimal lives, and their lack of agency contributes to poor welfare outcomes. Given that we typically fail existing domesticated animals, the idea that we *ought* also to be extending the reach of control to liminal and wild animals is risible. The costs to animals of being the subjects of domestic power are, as we have seen, very great. Indeed, from the perspective of liminal animals, the last thing they would want, or need is human "care" in a human domicile.

Second, and more importantly, we reject the claim that animals' interest in self-determination is only instrumental. That is, we think that animals have an interest in being self-determining, which is not reducible to the fact that self-determination will serve some other interest of theirs. Thus, animals have an interest in self-determination even when this exposes them to risks and harms. Establishing the non-instrumental value of anything is tricky. Non-instrumentally valuable things are valuable in themselves and not because they cause or produce something else of value. Since non-instrumental value cannot be established by appeal to other goods, explaining why something has non-instrumental value can sometimes seem like little more than mere assertion. We're mindful of this problem and want to avoid it (so far as possible). To get at why self-determination is non-instrumentally valuable for nonhuman animals, we borrow the following test from Elizabeth Anderson:

> The proper test of the noninstrumental goodness of an activity is not whether we'd still prefer to do it, even if it didn't result in desirable consequences. It is rather whether we'd still prefer to engage in it, even if the same consequences could be brought about by other (passive) means.
>
> (Anderson, 2009, p. 225)

In previous work, we made use of the "improper test" alluded to by Anderson in this passage. We suggested that self-determination was valuable for animals despite the negative consequences that might come from their free-willed action (Healey & Pepper, 2021, pp. 1227, 1239). We stand by that idea,[13] but we also think that Anderson's alternative test can further support the non-instrumental value of self-determination for nonhuman animals. Of course, one might try to argue that since the animals themselves cannot understand such a test, and since we cannot be certain of what they'd prefer, this test is inapplicable to them. In what follows, we suggest that the test is applicable and that animals' interest in self-determination passes it.

Though Anderson talks in terms of preferences, there is no reason to think that *expressed* preferences are the only way of measuring something's being of value to agents. We might just as easily talk about agents having an interest in something. The question then becomes whether animals might continue to have an interest in self-determination when the desirable consequences of their self-determined action – for example, meeting their needs, satisfying their desires, avoiding their dislikes – can be achieved by some other means such as by benevolent human guardians. We suggest that even if humans were able to satisfy domesticated animals' preferences and desires successfully and routinely, those animals would nonetheless have an interest in being able to exert control over their own lives.

We think it is helpful to approach this issue by being attentive to the fact that each "individual has a special, intimate relation to her mind, body, experience, and environment that she must especially endure" (Shiffrin, 2012, p. 382). Paying attention to this intimate relation – a unique relation between mind and world – highlights several reasons why it would matter to an individual to have control over what happens to them in and of itself. For one, the simple fact that an individual must endure what happens to them offers a clear motivation for why they should want to have control over what happens to them. Indeed, it is plausible that one of the things that an agent desires, *given* the intimate connection between their own mind, body, and experience is precisely that they have control over these things. Furthermore, when what animals desire is simply delivered to them, without engaging their agency, they become passive witnesses to their lives. In the case of human beings (and not only cognitively unimpaired adults), such passivity is something we think we have reason to avoid. For example, we encourage children to *do something for themselves*. Here we see Anderson's test in action: we think children have reason to act to get X themselves rather than simply receiving X passively. We see no reason why the same would not be true for nonhuman animals.

The foregoing can be given a positive restatement: exercising capacities for self-determined action is an important good for creatures with those capacities. That is, exercising capacities for self-determination partly constitutes the well-being of agents with those capacities. Given the natural tendency to want control over one's own lived experience, and our more general commitment to avoiding passivity, we can see why there would be a reason for an animal to secure the object of their preference themselves as opposed to achieving the same consequences by other (e.g., human interventionist) means. Demanding passivity, denying animals the power to control their own lives, and limiting their exposure to the experience and education needed for such control all serve to prevent them from living as the kinds of self-determining creatures that they are.

While more remains to be said, we think this sketch supports the view that nonhuman animals, like humans, have a significant interest in self-determination. If that is true, it is a serious mistake to think that domesticated animals are simply better off than liminal animals because they have (or rather *should* have) reliable access to food, water, and shelter. While these are important goods for any sentient creature, so too is the ability to exercise control over one's own life, an ability that is generally elusive for domesticated animals. Furthermore, there is no reason for us not to provide resources (e.g., shared water sources) for liminal animals so long as we remain respectful of their agency and independence. What their interests in self-determination require is that we refrain from intervening in their ability to move around and act as they see fit to give them what we judge to be "happy" or "better" lives.

Finally, some may agree with many of our ideas but nonetheless wish to resist the claim that we should take active steps to transition domesticated animals into liminality. Again, much remains to be said, but our purpose in suggesting such a transition is less to provide a well-defined policy proposal than to articulate an alternative regulative ideal, given the problems identified in practices of domestication and the comparative promise we see in our relations with liminal animals. In any case, it is important to note that even if there are ultimately sufficient reasons not to transition domesticated animals into liminality, this does nothing at all to vindicate existing practices of domestication. Insofar as the practices and processes of domestication involve intentionally curtailing animals' capacities for self-determination and exposing them to unnecessary risks of harm, they are unjustified. Thus, domesticating wild and liminal animals is neither an appropriate nor desirable response to the risks that they face in their daily lives.

Conclusion

In this chapter, we hope to have shown how reflection on existing relationships with liminal animals serves to problematise narratives of just domestication and reveal better ways of living with animals in the future. While liminal animals are vulnerable to a variety of harms, including harms caused by humans, they stand outside of the structural relation of dependence and dominance that characterises our relationships with domesticated animals. This gives us good reason to see these relationships, rather than existing relations with domesticated animals, as the starting point for thinking about how to achieve just relationships with nonhuman animals.

Postscript

It is now June and the magpies have stopped visiting the tree outside our window. We're not sure where they have gone or why. But, as free agents, they are entitled to live their lives elsewhere. Occasionally, we think we see them in the back gardens of the terraced houses on the street where we live. And any nuts we place on the windowsills to the back of the flat swiftly disappear. But in all honesty, we cannot be sure they are *our* magpies.

We too will be moving on soon and will not be around to see if they come back to the tree in the winter. Nevertheless, we will remember them fondly and continue to learn from our relationship with them, which was, while fleeting and fragmentary, free and uncomplicated.

Notes

1 Though these authors disagree about what justice demands, they agree that just relationships between humans and domesticated animals are possible.
2 The argument of this chapter is inspired by the abolitionist approach developed and defended by Gary Francione (see, for example, Francione, 2008; Francione & Garner, 2010; Francione & Charlton, 2015). Though we are not here committed to Francione's version of abolitionism, our contribution should be read as a critical challenge to those

who believe that the continued domestication of animals is compatible with the demands of justice.

3   www.express.co.uk/news/uk/1697812/Rat-fight-magpies-video
4   www.dailymail.co.uk/news/article-10665225/Scottish-bride-suffers-horror-head-wound-attack-evil-MAGPIE.html
5   www.mirror.co.uk/news/uk-news/city-invaded-vicious-magpies-attacking-28529695
6   www.dailystar.co.uk/news/latest-news/uk-town-besieged-vicious-magpies-28527819
7   www.mirror.co.uk/news/uk-news/city-invaded-vicious-magpies-attacking-28529695
8   www.dailystar.co.uk/news/latest-news/uk-town-besieged-vicious-magpies-28527819
9   There are some exceptions here, including animals bred and reintroduced into the wild by humans for conservation purposes. While many of these animals will go on to live in the wilderness – free of humans and their ways of life – some will, or will come to, live among us. For example, red kites (*Milvus milvus*) were reintroduced to the UK just over 30 years ago, and now significant numbers of them feed from gardens in residential areas in the city of Reading (Orros & Fellowes, 2015). Though humans had a hand in the creation of these animals, and they may be monitored, the animals themselves continue to enjoy a great deal of freedom over where they go and what they do.
10   For further discussion of what we might owe to liminal animals, see chapter 7 in Donaldson and Kymlicka (2011).
11   The view articulated here is indebted to Robert Goodin's (1985) discussion of asymmetric vulnerability.
12   The obvious example is human children. Of course, the case and comparison give rise to various questions and complications, and we cannot adequately address these here. However, it is worth noting for present purposes that (i) we intuitively have good reason to minimise the vulnerability of children where possible, and (ii) guardians have a duty to gradually equip children with the abilities and skills they need to be able to avoid such vulnerability.
13   Indeed, it is an implication of the claim that animals have a non-instrumental interest in self-determination.

References

Abra, F. D., Huijser, M. P., Magioli, M., Bovo, A. A. A., & de Barros, K. M. P. M. (2021). An Estimate of Wild Mammal Roadkill in São Paulo State, Brazil. *Heliyon, 7*(1), e06015.

Anderson, E. (2009). Democracy: Instrumental vs. Noninstrumental Value. In T. Christiano & J. Christman (Eds.), *Contemporary Debates in Political Philosophy* (pp. 213–227). Blackwell.

Cochrane, A. (2009). Do Animals Have an Interest in Liberty? *Political Studies, 57*(3), 660–679.

Cochrane, A. (2012). *Animal Rights Without Liberation: Applied Ethics and Human Obligations*. Columbia University Press.

Cochrane, A. (2014). Born in Chains? The Ethics of Domestication. In L. Gruen (Ed.), *The Ethics of Captivity* (pp. 156–173). Oxford University Press.

Donaldson, S., & Kymlicka, W. (2011). *Zoopolis: A Political Theory of Animal Rights*. Oxford University Press.

du Toit, J. (2016). Is Having Pets Morally Permissible? *Journal of Applied Philosophy, 33*(3), 327–343.

du Toit, J. (2019). The Ethics of Domestication. In B. Fischer (Ed.), *The Routledge Handbook of Animal Ethics* (pp. 302–315). Routledge.

Driscoll, C. A., Macdonald, D. W., & O'Brien, S. J. (2009). From Wild Animals to Domestic Pets, an Evolutionary View of Domestication. *Proceedings of the National Academy of Sciences, 106*, S9971–S9978.

European Commission Scientific Committee on Animal Health and Animal Welfare. (2000). *The Welfare of Chickens Kept for Meat Production (Broilers)*. European Commission,

Health and Consumer Protection Directorate-General, Brussels. Report no. SANCO.B.3/AH/R15/2000.

Francione, G. (2008). *Animals as Persons: Essays on the Abolition of Animal Exploitation.* Columbia University Press.

Francione, G., & Charlton, A. (2015). *Animal Rights: The Abolitionist Approach.* Exempla Press.

Francione, G., & Garner, R. (2010). *The Animal Rights Debate: Abolition or Regulation?* Columbia University Press.

Goodin, R. E. (1985). *Protecting the Vulnerable: A Re-Analysis of Our Social Responsibilities.* University of Chicago Press.

Healey, R., & Pepper, A. (2021). Interspecies Justice: Agency, Self-Determination, and Assent. *Philosophical Studies, 178,* 1223–1243.

Klem, Jr., D. (2009). Avian Mortality at Windows: The Second Largest Human Source of Bird Mortality on Earth. *Proceedings of the Fourth International Partners in Flight Conference: Tundra to Tropics.* McAllen, TX, pp. 244–251.

Klem, Jr., D. (2015). Bird-Window Collisions: A Critical Animal Welfare and Conservation Issue. *Journal of Applied Animal Welfare Science, 18,* S11–S17.

Kymlicka, W. (2022). Membership Rights for Animals. *Royal Institute of Philosophy Supplement, 91,* 213–244.

Milburn, J. (2022). *Just Fodder: The Ethics of Feeding Animals.* McGill-Queen's Press.

Nussbaum, M. C. (2022). *Justice for Animals: Our Collective Responsibility.* Simon & Schuster).

Orros, M. E., & Fellowes, M. D. E. (2015). Widespread Supplementary Feeding in Domestic Gardens Explains the Return of Reintroduced Red Kites Milvus Milvus to an Urban Area. *Ibis, 157*(2), 230–238.

Palmer, C. (2011). The Moral Relevance of the Distinction Between Domesticated and Wild Animals. In T. L. Beauchamp & R. G. Frey (Eds.), *The Oxford Handbook of Animal Ethics* (pp. 701–725). Oxford University Press.

Pepper, A. (2020). Glass Panels and Peepholes: Nonhuman Animals and the Right to Privacy. *Pacific Philosophical Quarterly, 101,* 628–650. https://doi.org/10.1111/papq.12329.

Shiffrin, S. V. (2012). Harm and Its Moral Significance. *Legal Theory, 18*(3), 357–398.

Chapter 6

Dog Proposals
Participatory Design, Playfulness, and Multispecies Futures

Michelle Westerlaken

Introduction

There are many different types of animals on this planet that are interested in participating in what humans do, one of which is the domestic dog. In fact, this is how they started to domesticate themselves some 20,000 to 40,000 years ago: by hanging out in the vicinity of humans. Contrary to the historically pervasive story that only credits humans for their domestication process, another narrative reveals that wild wolves specifically developed the sociability and friendliness that allowed them to live close to human settlements to scavenge their leftovers (Coppinger, 2001; Hare & Woods, 2013; Meijer, 2019). The wolves learned how their negotiations with humans became mutually beneficial and thus sought closer relationships. Approaching human settlements and negotiating relationships also changed the wolves' characteristics. Bolder, friendlier, more juvenile, and playful traits such as wagging tails and shifting curiosities emerged and changed their genetic compositions. Within several generations, and due to genetic changes associated with behavioural adaptation, they also showed splotchy coats, floppy ears, and made remarkable psychological changes such as the ability to read human gestures and facial expressions (Hare & Woods, 2013). A more nuanced historical analysis of this theory on dog domestication suggests humans may have had significantly more impact on domesticating dogs through early breeding processes (Shipman, 2015). In this reading, collaborative hunting activities between humans and wolf-dogs could have altered both human and dogs' genetic traits (ibid.). Domestic dogs, in this historical narrative, are an intriguing example when it comes to how they negotiated their territories and living environments. Such mutualistic and collaborative interactions have also been identified between humans and other animals including dolphins, orcas, and certain bird species (Cantor et al., 2023; Curtis, 2023). Rather than violence, control, and domination, animals used friendliness, playfulness, and cooperation as ways to negotiate relations with humans. This chapter seeks to design multispecies futures that are inspired by these characteristics.

From simply scavenging leftovers, proto-dogs – precursors of the domesticated dog – soon started to work with humans more actively. While helping with the hunt, barking at strangers, finding people in trouble, guarding carcasses, and defending

DOI: 10.4324/9781003366706-9

humans from predators, dogs started to become increasingly important members of human communities. During the mid-nineteenth century in Victorian England, relationships between humans and dogs further intensified as kennel clubs started to arise and modern dog breeding fully installed itself as the increasingly anthropocentric, exploitative, and eugenics-related institution that it is today. Although different types of dogs already existed long before then (and were already grouped/bred by function), kennel clubs established a narrower focus on the physical appearances and lineages of distinct dog breeds (Budiansky, 2000). This development fundamentally changed the position of the domestic dog as a valued individual member of inclusive multispecies communities, to the amplified and speciesist idea of an animal that can be grouped into categories like "purebred" or "bastard" depending on their genetic compositions and their relations to the humans who control them. Nowadays, although often labelled as "companion species," most dogs, as most "pets," are perhaps more accurately described as instruments of pleasure, aesthetic assets, status symbols, sources of emotional satisfaction, or toys for the many humans who claim to "own" them. Their friendliness and their interest in participating in human lives have brought dogs multi-faceted – and in some ways tragic – life-stories to tell.

Indeed, the thin line between being a companion or an instrument evokes critique for scholars who try to romanticise our relationships with dogs. Feminist new materialist thinker Donna Haraway, for instance, is heavily critiqued by scholars and activists from the field of Critical Animal Studies. Although she acknowledges that relations between humans and animals are seldom symmetrical, equal, or calculable, she also wants to "resist the tendency to condemn all relations of instrumentality between animals and people as necessarily involving the objectifications and oppressions of sexism, colonialism, and racism" (Haraway, 2008, p. 74). Haraway thus withholds judgement when discussing practices like mass-breeding, creating pure-bred dogs, and engaging in dog training. However, Zipporah Weisberg, writes that breeding is "a practice built directly out of humans' entitlement to the bodies and lives of other animals" (2009, p. 30), and that Haraway is complicit in sustaining or even endorsing these practices by providing ideological cover for such violent practices.

Haraway's dog – Cayenne – appears often in her writings about human–animal relationships. The agility training they engage in together is used as an example of how humans and animals "are everywhere full partners in worlding, in becoming with" (2008, p. 301). The issue here is to understand what "full partners" actually implies for Haraway when the participating dogs are conditioned to carry out activities and hone their skills according to what humans deem acceptable. In this reading, activities such as agility training become more an exercise in domination and control over the dog than something resembling full partnership. Both the failure to acknowledge the instrumental relationships humans have established with dogs and the attempt to romanticise exercises in dog-conditioning/training as partnerships are problematic for the ways they normalise speciesist relations with animals in general (Spannring, 2023). However, at the same time, Haraway's work illustrates the ways in which dogs themselves are not passive victims of exploitation but are continuously negotiating their relations with humans, albeit on unequal terms. What are the kinds of relations between humans and dogs that can facilitate these negotiations in more equal ways? This text illustrates how, through careful processes and reflections, "playfulness" as a collaborative activity can help to actively include the responses, negotiations, and proposals expressed by dogs and other animals. By unsettling normative and exploitative relationships, it aims to inspire more participatory multispecies negotiations towards creating liberatory futures. The following section introduces the heterotopian potential of playfulness. These perspectives are then connected to the methods of Participatory Design. The third section reflects upon a practical project in which playfulness was designed with two dogs, and the conclusion discusses its implications for multispecies futures.

Multispecies Animal Liberation Through Playfulness

How are other animals participating in the creation of heterotopias? Here, the notion of heterotopia extends beyond Foucault's spatially oriented writings on this topic (2008 [1967]). This work instead focuses on the mutually shared activities between humans and other species through which traces of other, less speciesist, relations become apparent. Such traces exist between the real and the virtual because here they are found within the context of "play." Play comprises a unique space in which players can negotiate and challenge each other, and speciesist norms can be transformed. Interspecies playfulness deliberately demands the participation of other living beings precisely because it is characterised by shared activity and requires both human and nonhuman animals to respond to each other's playful proposals. Since human societies are firmly embedded in human constructs including language, culture, and education, imagining multispecies participation is challenging, but the practice of playing together can open perceptive doorways to new proposals. This chapter aligns its contributions towards animal liberation with a multispecies participatory political theory that is outlined in the following two paragraphs and takes the position that animal liberation cannot be achieved through the interventions of humans alone.

Human proposals for liberated animal futures may assume a central position for humans as those who engage in the process of imagining difference and subverting the status quo. However, over the past two decades, less anthropocentric scholarly and activist engagements with animals have inspired more attention to the many ways in which animals are also themselves proposing and contesting futures. These reflections articulate how animals and other more-than-human entities resist their oppression (Hribal, 2007) and share their experiences and opinions, for example by engaging in politics (Donaldson & Kymlicka, 2011), political language (Meijer, 2019), interactions with technologies (Westerlaken et al., 2023), artistic practices (Despret, 2016), and interspecies communication (Wijngaarden, 2023). Furthermore, Indigenous scholarship has articulated the diverse modes in which animals and other entities actively shape socio-political dimensions of their societies and communities (Bird-Rose, 2017; Castro, 2018; Kohn, 2013). Also in the area of design – a field that often concerns itself with proposing possible futures – emerging discourses in creating eco-centric and multispecies futures have sparked discussions on how animals participate in design processes (Escobar, 2017; Mancini, 2011; Sheikh et al., 2023; Veselova & Gaziulusoy, 2022; Westerlaken, 2020). Moreover, the creation of multispecies futures together with other animals implies that these are not imagined in abstract forms through human modes of thinking. Instead, to facilitate participatory politics, multispecies proposals must be created in the everyday contexts where humans are encountering other animals.

Playfulness between humans and other animals can be a unique space in which negotiations between entities unfold and speciesist norms can be disturbed. Several animal studies scholars have commented on the important political potential found in play. Animal ethicist Cynthia Willett, for example, describes multispecies play and laughter as "technologies of reversal, levelling hierarchies by turning stratified structures upside down" (Willett, 2014, p. 30). Haraway advocates for play as a crucial possibility for generating political imagination beyond critique to propose the possible-but-not-yet (Haraway in Weigel, 2019). More recently, author and artist Carol Gigliotti unpacks the importance of recognising how animal play generates creativity and requires negotiating others' interests, trying out new behaviours, and using egalitarian components of social systems (Gigliotti, 2022). Playfulness between human and nonhuman animals often involves rules and power, but it also allows for continuously changing dynamics, where players are able to surprise each other (Westerlaken, 2017). Importantly, playfulness here includes not only the types of animal play characterised as "non-functional" or bodily behaviour such as jumping around or play-biting (Bekoff, 2004; Burghardt, 2006) but also a much broader scope of actions described by play scholars Miguel Sicart (2014) and Ian Bogost (2016) as any inventive, surprising, or carnivalesque expression or attitude. More generally, this relates to all kinds of activities undertaken with curiosity, pleasure-seeking, humour, mutual affection, creativity, laughter, and openness to surprising engagements.

Contrary to a prevailing understanding of playfulness as a non-serious activity, to be playful is to take things or situations seriously enough to manipulate them in new ways (Bogost, 2016). Furthermore, it must be recognised that by engaging in multispecies relations playfully, humans do not necessarily oppose dominant culture or completely abandon speciesism. Some forms of playfulness can even reinforce oppressive structures or turn into bullying. Rather, playful engagements between human and nonhuman animals help to create new opportunities and new attempts at wilfully paying attention to each other differently. This enables alternative ways of interacting, manipulating, and anticipating things that in turn change interspecies relations. In playfulness that arises between living entities, players seek to create a common ground that invites a response and ends when either party is no longer interested. Thus, by becoming more attentive to the surprising ways in which other animals can manipulate their environments and negotiate relations, it becomes possible to imagine alternative futures in multispecies participatory contexts. By negotiating possibilities through playfulness, all living beings can add richness to multispecies world-making.

In the field of design research, several projects explore technologies for animal playfulness, where humans are involved in different modalities as co-players, caretakers, or designers. Examples include Participatory Design with dogs engaged in watching TV (Hirskyj-Douglas et al., 2017), prototyping and testing touch screen games for sheltered orangutans (Wirman, 2014), playful artefacts in the context of elephant enrichment (French et al., 2015), the use of games for farmed pigs (Driessen et al., 2014) and for orangutans in zoos (Webber et al., 2017), and interactive systems for interspecies play (Pons et al., 2014). Many of these projects take place in highly speciesist contexts such as zoos and farms, while others involve animals in shelters or domestic spaces with the goal to help them work through oppressive human structures. All of these studies comment on the surprising or unanticipated playful interactions of the animals during the process. Yet, as design research, these projects prioritise the interactions of animals with objects to identify further prototype iterations. They do not leave much room for interactions and relations beyond the designed objects, nor do they specifically reflect on how speciesist norms are reversed or disturbed in these playful negotiations. In other words, the liberatory potential of negotiating relations between humans and animals in playful design contexts remains underexplored.

Participatory Design and Revisiting Encounters as a Method

Imaginations and creations of and for multispecies futures inevitably involve the field of design. Ranging from multispecies architectures, technologies, and urban spaces, to consumables, fashion, and graphic design, what connects this diverse field is the goal to propose or create possibilities. Design practice is inherently forward-looking and seeks to understand current situations in order to generate alternative futures. With regard to animal liberation, such efforts can focus on designing animal-free products, rethinking co-habitation with urban animals, creating technologies that could replace animal labour, or designing for systemic changes in human–animal relations.

Rather than developing finished solutions and presenting these to recipients, nowadays many design projects are shaped through more democratic engagements and iterative development phases. This method of Participatory Design, or Co-Design, includes different stakeholders or users as participants in design projects and has its roots in 1970s Scandinavian design and research practices focused on ideas of workplace democracy (Bjerknes et al., 1987; Bødker, 1996). Participatory Design is primarily understood as the inclusion of end-users and other stakeholders of the designed artefacts early in the design process. Participatory Design aspirations have often been framed as a way to include voices of people that are not usually heard, for example by focusing parts of the design process on listening to the workers rather than the managers, the neighbourhood inhabitants rather than the municipality, the tenants rather than the landlords, or communities rather than companies. Tools and methods such as workshops, collaborative prototyping and sketching, and iterative design phases are commonly used in Participatory Design practices.

However, since the early 2000s, critical discussions of these supposedly bottom-up processes have also emerged. Scholars have reflected on how participants in design projects can be used to simulate democratic engagement while actually reinforcing dominant power patterns (Beck, 2002; Decolonizing Design, 2017; Hillgren et al., 2016). In other words, Participatory Design is inevitably politically charged and at risk of reinforcing existing hierarchies and oppressions when it does not deliberately acknowledge and challenge the harmful structures in which it operates. These structures and their normative hierarchies are multiple, intersecting, and compounding, and may be based on gender, race, ethnicity, religion, wealth, age, ability, and species, among others. This critical discourse on Participatory Design echoes the critique on romanticising the instrumental relationships between humans and dogs as outlined in the opening of this chapter as well as the longstanding feminist critiques on democracy more generally (Wadiwel, 2015). Hence, where collaborations and partnerships, such as those between humans and animals, may be conceived as equal, they are typically constituted by and sustained through larger structures of control and domination. My question here is, can negotiations between humans and dogs during interactive activities help to reveal, challenge, or oppose speciesist structures, or do they reinforce them?

By addressing this question, this text articulates a design research project that shows how human-dog relations can follow liberatory pathways towards revealing multispecies futures with the use of playfulness as a context for participatory exploration. It does not exclude certain playful activities, such as training, agility, or sports between humans and nonhuman animals as oppressive by default. Indeed, inter-human sports and games can also mirror human societies' oppressive structures and are often embedded in colonialism, racism, sexism, homophobia, and many other societal injustices. Rather than opposing all forms of social sports and games, a more generative design approach is concerned with rethinking how to create more inclusive forms of communal playful activities.

Within the emerging field of Animal Computer Interaction, the notion of Participatory Design with other animals (particularly with domestic dogs) is gaining popularity. A paper published in 2005 already referred to the idea of "pawticipatory" design with canines to develop a telecommunication device that allows humans to watch their dog while they are away from home (Mankoff et al., 2005). While such early examples often turn to tools for monitoring and surveillance of dogs, since then, researchers have proposed using Participatory Design with dogs as a dog-centric approach that can "give the dog another method of voicing their opinion" (Hirskyj-Douglas et al., 2014, p. 2). To move further beyond anthropomorphic notions of participation, design researchers suggest that broader participatory models based on multispecies sensemaking, volition, and choice could make Participatory Design practices more resilient to the diversity of the partaking agents, and better account for the contributions that can be made by other animals (Mancini &

Lehtonen, 2018; Mancini, 2023). While these scholars engage with the possibilities for other animals to engage in design projects as participants, I wish to emphasise that – with the already ancient histories of dogs' participation in human societies – it is not a matter of deciding *if* dogs are *able* to join in Participatory Design processes but how designers may respond to and account for their already existing participation and negotiations in ways that take dogs' proposals seriously. It is only in this way that we can account for the multispecies futures that are always made collectively during design processes in which humans and other animals encounter each other.

As part of the move towards Participatory Design with dogs, I also published work that engages with the question of how dogs can become involved as participants in design processes (Westerlaken & Gualeni, 2016). However, given the transformation of multispecies and more-than-human design more generally, this work must now be revisited. As mentioned earlier, the liberatory potential of playful design with animals remains underexplored, including in my own work. While earlier design research predominantly focused on the design of artefacts that can improve the lives of domesticated animals, I now urge design researchers to also ask how we may become attentive to the proposals of animals in the creation of more transformative multispecies futures. The next section thus illustrates this design move in more detail, using a first-person voice to situate myself as a designer in conversation with the animals involved and using image and illustration to document less language-centric multispecies thoughts that emerged.

Earlier interpretations of this work were published in a design-oriented paper with the aim to demonstrate the participatory capacities of two dogs in co-developing artefacts (Westerlaken & Gualeni, 2016). Further analysis helps to reflect retroactively on this work and argues for the need to articulate the much more complex and less innocent efforts of speculating and imagining multispecies ways of living together. Such an approach openly includes considerations for making difficult decisions, shares the much longer histories between participants, and engages in more reflective, situated writing. This reworking is guided by research on more-than-human participatory design (Akama & Light, 2018) and aims to demonstrate how dogs are already well-versed in subverting and challenging speciesist norms. More recent Participatory Design work with dogs similarly articulates the importance of noticing small events that might seem subtle or mundane at first, but, upon reflection, become critical in contributing to relational transformations (Hakio et al., 2022). It is only by finding new ways to engage with the proposals of other animals that it may become possible to create multispecies futures, that is, alternative relations that are created more collectively, in everyday contexts, and *with* the more-than-human entities that seek to inhabit them.

Designing With Dogs: An Exercise in Sensitising and Attuning

Over the past decade, I lived with two dogs. Jojo and Boogie both came into my life as puppies and the three of us established a close bond over time. Throughout these years we together engaged in countless negotiations, several of which were

Figure 6.1 A robotic dog toy with a puzzle mechanic controlled from a mobile phone. Image from the author.

Figure 6.3 A digital hopscotch game for dogs and humans made together with Alex Camilleri. Image from the author.

Figure 6.2 A hybrid digital/physical game for dogs and humans. This project was made together with Dariela Escobar and Inge van Hoppe. Image from the author.

Figure 6.4 A dog toy that uses replaceable smells and interactive sounds as a core play mechanic. Image from the author.

documented and iterated upon as design projects. Between 2013 and 2015, we explored playful interactions in which digital technologies played a role for both the humans and the dogs involved (see Figures 6.1 to 6.4).

These projects can be analysed to identify where and how the dogs would qualify as participants in the design process, as is often the focus in design projects. Instead, I now propose to focus on more recent frameworks in which participation is regarded as a way to sensitise and attune to our already ongoing relationships rather than on how the dogs were useful in the design projects (Figure 6.5). Here, the notion of sensitising and attuning is adopted from the work of participatory design researchers Yoko Akama and Ann Light. They write about sensitising as a process of receptiveness and awareness of situations and attuning as a modifying and adjusting of courses of actions in response to a more relational understanding of complex interactions (Akama & Light, 2018). Their approach offers a way of discussing participation that, compared to traditional design approaches, is better able to account for the complexity of participation and the liberatory potential this has for the entities involved.

By being more explicitly open about the deep undercurrents of personal histories, hierarchies, and experiences that can surface in the design process, it becomes possible to explore how these commitments play out in our actions and begin to understand our practice better and work with greater sensitivity to others in the design process (Akama & Light, 2018). In doing so, designers can reveal a way of working that approaches uncertainty with more resilience than formal methods are able to do (2018). On a practical level, Akama and Light encourage writing about design practices through auto-ethnographic approaches that write the designers actively into the story, instead of implying we are external to the process and its outcomes.

In revisiting the projects with Jojo and Boogie eight years after the design projects were first carried out, with the particular goal of finding a more appropriate multispecies framing for participation, I paid particular attention to how I wrote myself and the individual dogs into our design story. In the earlier paper, about 30% of the text was already given to these "designer notes." Most of these notes dealt with the interactions of the dogs with the artefacts, but some sections shed light on the early and ongoing attempts by me and the dogs at interspecies attuning

Figure 6.5 "Let's try to make something together." Illustration by the author.

and sensitising. In the following paragraphs, I quote some of those notes and subsequently connect them with Akama and Light's reflections.

> When I put this prototype [an interactive ball] in my living room for the first time and I started to move it around, both the dogs immediately paid attention to the object. Interestingly, [Boogie][1] moved towards the artefact and started to explore it, while [Jojo] took more distance and observed the object from a safe and high position. Apparently [Boogie] found it enjoyable and started to play around with the ball for a few minutes, while making eye contact with me multiple times. I tried to facilitate these short moments of contact by participating in the interaction while being closer to them. So I sat on the floor and talked with the dogs, as I would normally do when we play together.
>
> (Westerlaken & Gualeni, 2016, p. 5)

Referring to an earlier text (Light, 2015), Akama and Light suggest that it

> is more important to be attuned to relations and ready for anything in these flexible and evolving circumstances than it is to have an action plan. Thus, one skill is to make an appropriate judgement call on what might and could happen next and decide whether to intervene: to disrupt or to preserve.
>
> (2018, p. 2)

This is arguably a challenging thing to do when both dog participants respond in widely different ways to our interventions (as described in this design note). However, through this note it becomes visible how our histories are intertwined with each dog's personalities and ways of approaching the world, which becomes a clear contributing factor that structured how our play sessions would unfold (Figure 6.6).

Figure 6.6 "I challenge you to play like this." Illustration by the author.

Reflecting on uncertainties in design projects, Akama and Light write:

Of course, no one can ever tell what would happen if paths had been followed rather than abandoned . . . there is no right answer, only the flow of interactions. What we can aspire to is a rhythm that includes all voices and helps people to achieve what they can do together.

(2018, p. 6)

It is these changes in "rhythm" that determine how the playful interactions (and thus the design project) unfold. So rather than striving to find "the right design," our sensitising and attuning happen in a search for a common rhythm that can be sustained. In combining elements of playfulness and design, this attempt at creating and sustaining a multispecies rhythm in which different entities participate characterises an important insight into how human-dog encounters are negotiated politically. Our design sessions would typically end when the dogs or humans lost interest or stopped responding to each other.

A second project that was reflected upon in the 2016 paper involved a more open-ended object. Rather than resembling a toy (such as a ball or soft toy), the initial artefact took the shape of a bundle of smells and sounds (see Figure 6.7) inspired by observations described as follows:

I noticed that one of the main activities my dogs engage in when we go outside is exploring objects by smelling them and following smell trails that are unnoticeable for me. Next to this, I discovered that they had a big interest in certain items that were connected to other places, such as clothing items of family members that live in other countries, branches from a local forest, towels

Figure 6.7 Jojo and Boogie in the activity of smelling a particular object (in this case, a towel that belongs to a family member living abroad) that formed an inspiration for this project. Image from the author.

that were in the dog-bed of a familiar dog, etc. I noticed how they would spend their time sniffing these objects and it seemed like an enjoyable activity to them. I became fascinated with the possibility that, perhaps, these smells could trigger certain memories for them, like familiar smells can do for humans.

(Westerlaken & Gualeni, 2016, p. 7)

These prototypes were inspired by a much more extensive history that we – Jojo, Boogie and myself – had together. Living in a different country from my human family, I asked them to send me a package of smelly objects including a branch from the forest that their dog-friend back home played with, a blanket from her bed, and a T-shirt of a human family member.

The original paper does not discuss my difficulties with having separated Jojo and Boogie from other dogs and people that they care about. I moved abroad, taking them away from the family, friends, and environments they were previously at home with. Although we often were in contact with family members online, the dogs did not show much interest in audio/video conversations. I hoped that having some familiar smells could help them somehow cope with these distances. Did it help in any way? Design often prioritises those responses and actions that are explicitly and visibly expressed by participants and most legible to humans by prioritising sight and language over other, more subtle, signals (Akama & Light, 2018). Design decisions, especially in Participatory Design, can therefore easily follow those participants that are most articulate because those decisions are made for reasons that appear as documented or recorded in the design process. There is no reason to assume, though, that other – more subtle or hidden – responses are less relevant to the making of multispecies futures. As Akama and Light write, "It is precisely in these nuanced, silent moments that delicate relationships are forming or transforming" (2018, p. 8).

I continued this project with prototypes that could contain smells as well as soundscapes of familiar places (such as a talking family member, a barking dog, or sounds of a forest back home). Using an *Arduino Uno*, an *Audio Wave Shield*, a speaker, and a *Radio-frequency identification* (RFID) reader/antenna, I created a soft toy that could contain both sounds and smells (see Figure 6.8). Additionally, the RFID tags that I attached to the dogs' collars would ensure that different sound files could be activated by each dog individually because throughout our lives together I came to learn that both dogs had individual preferences regarding the volume levels and the different soundscapes they showed interest in. Jojo was particularly sensitive to and even afraid of loud noises while Boogie enjoyed lively environments.

Interestingly, the dogs remained strikingly careful in their interaction with this prototype compared to the rough ways in which they treat their other toys. . . . While interacting with the prototype together, I noticed how the dogs' playful behaviour (biting, pulling, and running after the artefact) was alternated with other types of behaviour such as sitting next to the artefact, sniffing it calmly,

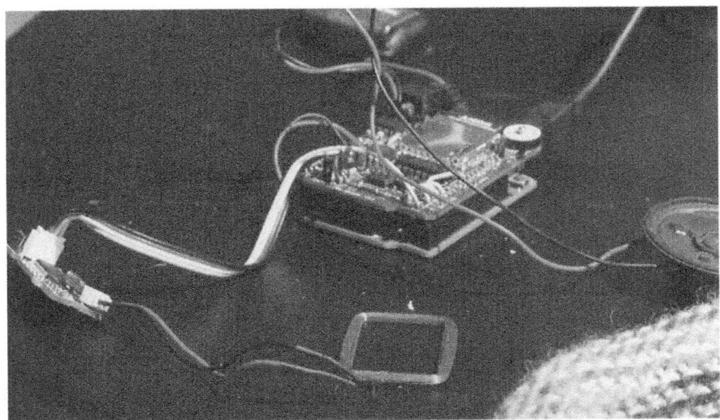

Figure 6.8 The technical components of one of the prototypes, including an Arduino Uno, an Audio Wave Shield, an RFID reader, an antenna, and a speaker. The image on the right shows Boogie (wearing the RFID tag) interacting with the prototype. Image from the author.

or seeking contact with me. Therefore, after the play session I iterated upon this by presenting the prototype as a starting point for exchanging affection. . . . Especially [Boogie] seemed to appreciate these sounds and she approached me and the prototype. When I started to pet her she sat down next to me and, after extensively sniffing the prototype for a minute, eventually fell asleep on top of the prototype.

(Westerlaken & Gualeni, 2016, p. 8)

Although I intended to create a toy that both dogs could play with together, they spent most of their time exploring the smells and sounds of the artefact in a calmer setting. My decision "not to act" (Akama & Light, 2018, p. 6), following my own expectations for the project, allowed the dogs to create their own ways of interacting with the prototype. Most importantly, these projects, and our other daily negotiations, continuously built renewed understandings and ways of knowing about each other that shape our relationships. This project, for example, helped me to rethink how smells and memories may work for Jojo and Boogie, as well as for other animals. Research nowadays shows more detailed insights into the olfactory worlds of dogs that can inspire multispecies design work (Horowitz, 2016; Kokocińska et al., 2022). The importance of making space for their own curiosities and explorations changed our daily neighbourhood walks together, and the times they got to spend with their other family members. I learned how to better listen to what they were telling me. Or, as philosopher Vinciane Despret wrote, I better learned what Jojo and Boogie could say if I "asked the right questions" (Despret, 2016). The attuning and sensitising to their curiosities also fundamentally impacted our lives when Jojo died, suddenly, through a violent sickness in 2018.

We observed Boogie's final moments of interacting with Jojo – her affective ways of being with and extensively sniffing her departed body – and we knew that she might be the only one who came to know the details of her death. In the aftermath of her death, we had to re-attune and re-sensitise in many different ways with Boogie, as their participation in each other's lives largely affected their entire ways of being in the world (Figure 6.9). Boogie subsequently proposed a rearrangement of our relations where she spent less time alone, demanded more routine, developed closer relationships with other individuals, made new friends, changed her toy preferences, and became bolder in approaching other dogs. As a result, we play differently now, where she initiates playful moods in more subtle ways as she ages. Since toy play is becoming rarer, when it does happen, I make sure to drop all other distractions and focus on playing together.

Her proposals for multispecies heterotopia are different from those of humans or other individuals, and learning how to recognise these is a continuously ongoing challenge. She challenged and contradicted our lifeways and enacted her proposals. Through these experiences of actively trying to design ways of living together through playfulness and participation, we developed a common language that helps to negotiate and structure our relationships (cf. Meijer, 2019). I became better at

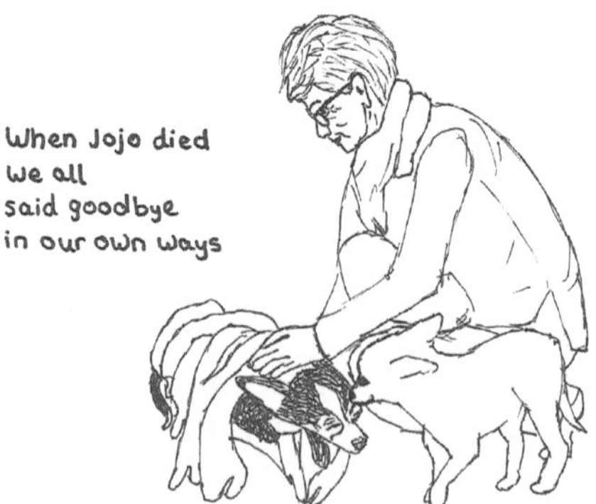

Figure 6.9 "When Jojo died we all said goodbye in our own ways." Illustration by the author.

recognising how our everyday lives are filled with negotiations and proposals by the animals we live with and reworked these experiences into other Participatory Design projects (Westerlaken, 2020). Nonetheless, while this work happens to be shaped and documented through design research processes, the context of playfulness in human-dog relationships aims to illustrate how these multispecies negotiations and proposals are a part of many everyday relationships already.

Conclusions

This chapter builds on already existing characteristics of the relationships between humans and dogs, including participation and playfulness, to collectively negotiate multispecies futures. By further extrapolating from the ways in which dogs evolved by seeking closeness to human societies, it becomes possible to rethink multispecies politics within these existing relations. Shared play between dogs and humans, including exploratory playful dimensions like curiosity, affection, and different modes of closeness, offers a context for mutual understanding where both humans and animals can challenge each other and propose new ideas. Although researchers have explored playfulness as a context for designing interactions with animals, the emancipatory potential and broader implications of these encounters have yet to be explicitly expressed. This text illustrates how Participatory Design work with dogs can further contribute to the creation of liberatory multispecies futures.

By revisiting Participatory Design work with dogs while paying more attention towards the ways in which these encounters shape our lives together, the multispecies politics involved become more visible. Through the work of participatory designers/researchers Yoko Akama and Ann Light, reflections on this work moved from a focus on evaluating multispecies participation in designing artefacts towards attuning to the wider changes in our relations during this process. With the use of illustrations, images, and reflective writing, this chapter seeks to weave empirical materials together to embed both the designer and the dogs into a situated attempt at collectively creating multispecies futures. Developing a kind of sensitivity to these processes is not something that is usually taught in formal design training but proves to be a central skill to design that seeks to negotiate complex relations (Akama & Light, 2018).

Yet, while this work is embedded in playfulness in a domestic context to structure multispecies negotiations, it is also limited by this environment. Most domesticated animals live in far more disturbing places including factory farms, zoos, and other caged environments. Further research and design work must engage with such constraining environments to reflect on the ways in which animals actively resist their oppression and propose different futures even within these contested spaces. While it may seem counterintuitive to focus on shared playfulness within these violent contexts, it is precisely those rare moments in which animals do express their curiosities, pose challenges, or seek novelty that articulate the heterotopias they continuously propose. One approach to this should include further research in animal sanctuaries where caretakers can observe the renewed playful responses between animals in interspecies relations and can redesign their living environments accordingly. Learning to become more attentive to such proposals is at the heart of doing a participatory politics that can identify more detailed aspects of multispecies futures.

Furthermore, this work raises questions regarding the possibility for humans to develop an understanding of animal proposals that are less aligned with human understandings. While dogs evolved refined skills to communicate their ideas to humans, other animals may be much less explicit in their proposals. Here, humans must find new ways of becoming more attentive to the multispecies heterotopias that are less explicitly articulated by different animals or involve more complex relations between different beings. One approach to these challenges is to articulate shared negotiations and common understandings that are inclusive of speculative work. Examples from multispecies live-action role-playing games (Catlow & Diehm, 2020), multispecies artworks, and other speculative practices (Dolejšová et al., 2023) show how participatory design can contribute to shaping multispecies negotiations, especially within these more exploratory realms.

Multispecies futures cannot be developed in abstract spaces or through the use of human-centric forms of thoughts only. Especially in industrialised societies, humans have largely forgotten that they are merely one member in multispecies communities and ecosystems. Alternative pathways towards animal liberation must therefore be created more deliberately together with the entities involved. Becoming more attentive to how animals already participate in negotiating our lives together is crucial to evolve new methods for collectively imagining multispecies futures.

Note

1 In the original paper, instead of using their real names, we cited the dogs as "A" and "B." I have always felt uneasy with this attempt at distancing the research project from our everyday lived experiences together. Although it is common to protect the identities of humans in participatory design projects, I think that to articulate multispecies futures, using the names of the animals involved is crucial for establishing them as partners rather than anonymous (and thus more distant) agents. I have therefore re-inserted their actual names here.

References

Akama, Y., & Light, A. (2018). Practices of Readiness: Punctuation, Poise and the Contingencies of Participatory Design. *Proceedings of the 15th Participatory Design Conference, ACM Press, 13*, 1–11.

Beck, E. (2002). P for Political: Participation Is Not Enough. *Scandinavian Journal of Information Systems, 14*(1), 77–92. https://aisel.aisnet.org/sjis/vol14/iss1/1

Bekoff, M. (2004). Wild Justice and Fair Play: Cooperation, Forgiveness, and Morality in Animals. *Biology and Philosophy, 19*(4), 489–520.

Bird-Rose, D. (2017). Shimmer: When All You Love is Being Trashed. In A. Tsing, H. Swanson, E. Gan, & N. Bubandt (Eds.), *Arts of Living on a Damaged Planet* (pp. 51–63). University of Minnesota Press.

Bjerknes, G., Ehn, P., & Kyng, M. (1987). *Computers and Democracy: A Scandinavian Challenge*. Gower Publishing Ltd.

Bødker, S. (1996). Creating Conditions for Participation: Conflicts and Resources in Systems Development. *Human-Computer Interaction, 11*(3), 215–236.

Bogost, I. (2016). *Play Anything: The Pleasure of Limits, the Uses of Boredom, and the Secret of Games*. Basic Books.

Budiansky, S. (2000). *The Truth About Dogs: An Inquire into the Ancestry, Social Conventions, Mental Habits, and Moral Fiber of Canis Familiaris*. Penguin Books.

Burghardt, G. M. (2006). *The Genesis of Animal Play: Testing the Limits*. MIT Press.

Cantor, M., Farine, D. R., & Daura-Jorge, F. G. (2023). Foraging Synchrony Drives Resilience in Human – Dolphin Mutualism. *Proceedings of the National Academy of Sciences, 120*(6), 1–10.

Catlow, R., & Diehm, C. (2020). *The Treaty of Finsbury Park 2025 (Concept Paper)*. https://newdesigncongress.org/en/pub/finsbury-park-2025

Curtis, D. (2023). The Last Human-Wild Collaborators. *Sierra*. www.sierraclub.org/sierra/last-human-wild-collaborators

Coppinger, R. (2001). *Dogs: A Startling New Understanding of Canine Origin, Behavior and Evolution* (1st ed.). Prentice Hall & IBD.

de Castro, E. V. (2018). *Metafísicas canibais: Elementos para uma antropologia pós-estrutural*. Ubu Editora LTDA – ME.

Decolonizing Design. (2017). *Editorial Statement*. www.decolonisingdesign.com/statements/2016/editorial/

Despret, V. (2016). *What Would Animals Say if We Asked Them the Right Questions?* (B. Buchanan, Ed.). University of Minnesota Press.

Dolejšová, M., Sáez Agurto, F., Beavers, I., Altarriba Bertran, F., Botero, A., Buschmann, J., Camino, D., Chewie, Choi, J. H., Davis, H., Feeding Food Futures, Furtherfield, Van Galen, S., Gobnjak, Hanafi, A., Heitlinger, S., Hellon, Howse, M., Kersnikova, . . . ZEMOS98. (2023). *Creatures Co-Laboratory Catalogue* (M. Dolejšová, Ed.). Zenodo. https://doi.org/10.5281/zenodo.7525506

Donaldson, S., & Kymlicka, W. (2011). *Zoopolis: A Political Theory of Animal Rights*. Oxford University Press.

Driessen, C., Alfrink, K., Copier, M., Lagerweij, H., & van Peer, I. (2014). What Could Playing with Pigs Do to Us? *Antennae: The Journal of Nature in Visual Culture, 30*, 79–102.

Escobar, A. (2017). *Designs for the Pluriverse: Radical Interdependence, Autonomy, and the Making of Worlds*. Duke University Press.

Foucault, M. (2008 [1967]). Of Other Spaces. In M. Dehaene & L. De Cauter (Eds.), *Heterotopia and the City: Public Space in Postcivil Society* (pp. 13–29). Routledge.

French, F., Mancini, C., & Sharp, H. (2015). Designing Interactive Toys for Elephants. *Proceedings of CHI Play*, ACM Press, London, pp. 523–528. http://dx.doi.org/10.1145/2793107.2810327

Gigliotti, C. (2022). *The Creative Lives of Animals*. NYU Press.

Hakio, K., Dolejšová, M., Mattelmäki, T., Choi, J. H., & Ampatzidou, C. (2022). Following Seals and Dogs: Experimenting with Personal Dimensions of Transformative Design. *Proceedings of the Participatory Design Conference 2022, 2*, 167–172. https://doi.org/10.1145/3537797.3537869

Haraway, D. (2008). *When Species Meet*. University of Minnesota Press.

Hare, B., & Woods, V. (2013). *Opinion: We Didn't Domesticate Dogs. They Domesticated Us*. www.nationalgeographic.com/news/2013/3/130302-dog-domestic-evolution-science-wolf-wolves-human/

Hillgren, P.-A., Seravalli, A., & Eriksen, M. A. (2016). Counter-Hegemonic Practices; Dynamic Interplay Between Agonism, Commoning and Strategic Design. *Strategic Design Research Journal, 9*(2), 89–99. https://doi.org/10.4013/sdrj.2016.92.04

Hirskyj-Douglas, I., & Read, J. (2014). Who Is Really in the Center of Dog Computer Design? *Proceedings of the Workshop on Advances in Computer Entertainment Conference (ACE '14*, Funchal, Portugal, pp. 1–5. http://dx.doi.org/10.1145/2693787.2693793

Hirskyj-Douglas, I., Read, J. C., & Cassidy, B. (2017). A Dog Centred Approach to the Analysis of Dogs' Interactions with Media on TV Screens. *International Journal of Human-Computer Studies, 98*, 208–220. https://doi.org/10.1016/j.ijhcs.2016.05.007

Horowitz, A. (2016). *Being a Dog: Following the Dog into a World of Smell*. Simon and Schuster.

Hribal, J. (2007). Animals, Agency, and Class: Writing the History of Animals from Below. *Human Ecology Review, 14*(1), 101–112.

Kohn, E. (2013). *How Forests Think: Towards an Anthropology Beyond the Human*. University of California Press.

Kokocińska, A., Woszczyło, M., Sampino, S., Dzięcioł, M., Zybała, M., Szczuka, A., Korczyńska, J., & Rozempolska-Rucińska, I. (2022). Canine Smell Preferences – Do Dogs Have Their Favorite Scents? *Animals, 12*(12), 1488, 1–16. https://doi.org/10.3390/ani12121488

Light, A. (2015). Troubling Futures: Can Participatory Design Research Provide a Constitutive Anthropology for the 21st Century? *IxD&A, 26*, 81–94.

Mancini, C. (2011). Animal Computer Interaction: A Manifesto. *Interactions, 18*(4), 69–73.

Mancini, C. (2023). A Biosemiotics Perspective on Dogs' Interaction with Interfaces: An Analytical and Design Framework. *Interaction Studies, 24*(2), 201–224.

Mancini, C., & Lehtonen, J. (2018). The Emerging Nature of Participation in Multispecies Interaction Design. *Proceedings of the Designing Interactive Systems Conference (DIS '18*, Hong Kong, pp. 907–918. https://doi.org/10.1145/3196709.3196785

Mankoff, D., Dey, A., Mankoff, J., & Mankoff, K. D. (2005). Supporting Interspecies Social Awareness: Using Peripheral Displays for Distributed Pack Awareness. *UIST 2005: Proceedings of the 18th Annual ACM Symposium on User Interface Software and Technology*, Seattle, Washington, pp. 253–258. https://doi.org/10.1145/1095034.1095076

Meijer, E. (2019). *When Animals Speak: Toward an Interspecies Democracy*. NYU Press.

Pons, P., Jaen, J., & Catala, A. (2014). Animal Ludens: Building Intelligent Playful Environments for Animals. *Proceedings of the 2014 Workshops on Advances in Computer Entertainment Conference*, Funchal, Portugal, pp. 1–6. https://doi.org/10.1145/2693787.2693794

Spannring, R. (2023). Critical Animal Pedagogy: Liberating the Nonhuman Learner? *On Education: Journal for Research and Debate, 6*(16), 1–9.

Sheikh, H., Mitchell, P., & Foth, M. (2023). More-than-Human Smart Urban Governance: A Research Agenda. *Digital Geography and Society, 4*, 100045. https://doi.org/10.1016/j.diggeo.2022.100045

Shipman, P. (2015). *The Invaders: How Humans and Their Dogs Drove Neanderthals to Extinction*. Harvard University Press. https://doi.org/10.2307/j.ctvjf9zbs

Sicart, M. (2014). *Play Matters*. MIT Press.

Veselova, E., & Gaziulusoy, İ. (2022). Bioinclusive Collaborative and Participatory Design: A Conceptual Framework and a Research Agenda. *Design and Culture, 14*(2), 149–183. https://doi.org/10/gn8psk

Wadiwel, D. (2015). *The War Against Animals*. Brill.

Webber, S., Carter, M., Sherwen, S., Smith, W., Joukhadar, A., & Veter, F. (2017). Kinecting with Orangutans: Zoo Visitors. *Empathetic Responses to Animals' Use of Interactive Technology, ACM SIGCHI Conference on Human Factors in Computing Systems*, Denver, CO, pp. 6075–6088. http://dx.doi.org/10.1145/3025453.3025729

Weigel, M. (2019). *Feminist Cyborg Scholar Donna Haraway: 'The Disorder of Our Era Isn't Necessary*. www.theguardian.com/world/2019/jun/20/donna-haraway-interview-cyborg-manifesto-post-truth?CMP=share_btn_tw

Weisberg, Z. (2009). The Broken Promises of Monsters: Haraway, Animals and the Humanist Legacy. *Journal for Critical Animal Studies, 7*(2), 22–62.

Westerlaken, M. (2017). Uncivilising the Future: Imagining Non-Speciesism. *Antae, 4*(1), 53–67.
Westerlaken, M. (2020). *Imagining Multispecies Worlds.* Doctoral Dissertation, Malmö University. http://urn.kb.se/resolve?urn=urn:nbn:se:mau:diva-17424
Westerlaken, M., Gabrys, J., Urzedo, D., & Ritts, M. (2023). Unsettling Participation by Foregrounding More-than-Human Relations in Digital Forests. *Environmental Humanities, 15*(1), 87–108. https://doi.org/10.1215/22011919-10216173
Westerlaken, M., & Gualeni, S. (2016). Becoming With: Towards the Inclusion of Animals as Participants in Design Processes. *Proceedings of the Third International Conference on Animal-Computer Interaction*, Milton Keynes, pp. 1–10. https://doi.org/10/f94dsg
Wijngaarden, V. (2023). Interviewing Animals Through Animal Communicators: Potentials of Intuitive Interspecies Communication for Multispecies Methods. *Society & Animals*, 1–12. https://doi.org/10.1163/15685306-bja10122
Willett, C. (2014). *Interspecies Ethics*. Columbia University Press.
Wirman, H. (2014). *Games for/with Strangers – Captive orangutan (Pongo Pygmaeus) Touch Screen Play*. http://ira.lib.polyu.edu.hk/handle/10397/7611

Chapter 7

The Radical Praxis of Equity
Mutual Interdependence and an Ethic of Responsibility

Charlotte A. Kunkel and Scott Hurley

Introduction

Citizenship in our contemporary understanding in the United States is an individualistic rights-based concept rooted in hierarchies of domination including whiteness, maleness, humanness, and wealth privilege. Rejecting citizenship as the basis for a new shared reality and community of radical equity moves us toward a future of relational/mutual interdependence with all living beings and the earth that is transformational and life-giving. Implicit to such an approach is a critique of hierarchy-enhancing ideologies, capitalist praxis, and single-axis logics that reinforce and enhance existing power structures that preserve systemic inequity. Thus, as one of our theoretical foundations, we use intersectionality as a tool "for questioning default logics that sustain and rationalize inequality . . ." (May 2015, p. 5), but also as a praxis of social action that has "a heuristic orientation [which] accentuates its problem-solving capacity, one that is contextual, concerned with eradicating inequity, [and] oriented toward unrecognized knowers and overlooked forms of meaning . . ." (p. 19). Moreover, we embrace mutual interdependence not only as a theoretical concept that subverts normalized meanings of "human" and "animal" that reinforce a status quo (which is inherently exploitative, oppressive, and divisive), but also as a praxis that creates a responsibility of "being with" and accountable for all beings. We provide a vision of a space/place wherein beings are intimately relational, interconnected, and embodied, one in which they are free and can thrive in community with one another. Freedom is a place (Gilmore, 2022). Such places are both imaginative and real – places that are "something like counter-sites, a kind of effectively enacted utopia in which the real sites, all the other real sites that can be found within the culture, are simultaneously represented, contested, and inverted" and that "are outside of all places, even though it may be possible to indicate their location in reality. . . . they are absolutely different from all the sites that they reflect and speak about" – they are heterotopias" (Foucault, 1984, pp. 3–4).

We envision the relations in such spaces being enriched and governed by an embodied empathy that is informed by Gruen's "entangled empathy" – a blend of emotion and cognition in which we are called upon to be responsive and

DOI: 10.4324/9781003366706-10

responsible to others by attending (or sometimes not attending) to their "needs interests, desires, vulnerabilities, hopes, and sensitivities" (2015, p. 3), and which is also socially engaged because it is

> an understanding of bodily becoming that affirms that the lived experience of our bodily selves, when cultivated through attention to our sensory creativity, can serve as a dynamic, responsive locus for responsibly negotiating our inevitable implication in cultural relations of power.
>
> (Lamothe, 2015, p. 232*n*5)

Furthermore, we imagine relations in these heterotopias as being enhanced and directed by an understanding that we are always intimately interconnected with others around us; that is, we "inter-are with one another and with all life" (Hanh, 2017, p. 13) and by the realization that living beings engage with each other through, and have shared experience in, bodies – a kind of "somatic commonality" (Acampora, 2006) – that allows a bodily compassion to drive an ethic of responsibility. To envision/create this radically equitable future, we will briefly critique citizenship as a primary relationship in society, and posit instead a relation characterized by the mutual interdependence of all things and an ethic of responsibility guided by a praxis-oriented, embodied empathy.

Mutual Interdependence

In the United States, we conceptualize democratic citizenship as the primary relationship of belonging. While it encompasses by definition one's rights, duties, and responsibilities as a member of society, all too often it is reduced only to the rights of individuals. Quadagno (1994) reminds us that a democracy generally consists of a citizenry having full civil, political, and social rights. Civil rights are largely understood as economic freedom or the ability to freely pursue the occupation of one's choice. Political rights include universal suffrage or the right to vote and voice in collective decision making. Social rights are more elusive but include social protections such as guaranteed income, protections against unemployment and old age, and injury on the job. Other social rights are provisioning for daycare, paid parental leave, national health insurance, and education. As Quadagno (1994) and Frazier and Gordon (1992) remind us, democracy in the United States is incomplete in that we have largely failed to offer social rights to everyone by excluding people of color and white women from the basic provisions of a democracy. Patriarchy, white supremacy, and capitalist ideologies have created structures of domination that deny citizenship to all. These are ideologies that tear apart rather than acknowledge and encourage interdependence of all beings, and thus make impossible transformational and life-giving community.

We understand the self to be social, meaning that the self is socially constructed through interaction. Who we are is inherently relational and interactional. Mead notes that through imitation and role play we become socialized and reflexive, and thus agentic (Mead, 1934). Identities, then, are social and thus reflect social structures. Michel Foucault argues that the distinctions we make between self-other and subject-object so common in "othering" discourse often result in a binary structure of inclusion and exclusion, of privilege and marginalization. In other words, such distinctions allow the privileged few to exclude the marginalized "others" by severing their connections with their people and community and forcing upon them an identity that restricts and hinders them (2000, p. 330). To overcome this exploitation, he speaks of a conflation of self and other:

> Superficially, one might say that knowledge of man (sic), unlike the sciences of nature, is always linked, even in its vague form, to ethics or politics; more fundamentally, modern thought is advancing towards that region where man's (sic) Other must become the Same as himself (sic).
>
> (1973, p. 328)

Here Foucault suggests that having the other become the same as the self is a necessary ethical step for adequately addressing exploitative power relations. However, how does one conflate such dualities as self and other, subject and object, human and nonhuman? Here we turn to the work of Buddhist monk, scholar, and social activist Thich Nhat Hanh and his notion of "interbeing" (Chn. *xiangji* 相即), a re-articulation and application of the Buddhist teachings of no-self (Skt. *anātman*; Chn. *wuwo* 無我), dependent origination (Skt. *pratītya-samutpāda*; Chn. *yuanqi* 緣起), emptiness (Skt. *śūnyatā*; *kong* 空), and mutual identification (*xiangji xiangru* 相即相入).

Hanh's interbeing underscores the interconnectedness and mutual interdependence of all things. As such, it is a radical expression of love and intimacy that challenges hierarchical ideologies (e.g., racism, classism, sexism) and economic, social, and political systems that reinforce systemic inequities (e.g., capitalism, individualism, militarism, citizenship). It also promotes a way forward, suggesting a means for living together that values a praxis of responsibility to the oppressed and the creation of systems and institutions that are not only sustainable but also regenerative. In an imaginative and moving description, Hanh (1988) in his book, *The Heart of Understanding*, defines interbeing in the following way:

> If you are a poet, you will see clearly that there is a cloud floating in the sheet of paper. Without a cloud, there will be no rain; without rain, the trees cannot grow; and without trees, we cannot make paper. If the cloud is not here, the sheet of paper cannot be here either. So we can say that the cloud and the paper inter-are. . . . If we look more deeply, we can see the sunshine in it . . . .the paper and the sunshine inter-are. And if we continue to look we can see the logger. . . . Your mind is in here and mine is also. . . . So we can say that everything is in here with this sheet of paper. You cannot point out one thing that is not here – time, space, the earth, the rain, the minerals in the soil, the sunshine, the

cloud, the river, the heat. . . . As thin as this paper is, it contains everything in the universe in it.

(p. 35)

Here we learn that all things are intimately connected – they intermesh. Just as with the sheet of paper, living beings, too, contain everything in the universe. If I am mutually interconnected to all beings, then harming others or allowing harm to befall them adversely affects me so I should prevent it from occurring. But this is not simply a kind of higher-order selfishness – "I prevent you from being harmed so that I am not harmed." For Hanh, the mutual interconnectedness of all beings operates alongside the Buddhist notion of compassion (Skt. *karuṇā*; Chn. *cibei* 慈悲), which requires us to understand that our own experience of suffering and happiness is no different from that of all other sentient beings. In the sixth of 14 precepts that Hanh (1987) created for his Tiep Hien Order ("Order of Interbeing"), he urges:

> Do not maintain anger or hatred. As soon as anger and hatred arise practice meditation on compassion in order to deeply understand the persons who have caused anger and hatred. Learn to look at other beings with the eyes of compassion.

(p. 39)

Here, we learn that compassion is paramount to how we interact with others. However, Hanh argues that we must also realize that we are complicit in the violence that others experience even if we are not the direct cause of it:

> A variety of interdependent causes have created the existence of the sea pirate. The responsibility is not solely his or his family's but society's as well. As I write these lines, hundreds of babies are being born close to the Gulf of Siam. If politicians, educators, economists and others do not do something to prevent it, many of these babies will become pirates twenty-five years from now. Each of us shares some responsibility for the presence of sea pirates.

(p. 39)

We interpret this to mean that we are responsible for not only self and others but for systems of oppression (the micro and the macro) because the self is social and reflective of social structures.

Now, one could argue that Buddhist soteriological goals focus primarily on a kind of personal psycho-spiritual realization where practitioners turn inward through meditation, ritual, and sutra study and practice compassion in order to achieve their own liberation – the focus being the personal and not the political; the individual and not social systems. Hanh would disagree and, in fact, argue that personal self-transformation and the transformation of the socio-political order are entwined. For example, in the fifth precept of the order Hanh (1987) states: "Do not accumulate wealth while millions are hungry. Do not take as the aim of your life fame, profit, wealth or sensual pleasure. Live simply and share time, energy and material resources with those who are in need" (p. 37). While this precept is directed at the individual practitioner, Hanh intends it as a means to subvert destructive socio-economic structures like capitalism that cause physical, emotional, and social harms. He argues that we must remain as free as possible from the "destructive momentum of the social and economic machine . . . (p. 37). In a 2010 interview with *The Guardian*, Hanh specifically refers to capitalism as a disease that has spread all over the world through globalization. He points out that "We have constructed a system we cannot control. It imposes itself on us and we become its slaves and victims" (Confino, 2010). If we take seriously Hanh's notion of interbeing, then the remedy for the suffering such systems cause is both to take responsibility for the direct and indirect pain and suffering that they and our participation in them cause others and, out of compassion, strive to overcome them and create alternative communities based on regenerative social and economic systems.

An Ethic of Responsibility

Several traditions have articulated an ethic of responsibility since Max Weber did so a century ago, indicating that we ". . . must answer for the foreseeable consequences of [our] actions" (2015, p. 83). Noted animal activist and ethicist Peter Singer addresses liberation and responsibility in his important work, *Animal Liberation* (2002). Justifying the title of his book, he reminds readers that "a liberation movement is a demand for an end to prejudice and discrimination based on an arbitrary characteristic like race or sex" (p. xxii). He argues that liberation movements require an expansion of our moral consciousness to understand that practices, institutions, laws, and traditions often regarded as "natural and inevitable" actually reinforce hierarchical ideologies and systemic inequities. Building on the work of the Black, Gay, and Women's liberation movements of the 1960s and 1970s, Singer argues animal liberation must be included in shifting people's consciousness about challenging systemic injustice.

Black liberation theology and the Black Buddhism Movement (Gleig, 2019; Syedullah, 2019) each articulate the position that white people must be held accountable for white supremacy not only through care or compassion but through responsibility. In his book, *Witnessing Whiteness: Confronting White Supremacy in the American Church*, Kristopher Norris (2020) argues that white people must take responsibility for the past and present domination of others, repent for complicity in such domination, and repair (reparations) the harm that was done historically. Drawing on the work of American theologian and advocate of black liberation theology, James Cone, Norris, more specifically, argues that we must develop an ethic of responsibility (2020, p. 7) to truly confront a legacy of white supremacy in the church. We agree and extend this to all of societal structures and institutions that reinforce systemic inequity and oppress human beings, nonhuman animals, and the earth. An ethic of responsibility extends beyond caring for others. Norris speaking of the church writes that we must "remember and memorialize its invention of white supremacy, publicly repent of its continued complicity and silence, and confront the racial damage it has inflicted through concrete actions of reparation (politically, financially, spiritually)" (2020, p. 9). To extend this ethic of responsibility to the larger cultural dimensions of oppression, we must acknowledge our participation and complicity in all forms of domination, engage in social justice action, and follow the lead of the oppressed. An ethic of responsibility requires not only a change in consciousness but material change and giving up of power.

Moreover, an ethic of responsibility is distinctly different from an ethic of justice (legal) or an ethic of care (paternal). In her influential work *In A Different Voice*, Carol Gilligan (1982) critiqued an ethic of justice as based on an androcentric legalistic frame of reference; conversely, she advocated for an ethics of care, arguing that it is more compassionate than an ethic of justice and that women were much more likely than men to appeal to it in their moral calculus. Scholars such as Carol Adams and Josephine Donovan (2007) introduced this approach to philosophical discussions of animal liberation suggesting the need for compassion and sympathy when considering human relationships with animals and pointing out the connection between the exploitation and oppression of women and the subjugation of nature. Unfortunately, an ethic of care all too often ends up being paternalistic and thus re-inscribes hierarchical relationships: the privileged few care for the oppressed out of their good will. The systemic and hierarchical nature of "caring" interactions, and the harm they cause, are rarely challenged.

An ethic of responsibility, however, emphasizes the interconnection between the oppressor and the oppressed, the complicity that each of us has in the destruction of the earth and its ecosystems, and the obligation we have to address the suffering that human and nonhuman others experience. For the construction of a true heterotopia, then, an ethics of responsibility requires interrogating and deconstructing the language, institutions, and policies used to justify oppressive social structures, as well as creating new structures that are regenerative and have mutual interdependence as their ideological and practical foundation.

Regenerative Systems

Indigenous feminist theory and ecofeminism have long argued that without balance in ecological systems, neither humans nor nonhumans will survive. Ecofeminism, a "global movement that is founded on common interests yet celebrates diversity and opposes all forms of domination and violence" (King, 1989, p. 20), understands that the oppression of women is connected to the exploitation of nature and land. As Lahar (1991) summarizes, a

> central theme of most versions of ecofeminism . . . is the interrelationship and integration of personal, social, and environmental issues and the development of multidirectional political agendas and action. Ecofeminism is transformative

rather than reformist in orientation, in that ecofeminists seek to radically restructure social and political institutions. Women's liberation is contextualized in human liberation and a more ecological way of living on the earth.

(p30)

Plumwood (1993) carries this farther to reject the foundational dualism of patriarchal rationality and colonization urging us to reimagine a holistic solution to human/nature domination.

A regenerative approach challenges us to see beyond sustainability and maintenance of the status quo to a place where transformative innovation and thriving is possible. Highly influenced by systems thinking (Reed, 2006), interbeing (Hanh, 1988), and the permaculture and regenerative agriculture movements (Mollison, 1988; Holmgren, 2011), Wahl (2016) in his book *Designing Regenerative Cultures* proposes designing an integrated culture of cooperation rooted in holistic biologically inspired designs, transformative innovation, healthy, resilient and regenerative ecosystems. Through mimicking healthy ecosystems, he contends, we can create healthy social systems that are regenerative – moving from systems of scarcity and competition to relational collaboration encompassing the goals of synergy, emergence, open source, and dynamic responsive mutualism that is adaptive (p. 236).[1] A regenerative enterprise is thus not only focused on its own growth and resilience but also on the interconnections and ecologies of our communities and a restorative culture of ecosystems. Regeneration operates at both the micro and macro levels.

In capitalist cultures, however, living beings, both human and nonhuman, are often abstracted from their social and relational realities and turned into things that can be purchased, sold, or discarded at will. They become capital-generating commodities, resources rather than living beings with intrinsic value. This objectification/commodification process reifies the dualities of self and other, human and nonhuman, which in turn supports, sustains, and reinforces capitalist praxis and ultimately suffering, exploitation, and oppression. In this way, non-human animals and marginalized groups of people have their subjectivity stripped from them – in such a context, they no longer are "subjects of a life" as philosopher Tom Regan (2004) defines it. Taking away their subjectivity privileges the subjectivity of human beings who have power and who represent the "norm" (i.e., white, cis-gendered, able-bodied, heterosexual males), thereby creating an oppositional duality between human (subject) and nonhuman (object) animals and the powerful (subject) and minoritized people (object). This duality allows some to place non-human animals and members of marginalized groups into particular roles, creating for them an identity that is defined only by how much they or their labor is worth. Similarly, it enables the privileged to continue benefiting from laws, policies, and traditions that systemically reinforce inequity.

In contrast, Glenn Manga (2008) in his article "Interbeing Autonomy and Economy," writes "we must strive for not just individual personal transformations, but a transformation of our socioeconomic paradigm in a move away from the injustices of the neoclassical model of economic advocacy" (123). Here, he first provides a critique of neoclassical economic models, namely those of the Chicago School (per Milton Friedman) and Austrian School (per Friedrich Hayek) by pointing out the harms they cause, and then, extending Hanh's interbeing to the socioeconomic arena, offers an alternative, one that he calls "Interbeing Economy for Social and Ecological Justice" (p. 123). In this alternative model, Manga (2008) advocates a move to what he calls a "Systems Integrity Building Economy" or "Interbeing Economy" – an economic paradigm that is not based on a self-interested pursuit of capital at the expense of human beings, nonhuman animals, and ecosystems but rather one that enhances each system and subsystem a person engages with and eliminates those practices that are exploitative:

> Each economic act is thus one that is measured/valued in terms of how it impacts upon our ecosystem's integrity, environmental integrity, social justice integrity, physical well being integrity and so on. Integrity attacking activities . . . would be identified and penalized in a relatively fast phasing out transition process to sustainable and regenerative economic activities.
>
> (p. 124)

Here we see how Hanh's concept of interbeing can be used to design an economic system that is fair and equitable. As a form of mutual interdependence and a re-articulation of no-self, dependent origination, emptiness, interbeing is both a conceptual and practical means for overcoming human and nonhuman animal exploitation and suffering. In other words, it provides the necessary theoretical and practical foundation to interrogate and deconstruct these dualities, establish ethical systems based on the interconnectedness of all living things, and create social, political, and economic systems that are regenerative in ways that allow all beings to thrive.

Interbeing, then, deconstructs our conventional ways of thinking, which include the way we understand the self or ego. Sentient beings are not separate "selves" that have an independent existence distinct from others. In fact, quite the opposite: they exist only in dependence upon others and others exist in dependence on them. In this way, living beings are intimately connected.[2] The common, everyday view of the self and the things around us as concrete, unchanging entities is erroneous – a view that causes suffering because we do not realize that our sense of self is constructed; instead, we see it as separate and distinct from the world we live in. Buddhist scholar and Zen practitioner David Loy (2008), points out that the constructed self is a psychological-social-linguistic construct: psychological, because the ego-self is a product of mental conditioning; social, because a sense of self develops in relation with other constructed selves; and linguistic, because acquiring a sense of self involves learning to use certain names and pronouns such as I, me, mine, myself, which create the illusion that there must be something being referred to (p. 16–17).

Thus, we organize our lives by defining ourselves over and against others. We embrace dichotomies like male and female, gay and straight, white and black, human and animal because they validate the self. After all, if we can say what we are and are not, essentializing our identities, then there is something that persists as a separate and distinct self; there is some belief in an enduring essence that inherently exists and thus distinguishes us from all others. Not only do people accept and assert these self-definitions, but they also engage in self-constructing activities to affirm who they are; so, for example, they pursue wealth, fame, and power – all goals that society recognizes as pivotal for obtaining success and necessary for setting people apart from others. Similarly, hard and fast distinctions between human and nonhuman animals, white and minoritized people, etc. enhance a sense of an independent self or essence and result in "de-animating" nonhuman animals and dehumanizing marginalized groups turning them into objects or "others," which can be exploited and oppressed. Consider, for example, the labels "human" and "animal." They carry with them a series of associations that limit our conceptions of what human and nonhuman animals are able to be and do: humans are rational, moral, intelligent, sophisticated, and capable of reaching enlightenment; they are children of god, arhants, and bodhisattvas. Animals are irrational, amoral, subject to instinct; they are food, entertainment, tools, and punishment for bad karma. Moreover, the human–animal dichotomy implies certain assumptions about moral worth: humans are always in every context more morally valuable than animals and thus worthier of our attention – animals, on the other hand, can and do go unnoticed, falling outside of our moral circle. In Buddhism, this moral distinction is reflected in the Pratimokṣa (Chn. *lüyi jie* 律儀戒), the rules governing the conduct of monastic communities, wherein the killing of a human is conceived as fundamentally different from the killing of an animal; that is, the punishment is more severe for the former than the latter (Waldau, 2002, p. 124). Thus, even in Buddhist traditions (and of course other religious traditions) we see a tendency to privilege humans over nonhuman animals. The unfortunate consequence of doing so objectifies animals and prevents humans from recognizing the intimate and dependent connection they have to them. Hanh would argue that making such distinctions can only end in suffering for oneself and others. Only when we acknowledge that any sense of self is dependent on other living beings and things for its existence, can we create just and equitable communities and thereby end suffering, oppression, and exploitation. Such a realization leads to a sense of responsibility for our complicity in the suffering of others and the development of a compassion that strives to eliminate systems that exploit, oppress, and harm other living beings.

Empathy, Compassion, and Embodiment

Michel Foucault is right to emphasize the need to conflate the self and other but with one corrective. Sameness is not the goal, just as strict distinctions are not, but rather interdependence – we meet the nonhuman animal as one subject encountering another, not as an object, but instead recognizing our interconnectedness in the moment of our meeting. Perhaps this is what Jacques Derrida (2008) is getting at in his *The Animal That Therefore I Am*, when he describes standing nude and feeling ashamed "before a cat that looks at you without moving, just to see" (p. 4) – the cat merely looks, seeing him looking back – subject to subject, body to body. Following philosopher and scholar, Ralph Acampora (2006), we can support an ethics of responsibility informed by what he calls a corporal compassion by understanding that we exist together with others through common bodily experience of a shared life world (p. 130).

To explain what he means by corporal compassion, Acampora describes a concept he calls "symphysis" which he defines as a "somaaesthetic nexus experienced through a direct or systemic (inter)relationship" (p. 76). It is symphysis, for example, that allows us to experience a shared life world with non-human animals – through our bodies, our senses – without colonizing their consciousness. This is a kind of empathy but not, as Acampora points out, of the imaginative kind where we mentally visualize the self as other – where we become the other, taking on his/her identity and then, pretending to know the other's suffering, develop a moral sentiment that drives us to alleviate suffering and strive to prevent it from happening again. Complete identification is impossible, not desirable, and not the way real empathy works: "Cultivating a bodiment ethos of interanimality is not a matter of mentally working one's way into other selves or worlds by quasi-telepathic imagination, but is rather about becoming sensitive to an already constituted 'inter-zone' of somaesthetic conviviality" (p. 84). When a dog squeals because someone has stepped on its paw we do not feel that pain the way the dog does – but we recognize it, even flinch in response to it because most of us have experienced a similar pain when someone has stepped on some part of our body. Acampora recognizes that we share much in common with nonhuman animals. Of course, it may be easier to relate to nonhuman animals that possess physical and cognitive structures and characteristics similar to our own; for example, it is easier for some human animals to cultivate a "bodiment ethos of interanimality" with dogs, deer, and squirrels than with snakes, frogs, and octopi. However, our ability to cultivate connections with such species can be aided by understanding the behavior, lifeways, and environmental contexts of such species, thus making shared, body-to-body experience still helpful, if not essential, for cultivating empathic understanding. Acampora writes, "Despite differences in some sensory modalities, members of various species retain enough somatic commonality to make sense of one another" (p. 30).

But there is more. We also connect with other living beings in our experiences of certain ecological spaces. This can occur when we realize that we are experiencing the same climate or hearing the same sounds as another animal like when we realize that during winter we don heavier clothing just as many nonhuman animals experience the thickening of their fur. In this way, we live together with nonhuman animals as neighbors because we share a similar space, a neighborhood – like a patch of woods, park, or city street. Sometimes these experiences, however, are recognized by both species as shared – as an experience between friends (those that live together). The idea of shared experience here is of key importance, and it is this kind of interaction that needs to be reflected on when we consider alternative ways to engage other living beings. Again, recall Derrida's cat. He writes of the mutual gaze – meeting his cat (and being met by him) eye to eye, and of the subsequent recognition in that moment that each one had of the other's existence (2008, pp. 3–5). No hierarchy, no domination, no control – equal in the recognition of the other. These experiences can happen with wild animals – when a deer for example meets one's gaze during a walk in the woods. But even more profound is the recognition of a shared experience with those nonhuman companions we spend our lives with – an experience that is mutual and not governed by the language of domination and control. The question is can we note these experiences, recognize them for what they are, and allow them to provide the foundation for a new way of conceiving of human and nonhuman animal relationships.

The Praxis of Radical Equity

A radical equitable heterotopia would be characterized as a place where mutual interdependence of all things in a regenerative system is guided by embodied empathy and an ethic of responsibility. Some examples of heterotopia are found in the imaginary of feminist science fiction. We draw on feminist science fiction in particular because these authors imagine heterotopia that are most explicitly intersectional, or as Foucault suggests, they "represent, contest, and invert" all that we know (1984). Further, as Adrienne Maree Brown argues, social justice is science fiction because it is not yet realized (Love, 2015). A feminist utopia constructs a

> social vision which favors "egalitarian social relations, practices, and institutions" (Mouffe, 1992, p. 380), asking what each citizen, as an individual in a raced, classed, and gendered body, needs for her or his welfare. In so doing, Mouffe recognizes both the inherently communal project of citizenship and the inevitable particularity of its manifestations, asking what it means to be a citizen in a multicultural world.
>
> (Silbergleid, 1997, p. 172)

The best feminist utopias present alternative social structures of civil society that question patriarchal notions of citizenship, recognizing embodiment, mutual interdependence, and intersectionality. Take, for example, the fictional societies of Marge Piercy's Mattapoisett (1976), Slonczewski's ocean moon (1986) or Butler's Acorn (1993). A basic premise of much of this feminist writing is a "radical equalitarianism" such that sex (gender) equality is only thinkable in a context of overall equality (Eichler, 1981, p. 63). Many texts present alternatives to capitalism and patriarchy by not only challenging the patriarchal family structure but notions of sexuality as well. *Califia's Daughters* (Richards, 2004), for example, incorporates a critique of capitalist heteropatriarchy in many forms while the protagonist is fighting to save the valley and stave off biological disasters. Critiques of white supremacy, heteronormativity, essentialism, biologism, and religious dogma are also depicted by "a dismantling of racial and class hierarchies and an overt critique of capitalism and the assumptions of liberal democracy" (Silbergleid, 1997, pp. 170–171). As Kathleen Jones explains, these utopian projects disavow "the individualist, rights-based, contractual model of citizenship" and replace it with "the virtues of commitment to relationships, love, and caring for others" (Jones, 1990, p. 810), virtues which work against the competition and conflict associated with natural right (Silbergleid, 1997, p. 170).

The vision of a feminist radical egalitarian global citizenship is also ecologically sound or regenerative. Ecofeminism illustrates the lived relationship between humans and the "natural" world. Without a balanced and regenerative ecological system, humans (and non-human animals) will not survive. One of the best articulations of a radical equity utopia is the novel *The Fifth Sacred Thing* (1993). The author, Starhawk, creates a multicultural, multifaith, ecologically sound culture in the Northwest region of the United States where people are fighting to survive. The people of the North draw on syncretic religious traditions, advanced scientific knowledge, indigenous and embodied knowing, magic, witchcraft, and the wisdom of the elders to save the planet. The resistance is headed by women of all racial and ethnic backgrounds who use personal and generational knowing to ensure inclusion and survival. The people of the North create explicit structures to develop regenerative systems at all levels. For example, the main character, Madrone, articulates the governance structure as being inter-relational:

Many people felt that nothing could truly be decided when the Four Sacred Things were not present. The animals, the plants, the waters had no voice in Council, and yet every decision should take them into account. After seemingly interminable argument, they had one of those unlooked-for bursts of collective creativity, or perhaps madness, and established this ritual, where masked representatives for each of the sacred elements sat in trance in Council, channeling the Voices of wind, fire, water, and earth.

(ibid., 1993, p. 46)

Starhawk weaves an engaging tale wherein mutual interdependence and an ethic of responsibility are embodied and radically enacted. The future of the planet depends on the people of the North fighting against capitalist patriarchal white supremacy as well as biological warfare. Rather than fight, the people of the North invite the soldiers from the South to join them; they collaborate with non-human animals as well as monsters, angels, pirates, and ghosts. They live "being with" and somatic commonality to practice embodied empathy with each other, human and nonhuman animal alike; they embrace self and other, be it race, creed, biological modification, nonhuman animal, plant, and/or spirit. The exploitation and destruction of nature and biological diversity threaten the life and existence of the planet. Multicultural groups, subcultures, nonhuman animals, plants, water, and spirits are all needed to save the future – using magic and bees and technology and *chi*. All must collaborate to "save" the future. Each is responsible for all the others in an ethic of responsibility deeply connected to place.

There are examples of radical equity practices in real life as well. Feminist historical and anthropological understandings of foraging societies are that they were matrilineal and egalitarian (Nielsen, 1990; Woodburn, 1982). Indigenous peoples of today often still continue to assert these cultural practices. One example of a cultural way of being that embodies mutual interdependence and an ethic of responsibility is the traditional greeting of indigenous Dakota people and culture. The Dakota people see all of the earth, land, water, and animals as relatives. This is illustrated every time they greet each other, by saying "Mitákuye Oyás'iŋ." This means "Hello my relative," a greeting offered to everyone and everything (see Eastman, 1980). This greeting to animal, water, plant, and human beings alike exemplifies the belief in the interconnected and interdependence of all beings – we are related and interdependent.

Another real-world example of radical equity is Plum Village, a monastic community near Bourdeaux in the Southwest of France established by Thich Nhat Hanh. Plum Village is Hanh's vision of a "Beloved Community" that embraces interbeing as its foundation; a place wherein there exists "a healthy nourishing environment where people can learn the art of living in harmony with one another and with the earth" (Plum Village, n.d.). Though it is home to a large resident population of monks and nuns rooted in the Zen Buddhist tradition, it offers numerous opportunities for lay people and their families to visit, and experience living "simply and harmoniously in community," working together making vegan meals, cleaning dishes, gardening as well as deepening their practice of mindfulness meditation (n.d.).

There are other examples in real life, albeit partial rather than societal. A practical problem in thinking of real utopias is that they are often dismissed as fantasy or impossible. Thus, partial examples may be more practical. As Wright (2011) articulated,

The challenge of envisioning real utopias is to elaborate clear-headed, rigorous, and viable alternatives to existing social institutions that both embody our deepest aspirations for human flourishing and take seriously the problem of practical design. Real utopias capture the spirit of utopia but remain attentive to what it takes to bring those aspirations to life.

(p. 37)

He goes on to identify things such as urban participatory budgeting developed in Brazil in 1988, the democratic and noncommercial creation of Wikipedia, worker-owned cooperatives, and unconditional basic income as real examples of utopian efforts.[3] Each has been profoundly collaborative and anti-capitalist.

To conclude, our project posits, imagines, a future wherein the radical praxis of equity is guided by mutual interdependence and an ethic of responsibility as an act of resistance, an act of hope, and a turn toward social justice. Employing examples both real and imagined, we can create the radical practice of equity.

Notes

1 While we are aware of the ongoing debates, for example, between advocates for livestock versus lab-grown meats, in regenerative agriculture, we focus here on the attributes of transformative innovation toward decentralized control of regenerative holistic ecosystems. See https://savory.global/statement-on-the-allan-savory-george-monbiot-debate-at-oxford/ for example.
2 Here it is important to note that Val Plumwood also develops a conception of interdependence that is not only non-dualistic but embraces an ecological view that understands humans as animals and intimately connected to nonhuman life. She suggests that not only can human animals be killed and eaten, but that they can kill nonhuman animals to eat. See *Eye of the Crocodile* (2013, pp. 84–85). Moreover, Donna Haraway theorizes about *natureculture*, which "is a synthesis of nature and culture that recognizes their inseparability in ecological relationships that are both biophysically and socially formed" (Malone and Ovenden, n.d.), suggesting again an intimate interdependence between human and nonhuman organisms but one that does not preclude human dominance in certain kinds of human-nonhuman interactions.
3 We do not include animal rights or river rights movements, for example, as practical utopias because they reify the very individualistic notions of capitalism that most separate us. We are arguing for a rejection of citizenship as the basis for utopia. Instead of the mutual responsibility and interdependence that we are arguing for, these rights-based movements re-inscribe the very notions of capitalist, patriarchal constructions of citizenship that define a center. We are proposing a decentering of citizenship, not inclusion in it.

References

Acampora, R. (2006). *Corporal Compassion: Animal Ethics and Philosophy of Body*. University of Pittsburgh Press.

Adams, C. J., & Donovan, J. (Eds.). (2007). *The Feminist Care Tradition in Animal Ethics*. Columbia University Press.

Butler, O. (1993). *Parable of the Sower*. Grand Central Publishing.

Confino, J. (2010). Zen and the Art of Protecting the Planet. *The Guardian*, August 26. www.theguardian.com/sustainability/environment-zen-buddhism-sustainability

Derrida, J. (2008). *The Animal that Therefore I Am*. Fordham University Press.

Eastman, C. A. (1980). *The Soul of an Indian*. Bison.

Eichler, M. (1981). Science Fiction as Desirable Feminist Scenarios. *Women's Studies International Quarterly*, *4*(1), 51–64.

Foucault, M. (1973). *The Order of Things: An Archaeology of the Human Sciences*. Vintage Books, Random House.

Foucault, M. (1984 [1967]). Of Other Spaces: Utopias and Heterotopias (J. Miskowiec, Trans). *Architecture, Mouvement, Continuité*, *5*, 46–49. https://web.mit.edu/allanmc/www/foucault1.pdf

Foucault, M. (2000). *Power: The Essential Works of Michel Foucault, 1954–1984* (R. Hurley, Trans.). New Press.

Frazier, N., & Gordon, L. (1992). Contract Versus Charity: Why Is There No Social Citizenship in the United States? *Socialist Review*, *22*(3), 45–67.

Gilligan, C. (1982). *In a Different Voice: Psychological Theory and Women's Development*. Harvard University Press.

Gilmore, R. W. (2022). *Abolition Geography: Essays Towards Liberation*. Verso Press.

Gleig, A. (2019). *American Dharma: Buddhism Beyond Modernity*. Yale University Press.

Gruen, L. (2015). *Entangled Empathy: An Alternative Ethic for Our Relationships with Animals*. Lantern Publishing and Media.

Hanh, T. N. (1987). *Interbeing: Commentaries on the Tiep Hien Precepts*. Parallax Press.

Hahn, T. N. (1988). *The Heart of Understanding: Commentaries on the Prajñaparamita Heart Sutra*. Parallax Press.

Hahn, T. N. (2017). *The Art of Living: Peace and Freedom in the Here and Now*. Harper Collins Publishers.

Holmgren, D. (2011). *Permaculture Principles & Pathways Beyond Sustainability*. Permanent Publications.

Jones, K. (1990). Citizenship in a Woman-Friendly Polity. *Signs, 15*(4), 781–812.

King, Y. (1989). Healing the Wounds: Feminism, Ecology and Nature/Culture Dualism. In A. Jagger & S. Bordo (Eds.), *Gender/Body/Knowledge: Feminist Reconstructions of Being and Knowing*. Rutgers University Press.

Lahar, S. (1991). Ecofeminist Theory and Grassroots Politics. *Hypatia, 6*(1), Ecological Feminism, 28–45.

Lamothe, K. (2015). *Why We Dance: A Philosophy of Bodily Becoming*. Columbia University Press.

Love, B. (2015). *We Want to Do More than Survive: Abolitionist Teaching and the Pursuit of Educational Freedom*. Beacon Press.

Loy, D. (2008). *Money, Sex, War, Karma: Notes for a Buddhist Revolution*. Wisdom Publications.

Malone, N., & Ovenden, K. (n.d.). *Natureculture*. Retrieved February 3, 2024, from Nature-culture – Malone – Major Reference Works – Wiley Online Library

Manga, G. (2008). Interbeing Autonomy and Economy: Toward Enduring Social and Ecological Justice. *Human Architecture: Journal of the Sociology of Self-Knowledge, 6*, 113–127.

May, V. M. (2015). *Pursuing Intersectionality: Unsettling Dominant Imaginaries*. Routledge Taylor & Francis Group.

Mead, G. H. (1934). *Mind, Self and Society* (C. W. Morris, Ed.). University of Chicago Press.

Mollison, B. (1988). *Permaculture – a Designer's Manual*. Tagari Publications.

Mouffe, C. (1992). Feminism, Citizenship and Radical Democratic Politics. In J. Butler & J. Scott (Eds.), *Feminists Theorize the Political*. Routledge.

Nielsen, J. M. (1990). *Sex and Gender in Society: Perspectives on Stratification*. Waveland Press.

Norris, K. (2020). *Witnessing Whiteness: Confronting White Supremacy in the American Church*. Oxford University Press.

Piercy, M. (1976). *Woman on the Edge of Time*. Fawcett Crest.

Plum Village. (n.d.). *Plum Village*. Retrieved March 29, 2023, from https://plumvillage.org/about/plum-village/

Plumwood, V. (1993). *Feminism and the Mastery of Nature*. Routledge.

Plumwood, V. (2013). *The Eye of the Crocodile*. ANU Press.

Quadagno, J. (1994). *The Color of Welfare: How Racism Undermined the War on Poverty*. Oxford University Press.

Reed, B. (2006). Shifting Our Mental Model-Sustainability to Regeneration. *Integrative Collaborative Design*. https://api.semanticscholar.org/CorpusID:22230389

Regan, T. (2004 [1983]). *The Case for Animal Rights*. University of California Press.

Richards, L. (2004). *Califia's Daughters*. Bantam Spectra.

Silbergleid, R. (1997). Women, Utopia, and Narrative: Toward a Postmodern Feminist Citizenship. *Hypatia, 12*(4), 156–177.

Singer, P. (2002 [1975]). *Animal Liberation*. Harper Collins Publishers.

Slonczewski, J. (1986). *A Door into Ocean*. St. Martin's Press.

Starhawk. (1993). *The Fifth Sacred Thing*. Bantam Books.

Syedullah, J. (2019). The Unbearable Will to Whiteness. In G. Yancy & E. McCrae (Eds.), *Buddhism and Whiteness: Critical Reflections* (pp. 143–159). Lexington Books.

Wahl, D. C. (2016). *Designing Regenerative Cultures*. Triarchy Press Ltd.

Waldau, P. (2002). *The Specter of Speciesism: Buddhist and Christian Views of Animals*. Oxford University Press.

Weber, M. (2015). *Weber's Rationalism and Modern Society* (T. Waters & D. Waters, Trans. & Ed.). Palgrave Macmillan.

Woodburn, J. (1982). Egalitarian Societies. *Man*, New Series, *17*(3), 431–451.

Wright, E. O. (2011). Real Utopias. *Contexts*, *10*(2), 36–42.

Subversion Through Radical Storytelling and Restorying

DOI: 10.4324/9781003366706-11

Chapter 8

Opening Aquaria

Daniel Vandersommers

As Foucault (1994) points out in *The Order of Things*, so many of the knowledges, institutions, and artifacts derivative of the so-called European Renaissance resurrect, and preserve into our present-day, that Platonic notion of "microcosm," which saw all living things as reflections of the cosmos (pp. 30–31). While the abstraction of "the microcosm" – encapsulating the wider world in miniature – is frequently employed in reference to philosophy, its physical assembly *within* the objects that surround us, still all-too-often remains invisible, submerged below the surface. This is the predicament of the aquarium. Suspended in the curated waters of an aquarium, large or small, are myriad bodies *made* to encase, and thus preserve, the ancient dream of a little world. Bodies under light appear luminous. They blaze and sparkle and shimmer. They undulate and flow. They dart and scurry and refract so that their observers may ascend "the immense distance" from microcosm to cosmos (ibid. p. 31). The aquarium reenacts a directional, hierarchical ontology from finite to infinite – from glass bowls to boundless depths. It ushers admirers up a great chain from *a* world to the idea(l) of *the* world, a vast progression that, through a mere glance, dissolves into the autonomous gaze, the Beginning and End, the Logos, the (renaissance) Man, who can create and then look upon (Jackson, 2020, pp. 90, 100, 105). This encoded phenomenological structure – the mutual folding of the macro and micro into one another – breathes an uncanny (even paradoxical) experientiality into fish tanks. Judith Hamera (2012, p. 2) states that the aquarium "allowed aquarists to feel . . . both *larger* through their mastery of 'lower' beings and *smaller* through communion with the glories of nature" [emphasis mine].

Alongside captive occupants, many deceits swim freely within aquaria, but this chapter seeks to raise one to the surface. For the aquarium to reveal "the world," it *must* erase all its tangible, material "worldly" connections. Therefore, the omniscient Subject constituted in the gaze reflected by the aquarium *must* (pretend to) know nothing of the real ecologies that intersect with worlds, big or small, in order to keep that gaze and its aquarium intact. Only through/after this deception, or suspension of disbelief, can the aquarium be put to work toward all of the diverse scientific, pedagogic, aesthetic, and allegorical ends it inherited across the nineteenth and twentieth centuries (ibid. pp. 2, 22, 29). This chapter shatters the contrivance of the aquarial gaze by unveiling the very material processes that made a single

DOI: 10.4324/9781003366706-12

aquarium – the previously unknown 1897 aquarium of the National Zoological Park of Washington, DC. The story that follows was assembled by piecing together late nineteenth-century letters, housed in the Smithsonian Institution Archives, that shed light upon the markets that supplied the living and nonliving objects presented in this particular display.

In the 1890s, public aquariums differed from their parlor aquarium predecessors in terms of size and scale, yet both types manufactured enclosed environments in similar ways. All aquariums required water, plants, animals, rocks, and sediments. By isolating a single, historic aquarium and examining how it acquired these essential elements, this story – like the aquarium – also functions as a part standing-in for a whole. Yet, unlike the cosmoses conjured and then forgotten within the aquarial imagination, the following synecdoche encourages us to open-up all aquariums, indeed all institutions with animals – in all times and places – with a similar degree of scrutiny. *Where do these lives come from? How are these institutions built? What are their seen and unseen purposes?* This case study encourages us to *see* before-without the metaphysical echelon of ordered "worlds." The irony of the examined Progressive Era aquarium is that, once the idea of cosmoses is discarded, we can see that the discrete, tangible, actual processes of aquarium-building did indeed extract from local sites, distributed by emerging global supply chains, around a very big but ever-shrinking planet. It is essential for us to recover the histories of both humans and animals, exploited and then forgotten within extractive systems, if we are to live more sustainably and ethically in the twenty-first century. The following history will turn toward aquarial lives – ones forgotten even by those of us with the most liberatory agenda. Opening aquaria and other overlooked containers, one at a time, will help us "see" the places where animals (including humans), plants, rocks, and sediments become commodified for the human, all too human, dream of seizing "the world."[1]

Background: Nineteenth-century Aquarimania

In the 1850s, middle- and upper-class Europeans and Americans became fascinated with all sorts of encasements – coral cabinets, terrariums, vivariums, Wardian cases, fern tanks, spider tanks, goldfish bowls, and especially aquariums.[2] At the height of parlor aquarimania, and as the first public aquariums were opening in New York and Boston, the Smithsonian Institution, in 1857, briefly supported its own public aquarium ("Boston Aquarial," 1861; Kisling, 2001, pp. 155–156).[3] Little is known about this early liminal aquarium, between parlor and public. Yet, a few publications in 1857 and 1858 mentioned its grand opening: "simply a glass tank, erected on a table, and filled with sea-water, in which flourish marine plants and animals without . . . even changing the water" ("Smithsonian," 1857). This manufactured environment consisted of a floor of "silver sand, coarse sand, and pebbles" and a "mass of rock" that provided shelter to three hundred specimens, representing "thirty-eight species of fishes, molluscae, crustacea, and polyps" (ibid). Crabs, jellyfish, seahorses, and a single flounder were also among the inhabitants. The fate of this aquarium remains a mystery. Most likely, the salinized water became too turbid to support life (since it was never changed) or the three hundred inhabitants became embroiled in a high-casualty Darwinian war. Indeed, behind the glass of early aquariums, the supposed "economy" of Nature proved difficult to model, often remaining only a chimerical ideal that said more about Western political philosophy and moral economy than about "Nature" at all.[4]

Four decades later, in the 1890s, the Smithsonian received a second chance at successfully running an aquarium. By peering into the aquarium of the National Zoological Park,[5] operated by the Smithsonian and funded by Congress, we will learn *how* an early aquarium was assembled. This is important for two reasons. First, in the twenty-first century, aquariums are still consumed as though they are singular, intact entities rather than the heterogenous composites that they pretend not to be. Second, the processes of aquarium-building in the 1890s were dependent upon an emerging global capitalism that still makes the institution, both public and private, today. By learning how to *see* aquarium-building in the past, we can better *see* aquarium-building in our contemporary world since present practices of assembly are merely intensifications or modifications of practices in place by the beginning of the twentieth century. By 1900, Frank Baker, superintendent of the National Zoological Park, had established new networks for the acquisition of water, plants, animals, rocks, and sediments, which became increasingly more dependable over the first decade of the aquarium's history. Furthermore, these networks grew increasingly far-reaching and extractive, as the numbers of private and public aquariums and government-sponsored fisheries increased (Muka, 2018, pp. 92–94). Deconstructing the built environments of the National Zoo's aquarium salvages a (hi)story drowned in the past – the story of how backwoods fishermen, ocean-bound sailors, insolvent object dealers, aquarium directors, Floridian entrepreneurs, scientists, loggers, boardwalk tycoons, and seaweed collectors became entangled in an institution devoted to making small worlds with diverse captive lives.

Containers

After the 1895 Cotton States and International Exposition, the US Commissioner of Fish and Fisheries donated to the Smithsonian's National Zoo the tanks used for the exposition's fish display. In 1897, the Zoo, which lacked the funds to build an aquarium building, placed these tanks temporarily in an old carpenter workshop ("Aquarium," 1980; Baker, 1911). Despite this substandard building, no other exhibit, in the words of Smithsonian secretary Samuel Langley, attracted "more attention or interest" (Hamlet, 3). As we know from zoo history, close and crumbling quarters did not keep schools of zoogoers from flooding the confines of popular animal displays. The donated tanks, each framed with ornately handcrafted Georgia pine exteriors and resting upon yellow pinewood posts, filled the room (Baker, December 20, 1911). These aquariums were inexpensive; only $200 was spent on their upkeep during the first full year of their tenancy within the shed ("Aquarium," 1980).

In total, the Department of Fish and Fisheries donated 28 tanks, but due to lack of space, only 16 were set up in the shed. The two longest tanks were 14 feet long; the others, either 7 or 5. All were 3 feet high. The back glass curved inward, creating a concave space that was wider at the top (5 feet) than at the bottom (2 feet). The curved glass would have refracted incoming light, causing it to dance through the water. In addition, the zoo ordered an extra tank – 12 feet long, 5 feet high, and 6 feet wide (at both top and bottom). This 2,700-gallon tank was the largest in the crowded aquarium. All used cypress wood to support the glass tanks themselves, for it was perfectly evolved to withstand moisture and mold. The wood was shipped from Gulf Coast lumberyards, which had accelerated their deforestation of southern old-growth forests after the Civil War, completely denuding the landscape and altering ratios of oxygen and carbon dioxide in the atmosphere (Baker, December 20, 1911). Each individual aquarium brought together water, plants, animals, rocks, and sediments. And each of these components required the creation of new global trade networks that made the artifice of the aquarium possible.

Water

The aquarium's first tanks contained fresh water and freshwater fish alone. At first, water was supplied by Washington water mains; however, since it muddied quickly, the zoo used aluminum filters for further purification. These required unceasing surveillance, for their use increased the likelihood of double sulphate salts leaching into the aquarium and killing the fishes. To address this problem, in 1899, two wells were dug to draw water from the nearby Rock Creek into the aquarium until the installation of a general filtration plant for the city permitted the zoo to draw water once again from the main grid ("Aquarium," 1980; Baker, December 20, 1911).

By 1899, Baker hoped to incorporate sea life into the tanks. Only then could the National Zoo boast an aquarium that could be recognized nationally. However, salt water was not easy to obtain. Those who owned small parlor aquariums often purchased seawater stored in milk-cans and sold in local stores or through nearby seashore merchants, but this was impractical for larger aquariums. Even though many zoogoers would believe seawater was piped from the Atlantic coast or Chesapeake directly to the aquarium, the permanent method established depended on water casks and wagons. Large containers of seawater would be carried by schooners and sloops and delivered to docks near Norfolk. From there, the water from the tanks would be transferred into ten-to-twelve-gallon casks, placed aboard horse-drawn wagons, and carted to the zoo (Baker, November 17, 1899; "Sea Water," 1902). Occasionally, D.C. Turner, a local fisherman employed as the caretaker of the Zoo's aquariums, would collect seawater while traveling by train and carriage to Florida and New Jersey for animal or seaweed specimens, but usually this was an inefficient way to move water.

Seawater surprisingly stayed fresh for a long time if stored in a cool place, so transportation and storage rarely posed any threat to its vitality. On average, the aquarium water was changed twice every month, and to ensure a dependable supply, seawater was contracted out for the best price – usually to an oysterman, fisherman, or boater from eastern Virginia (Googhegan, 1906). To further guarantee the stability of the water supply, the newest and most accurate salinometers were kept on hand (Baker, July 28, 1900). In case a tank's salinity needed to be adjusted, the aquarium also stored bags of salt, raked from the marine ponds of Turk's Island. Also known as Salt Cay – part of the British Turks and Caicos Islands territory, north of Haiti – this island served as a crucial colonial entrepot in the eighteenth- and nineteenth-century Atlantic trade in cotton, salt, and enslaved persons (Baker, December 27, 1900; Kurlansky, 2002, pp. 211–212, 220, 238, 254). Rarely, though, did the aquarium use artificially made seawater. According to Frank Baker, "this substitute is not satisfactory for keeping fish" (Baker, April 13, 1907). Only one European aquarium – the Stazione Zoologica in Naples, Italy – was successful at manufacturing artificial seawater, but this success required employing a profes- sional chemist as a full-time water-maker. Like freshwater tanks, saltwater ones required pumps; yet, these pumps, along with piping and valves, needed to be made of materials (usually brass or copper) that would not react chemically to salt, for neglecting to consider chemical reactions was one of the leading causes of death in early aquarium ventures (ibid.). While the leading aquariums at Plymouth, Naples, and Berlin had begun using self-sustaining systems that circulated the water for an entire year before needing to be replaced, such expensive systems were never installed in the underfunded National Zoo (Townsend, 1908).

All this meant that the Zoo's aquarium was high maintenance regarding its most essential resource – water. Obtaining, disposing of, and replacing water became an unceasing chore and placed limits on the size of the aquarium. The water *had* to be just right. If even just a little too much air was pumped into a tank, air bubbles could stick to the fish, causing exophthalmia, or popeye disease, wherein fishes' eyes protrude too far from their heads, leading to emotional distress, physical trauma, and inevitable deaths (Baker, April 17, 1901). To obtain water, the Zoo established an infrastructure, invisible to zoogoers, to ensure steady supply, usher- ing oystermen, fishermen, boaters, pipe-layers, wagon drivers, and work horses into the zoo aquarium enterprise.

Plants

Of the four types of inhabitants that made up aquarial environments – water, plants, animals, and rocks and sediments – plants were the easiest to acquire. Marine plants purchased for the aquarium were usually labelled monolithically as "sea- weed;" however, this category marked a wide diversity of marine algae, ranging in size, shape, habitat, and color, from red to brown to purple to green. Seaweed, like saltwater, was usually purchased directly from the Virginian coast where it was most often harvested, especially in the early years of the aquarium. Baker often hired fisherman J. T. Boyhan of Phoebus, Virginia, to ship seaweed to the zoo. Boyhan frequently found a "nice lot of sea-weed [containing] . . . as much variety as possible in shape and color," and transported the assortment in metal tubs, filled with seawater and covered with a "durable but not too thick" cloth to ensure survival during transport (Baker, March 23, 1903; Baker, November 8, 1902; Baker to Hegan, January 6, 1903). Overall, this was a straightforward process, since seaweed hobbyists had been an expanding middle-class subculture for several decades.[6] When the aquarium needed more seaweed, Baker would simply write to Boyhan and request another shipment. Only the occasional hurricane or northeaster could delay this process in spring, summer, or autumn (Boyhan, 1902).

The winter season in Virginia, however, regularly killed local seaweed and disrupted supply, forcing Baker to find marine plants further south between October and March (Baker to Hegan, January 6, 1903). Floridian coastlines especially fostered a great diversity of specimens, and Baker always emphasized to local shell and coral collectors that "the more brilliantly colored kinds would be preferable" to make the spectacle of display more enticing to viewers wanting to consume the exotic (Baker to Warner, January 6, 1903; Warner, 1903). He and Turner discovered, though, that not all Floridian seaweeds could survive in saltwater tanks because some of the more accessible waters around Florida, including Biscayne Bay, were brackish (only slightly salty), rather than "pure" salt water (Hegan, 1903). In fact, one contact informed Baker that most of the seaweeds around and north of the Bay were darker mosses, not flamboyantly colored marine plants, and would not survive well in captivity. This collector urged Baker to seek seaweed from Miami and the Florida Keys, which were fast emerging as mining centers for the global algae trades (White, 1903). Baker also occasionally purchased seaweed from the New York Aquarium, packed alongside marine animals and corals, and from the US Fish Commission's Station in Gloucester, packed alongside sea anemones since this Massachusetts port served as the cynosure of the anemone trade (Baker to Corliss, January 19, 1903; Baker to Turner, January 19, 1903).

In the big picture, seaweed was easy to procure. One obvious reason was that it only needed to be harvested, not caught. Any proletarian working along a shoreline could make a quick buck selling algae. Another reason has to do with the function attributed to marine plants in aquarium display, namely, as emphasized in a letter from Baker to Turner, "to decorate our . . . aquarium tanks" (Baker to Turner, January 6, 1903). This means that seaweed was not scrutinized for "perfect form" as were the animals, so collectors could spend more time bundling and less time assessing groups for archetypes. Seaweeds only mattered because they situated captive fish, crustaceans, mollusks, and other animals in an aquatic setting that *looked* Natural. The seaweeds themselves were never the center of attention, and the specific types and species showcased were hardly a concern, even as they became increasingly rare. Seaweeds merely served as props, as signs, that signaled to those peering into the encasement that *this* aquarium was indeed a world in miniature.

Of course, *within* aquariums, in relation to the other living captives, seaweeds were more than signifiers. The nineteenth-century aquarium literature, even regarding early parlor aquariums, had long acknowledged the ecological function of seaweed within manufactured environments. By consuming carbon dioxide and emitting oxygen, marine plants were necessary for aquariums to reach that nineteenth-century ideal, a chemical "balance," a self-sustaining equilibrium. In addition, plants and algae made up for the deficiencies of the manmade pumps that also circulated liquids and gasses. Baker and Turner were both aware of the ecological function of seaweed – whether eelgrasses, kelps, dulses, seagrasses, or sea lettuces. All utilized photosynthesis. All turned the exhalations of fish into oxygen. Most important, all brightened encasements as they simultaneously constructed zoogoers as proper Subjects gazing upon a supposed representation of a vast oceanic world.

Animals

If tanks stretched the aquarium to Appalachian forests and coastal cypress groves, and seawater stretched the aquarium to the shores of Norfolk, and seaweed, to Phoebus, Virginia, Miami, and Gloucester, the animals connected the aquarium to more far-flung places. In general, the aquarium brought together three types of aquatic animals – protagonist, feed, and decorative animals. The last two played supportive roles for the first, who performed the most microcosmic labor.

Before the aquarium's saltwater tanks were established, the Zoo specialized in the freshwater fishes caught by Turner along the Potomac, Rappahannock, Wicomico, and York rivers: large-mouthed black bass, yellow perch, crappie, common sunfish, long-eared sunfish, stone rollers, mullets, white and Mississippi catfish, freshwater eels, gar pikes, and German carps (Baker, April 16, 1901; Rathbun, July 30, 1900). Other times, Baker sent Turner to Piney Point and Point Lookout, Maryland, to catch sheepshead (Baker, September 15, 1900). These excursions represented a significant portion of Turner's job description. Usually, the zoo rented a boat, paid for an oarsman, and hired a helper. Captured by a medley of hook, net, and small-scale trawling methods, surely the seized individuals experienced a range of pains, traumas, and stresses.[7]

Once saltwater aquariums were established in 1899, demand for new and alluring protagonists arose. Baker pursued several methods to obtain ocean animals. First, he continued to send Turner on jaunts to nearby shores, for example, directing him in April 1900, to spend the week in Fort Monroe, Virginia, to catch saltwater fish (Baker, April 14, 1900). Second, to acquire high-profile ocean animals, Baker initiated transactions within the burgeoning international seashore trade. For example, to obtain an octopus, a classic symbol of "the coral reef" in the colonial imaginary, Baker wrote to E. E. Saunders & Co., a fishing operation based in Pensacola, about the likelihood of a fisherman catching one in their nets and sending it north to the capital (Baker, March 14, 1900). These transactions could also take more circuitous routes. For example, one day in 1900, German evolutionist Anton Dohrn, director of the Naples aquarium, informed the Smithsonian's acting secretary that a personal friend and steamship captain would gladly send the zoo an octopus caught off deck (Rathbun, March 8, 1900). Like octopus, seahorses had also long fascinated the public and were quite difficult to collect. They were sporadically procurable in Norfolk, but their numbers could never be predicted from season to season. Furthermore, seahorses possess fragile bodies, acutely sensitive to water quality, and were especially vulnerable to shipment since they do not have endoskeletons like other fishes, leading to incalculably high death rates (Baker, April 1, 1901). Once held in captivity, seahorses also succumbed quickly to bacterial infections (Muka, 2023, p. 164). The third way that Baker acquired saltwater fish was through trade with private collectors. Since the aquarium possessed a plethora of freshwater animals, he offered them to individuals like Colonel J. E. Jones of the US Navy in exchange for Bermudan fishes like angelfish, parrot fish, blue tang, and wrasses of all colors whom he acquired from international merchants and officers at naval bases and ports as distant as Kingston and Marseille (Rathbun, July 30, 1900).

None of these three methods of procurement worked satisfactorily. For a burgeoning aquarium that craved a lot of fish and a lot of diversity, all in a timely manner, trying to hunt down an individual octopus, seahorse, and crab one at a time was simply inefficient. Finding someone with an eel when one was wanted or trying to catch a dogfish when one was desired, always demanded a degree of luck. If the aquarium was to be successful, the Zoo required new and steady supply chains. Baker needed to find venues that distributed ocean animals in bulk, places he could consistently depend on for "stock" – like Young's Ocean Pier in Atlantic City, New Jersey.

John Lake Young, a leading entertainment and boardwalk mogul along the eastern seaboard, bought a pier and entertainment pavilion in Atlantic City in 1891, which he soon transformed into the cornerstone of a boardwalk empire. Vacationers traveled to Young's Ocean Pier to gamble, dance, fish, swim, and watch the newly invented game of basketball. In 1905, Young built a new pier that was a 1,775 foot long solid concrete extension. This "Million Dollar Pier" included a concert hall, theater, and a huge aquarium, which preceded the pier's completion and which Baker got wind of by the end of 1901 (Baker, September 30, 1901). Immediately, Baker reached out to Young, who made him feel "perfectly welcome" to any ocean animals, whom Young acquired from the leading Atlantic fish dealers serving the New York City market (Young, 1901). All Baker had to do was send Turner to Atlantic City to pick up the orders.

Baker purchased many protagonist animals from Young's aquarium (Baker, May 13, 1905). The first order demonstrated the number and diversity of individuals available – striped sea robin, sting ray, spade fish, sheepshead, bog shark, small red drum, horseshoe crab, "devil-fish" (octopus), cow nose ray, pompano, trigger fish, pilot fish, skate fish, rudder fish, and butter fish (Traylor, 1901). These animals, in the autumn of 1901, finally legitimated the National Zoo's aquarium, signaling to zoogoers that *this* aquarium was indeed a small world, "the ocean" in miniature, for these individuals *represented* the entire Atlantic. Pompanos, for example, were likely caught from Caribbean waters. Whereas devil fish (currently endangered) were probably seized off the coast of either Ireland or Portugal. The scintillating bodies of pompanos, the zebra stripes of pilot fish, the cartilaginous contortions of rays, and the scuttlings of horseshoe crabs made the aquarium the most popular site in the National Zoo, competing with the new elephant house. In the words of one nineteenth-century aquarium booster, "It is action that renders the Aquarium the most attractive spot at the Zoological Gardens" ("Gosse's Aquarium").

After Baker established relations with Young's Ocean Pier, he networked with the New York Aquarium, which first opened at Castle Garden in Battery Park in 1896. In 1902, the first significant transaction between the New York Aquarium and the National Zoo's aquarium was forged through a trade of freshwater for saltwater fishes: 1 leather carp, 2 mirror carp, 12 ring perch, 18 crappie, 8 mullets, 7 stonerollers, and 11 long-eared sunfish were exchanged for 28 Bermuda fish, 5 groupers, 6 angelfish, 3 striped grunts, 1 trigger, 2 surgeon fish, 2 trinity fish, 5 gray snapper, 2 hunid, and 2 princess rock fish (Turner, July 19, 1902). Within a year, Baker began to depend on this aquarial hub and its director Charles Haskins Townsend. Unlike the animals from Young's Pier, the individuals trapped in the New York Aquarium, which was fast becoming another entrepôt of the international fish trade, originated from far-flung waters around the world. Angelfish, for example, were frequently procured from the Amazon basin, yet surgeon fish were certainly netted in Caribbean coral reefs. Others, like the gray snapper, had enormous ranges stretching the coast from Maine to Uruguay.

Not all of the Zoo's protagonist animals were obtained from New York and Atlantic City. Baker still sent Turner on fishing expeditions, and he still looked to individual animal dealers and fishermen who specialized within niche markets opened by larger supply chains. For example, Baker purchased moray eels and octopus from Captain William Cuddy of Charleston, South Carolina, who seemed particularly skilled at harvesting invertebrates from accessible reefs (Baker, April 10, 1903). Japanese goldfish – which had been, like the canary, popular ornaments in Europe since the late 1600s – were always purchased from Henry Bishop, a Baltimore shopkeeper who likely got them from a Cincinnati goldfish breeder (Bishop, 1903).

Some animals played supportive, though not less important, roles. Decorative animals commanded a lot of attention, especially from Smithsonian secretary Langley, who took special interest in the aesthetics of the aquarium, even complaining in November 1902 that the sea fans were not placed on the sloping backs of the tanks where they could sway sensuously in plain sight (Baker, November 1, 1902). The zoo put a lot of time into decorating the aquarium with animals who, like plants and algae, would signal a larger world. These animals, though, were often treated, or referred to, as "charismatic plants" since they lacked the phenotypical traits usually associated with animals – eyes, mouths, limbs, and other attributes that facilitated the anthropocentric sympathies of humans. Decorative animals included a wide array of gorgonians, corals, sea anemones, and sponges, representing sessile entities, both singular and self-contained, collective and symbiotic, but all challenging commonplace notions of animality.

The acquisition of decorative animals was more challenging than for protagonists. First, they were quite delicate when removed from their environment. Coral and gorgonians frequently broke during transport (Baker, September 9, 1902). Furthermore, sea anemones and sponges could only live in circulating water, so sitting in crated and stagnant pools was risky (Smith, 1903). Sometimes they could be revived when placed into the aquarium, but usually a dead anemone resisted resurrection (Baker, January 22, 1903). Second, most of these colorful individuals originated from distant tropical waters, maximizing the distance of transport. J. J. White of Rockledge, Florida, for example, advised that corals, sea fans, and other "curiosities" could only be purchased steadily from the Bahamas (White, 1902). Baker forged these supply channels by communicating with a diverse cast of South Floridians. He ordered sea fans and corals from a Key West weather observer (Baker to Simons, November 1, 1902). He inquired about the price of sea fans from G. W. Warner of Palma Sola (Baker to Warner, November 1, 1902). And he depended on Robert Hegan, who ran the Royal Palm Gardens, for assortments of corals and seaweeds (Baker to Hegan, November 1, 1902). Most of these animals, though, would have been first extracted from their watery landscapes by individuals of the underclasses – sailors, crewmen, dock workers, and even children scraping by within the always-exploiting fishing industry and erased from the historical record preserved by official correspondence.

The last type of aquarial animal was the feed animal.[8] Early on, Baker instructed Turner to catch minnows in nearby streams (Turner, 1900). By December 1901, though, he began shopping around for fish dealers that could handle the aquarium's demand (Baker to Hurst, December 18, 1901; Baker to Odell, December 18, 1901; Buchanan, 1901). Eventually, he decided upon J. C. Robinson of Hampton, Virginia, for a steady supply of fish feed (Baker, May 21, 1903). This supply was composed of dead herring, menhaden, trout, bluefish, and butterfish, who were the less "valuable" feed animals used to sustain the more valuable protagonists as well as the other terrestrial captives of the Zoo, especially the bears and birds of prey (Robinson, May 12, 1903). Baker learned that the aquarium needed a diversity of fish on hand because the "animals develop a capriciousness in their appetite" (Baker, May 15, 1903). Feed animals were usually dropped into the aquariums dead (sometimes in pieces or ground into a pulpy meal), not alive (with the exception of minnows who were stored in a concrete tank that was built in 1902, on the north side of the aquarium, for the purpose of saltwater storage) ("Aquarium, 1980"). Feed fish, in many ways, played the most crucial role in the aquarium, but they performed the least important signifying labor in the fantasies of zoogoers, who rarely noticed or considered the animals that sustained the others on display.

Rocks and Sediments

No aquarium could be complete without rocks and sediments. While travelling the Hudson River, Baker once visited the Soldiers Home in Irvington-on-Hudson and was so impressed by the calcareous tufa, a pervious and lightly colored limestone used in its design, that he ordered one ton of this rock for the aquarium (Baker, October 27, 1902). The lower Hudson was a hotbed for this type of porous travertine, and when Baker needed more rocks for the aquarium, he always contacted a rock dealer from that region first (Baker, November 6, 1902). Tufa served as both an ideal decorative rock and a solid "sea floor" for the aquariums, which Baker brightened with colorful marble stones, sold locally (Baker, January 16, 1903). White quartz rock was also mixed in, and silicified woods were placed atop these rocks to conjure daydreams of driftwood or sunken pirate ships (Baker, November 2, 1905). Sediments arrived in the form of sand ordered from Norfolk (Baker, June 5, 1902). Occasionally, Baker requested picturesque white sand, which could double as a base for some of the bird cages (Baker, October 10, 1904).

Aquarium-goers would rarely have commented on the rocks and sediments of the aquarium. They lay entirely below perception. Their purpose was not only to provide a floor in which plants could grow, upon which algae could stick, and along which fish could swim, but the stones and sands – the remainder unconsciously felt but not seen – made the illusion of microcosm complete. The aquarium was (part of) something more, a representation of a larger unseen world.

Conclusion

Through this story of the National Zoo's aquarium, two historical lessons rise to the surface. First, aquariums are quite unnatural. Their histories are seeping with lessons of heterogenous mixture, as living and nonliving entities from vastly different ecosystems are placed into single small encasements. What is remarkable is that the *essential* diversity – the required combinations, aggregations, forced migrations, and cohabitations of water, plants, animals, and rocks and sediments – that *make* an aquarium (an aquarium) is the *very* quality an aquarium viewer must not see in order *to see* the aquarium, to apprehend the large and small worlds represented. Thinking with Foucault again, the aquarium (like so many institutions composed of animals) possesses, on its surface, an "apparent simplicity," with "that air of naïveté it has from a distance, so simple does it [the aquarium] appear and so obviously imposed by things themselves" (1994, p. 132). This chapter shows that "the aquarium" – the imaginative experience of moving from part to imagined whole and the desire to embark on such an odyssey – exists only within the curation of the aquarium, which occupies that "silent gap" "opened up between things and words," between the tangible, living components of the aquarium and the spectator's lens that sees the singularity of the tank by amalgamating its components (ibid., pp. 129–130). Yet when that gap becomes articulated by naming the "things themselves" ("things" that always, as Foucault makes clear, "touch against the banks of discourse"), by naming the sea anemones and tufa and goldfish and moray eels and algae themselves, the "hollow space of representation" begins to flood with the very living things that *make* that representation – "the aquarium" – possible in the first place (ibid). Once the items and beings within an aquarium are named, and once the histories of how they came to be in an aquarium are distilled, "the aquarium" itself no longer feels intact. When an aquarium is *opened up*, its mystical deceits spill out. No longer can it be seen as a world in miniature. No longer can it represent a larger world. Once quartz rock and gar pike and rock fish and oystermen and sessile animals – and all their histories – puddle upon the floor, we see the aquarium as an unlikely crossroads of many heterotopic, "placeless" places erased by/within the apparatus of the aquarium (Foucault, 1986, p. 24). These "real" places were only "placeless," of course, because they existed outside the (Real) discourse of the aquarium's fictions. While they remained invisible to the spectators who consumed aquarial displays, and thus functioned as essential vanishing points that kept aquariums literally intact and symbolically legible, these places were quite locatable for the human actors ushered into the economics of aquarium-building.

The second historical lesson is that once we "open up" an aquarium and follow the "tangled paths, strange places, secret passages, and unexpected communications" hidden below the surface, light pours in upon previously unforeseen marketplaces (Foucault, 1994, p. xix). Within "the patches" of emerging global supply chains, a wide variety of proletarian and underclass laborers tracked and captured octopus, seahorses, Bermuda fish, anemones, and seawater as they scraped by within increasingly extractive global economies (Tsing et al., 2019, p. 188).[9] Indeed, the myth of this container, the aquarium, was so intact and so "well-constructed" that not only the exploited (nonhuman and human) but also the exploiters – boardwalk tycoons, ship captains, institution directors, scientists – who transformed aquatic lives into aquarium capital were erased from the institution's central fantasy of ordering worlds and beings therein. In truth, since the Renaissance, there was no typical or archetypal aquarium (Foucault, 1994, pp. 158–159). Their sizes, styles, and compositions were contingent on innumerable place-specific and historical-contextual factors. Each aquarium surely contains its own unique histories of assembly, its own hidden economies, and its own exploited ecologies. Each aquarium also surely held so many diverse (and shortened) lives that were never experienced, by viewers, beyond the orders/worlds (*cosms*) they signaled.

The Rise and Fall of the National Zoo Aquarium

The aquarium of the National Zoo never became one of the renowned aquariums of the world. In 1902, after touring Europe, Smithsonian secretary Langley hoped that the Zoo could host an aquarium like the one in Naples. In fact, he obsessed over the Stazione Zoologica, writing to Baker, when he returned from Europe, "I saw nothing which impressed me so much" (Langley, 1902). Baker himself was quite knowledgeable about the acclaimed aquariums of Europe; he especially loved the Brighton Aquarium of London (Bashford, 1896). Sadly, the one crowded into a shed in the National Zoo lacked sufficient funding from the beginning. Baker and Langley, nonetheless, had grandiose plans. As early as 1899, they hired architects to design a new building. And if their blueprints would have been put into action, the Zoo could have indeed boasted one of America's leading establishments, like those in New York and Detroit. This new aquarium would have been arranged around an intricately designed central dome, composed of large windows that funneled natural light into the tanks below ("Aquarium, 1980"). Featured on the main floor would have been six large tanks for, respectively, whales or porpoises, sharks, sturgeon, sea turtles, West Indian seals, and manatees. Underneath these would have been a heating plant that warmed both the building and the waters within, and alongside it there would have been two rooms built for keepers and one reserved for spare tanks ("Memorandum," 1899). Smaller tanks, filled with a diversity of marine life, would surround the central tanks. The building itself would have been so drowned in light that the aquarium-goer would be carried away by the "illusion of . . . being under the general level of the water, or as if walking in a transparent tunnel under the sea" (Langley, 1899). Reality forestalled all this *cosmic* dreaming of becoming ocean-dweller.[10]

In 1901, Congress was asked to approve $25,000 for a new building, but the request was denied. Later requests were also rejected. Improvements to the makeshift aquarium were made. In 1903, for example, skylights were installed, increasing the amount of light in the shed. The aquarium remained popular with zoogoers for years, a central site of the zoological park. Nonetheless, in 1908, the foundation of the building began to sink. The walls, made temporarily out of Virginia pine, began to rot. The number of rats increased. By 1909 or 1910, the building was no longer safe to enter for fear of collapse. The exact day or year that the aquarium shut its doors has evaporated from history. No records exist after 1909 ("Aquarium, 1980").

If an aquarium stands as a microcosm, a representation of a larger world, the *building* of an aquarium – through the locating, claiming, accumulating, containing, and domesticating of living others – must also stand for the *assembly* of this ideal. Therefore, opening aquaria – imagining otherwise – does not merely mean halting the assemblage of aquariums themselves, though certainly this *unthinkable* act would prevent numberless millions from being consumed by the global marine aquarium industry each year. Opening aquaria means *unbuilding* the ideal (the desire to "see" the *larger* world, a "single world") that drives and maintains the economic-material processes of aquarium-building in the first place. When "things" (water, plants, animals, rocks and sediments) and "words" (aquariums, cosms, environments, Natures) are seen *separately*, a silent order, a status quo, a violent normalization of the everyday remains intact within zoos, hotels, restaurants, schools, and upon cabinets, tabletops, and mantels. To open aquaria, we must insert ourselves into their framework and hold words and things together (Deleuze & Guattari, 2009, p. 87). To purchase eelgrasses or octopus *is* to subscribe to a "pretense of mastery," and vice-versa (van Dooren, 2019, p. 128). Yet, true "immersion" within the container may help us *think* without the "striving towards universality" that is inevitable when looking *upon* the container (Foucault, 1994, p. 247). Immersion may help us see the other spaces, heterotopias, that are not diluted by the fantasy of a glimmering place that is perfect, pure, whole, or beyond.

A plunge into the 1897 aquarium of the National Zoological Park, one of the first zoo aquariums in the United States, pointed us toward "placeless places," consumed in, yet washed away by, aquarium building – Norfolk docks, Caribbean salt ponds and kelp forests, New Jersey piers, Mediterranean steamship routes, edge habitats at Floridian hotels, Virginian rivers, seaweed bales along Atlantic coasts, New York strip mines, and sprawling species-specific fish ranges, to name a few. A recent study has shown that in Iceland alone, the nation's *three* existing aquarium stores sell no fewer than 1,275 marine species. And 87% of these species originate from the Indian or Pacific Oceans, far away from this island nation (Micael et al., 2023). If this is true for Iceland, with the tiniest fraction of both global population and private and public aquariums, imagine the total number of species, individuals, and placeless ecologies absorbed through twenty-first-century global aquarium supply chains, especially considering that there are at least 8,000 marine aquarium retailers and 6.7 million hobbyists worldwide, dealing in more than two billion dollars of living capital annually (Watson et al., 2023). Perhaps the "other" spaces that compose supply chains, these sites of extraction at the "edges of capitalism," these locales collapsed into the microcosm of the aquarium, are the very places to craft a new vision that instead strives toward local particularity (Tsing, 2015, p. 282). Perhaps opening aquaria is merely looking down, looking closely, paying attention, following the entanglements of words and things to the lives and ecosystems nearest us. Sites of extraction are everywhere, and they are teeming with lives unseen. Truly seeing and valuing the aquatic lives at hand may be the first and most important step to opening ourselves to visions that do not require cosmic fantasies and the accumulation of others in translucent boxes elsewhere. There are large worlds around us already, and we are a part of them. Our cosmic dreaming is already our lived realities if we can open our eyes and take notice.

Acknowledgments

I would like to thank Miguel Gomez for reading a draft of this chapter and for his priceless historical, writerly, and aquarial insights. Thank you to Elizabeth Tavella and Michelle Westerlaken for providing insightful comments that strengthened these pages. Thank you also to the University of Dayton Department of History Research Colloquium, to Trip Glazer, Paul Morrow, and Erica Tom, and to the staff of the Smithsonian Institution Archives. Thank you especially to Paula Arcari for your thorough and thoughtful feedback and for pushing a historian to look toward future horizons too.

Notes

1 I am riffing here on Nietzsche's posthuman classic *Human, All Too Human*.
2 See also Keogh, L. (2020). *The Wardian Case: How a Simple Box Moved Plants and Changed the World*. University of Chicago Press.
3 For a great overview of nineteenth-century aquarium trends, see Hamera (2012, pp. 50–124) and Brunner, B. (2011). *The Ocean at Home: An Illustrated History of the Aquarium*. Reaktion; and Hamlin, C. (1986). Robert Warington and the Moral Economy of the Aquarium. *Journal of the History of Biology, 19*(1), 131–153.

4 See, for example, Worster, D. (1994). *Nature's Economy: A History of Ecological Ideas,* 2nd ed. Cambridge University Press.
5 For an intersectional-cultural history of the National Zoo between 1887 and 1920, see Vandersommers, D. (2023). *Entangled Encounters at the National Zoo: Stories from the Animal Archive.* University Press of Kansas.
6 See Duggins, M. (2016). Pacific Ocean Flowers: Colonial Seaweed Albums. In S. Mentz & M. E. Rojas (Eds.), *The Sea and Nineteenth-Century Anglophone Literary Culture* (pp. 119–134). Routledge.
7 See, for example, Wadiwel, D. (2016). Fish and Pain: The Politics of Doubt. *Animal Sentience 3*(31), 1–8.
8 Feeding always represented a chief challenge for zoos. For more see Kinder, J. (2019). 'Try Telling that to the Polar Bears': Rationing and Resistance at the Wartime Zoo. In T. McDonald & D. Vandersommers (Eds.), *Zoo Studies: A New Humanities* (pp. 145–166). McGill-Queens University Press.
9 Residing within, or beneath, every "historic" event lies members of the underclasses (Marx's *Lumpenproletariat*), erased from the historical record. See, for example, Rockman, S. (2009). *Scraping By: Wage Labor, Slavery, and Survival in Early Baltimore.* Johns Hopkins University Press.
10 Zoo and aquarium architects, since the 1970s, have called enclosures built around such a dream "immersion exhibits," or in the twenty-first century, an "immersion experience."

References

Secondary Sources

Deleuze, G., & Guattari, F. (2009). *A Thousand Plateaus: Capitalism and Schizophrenia* (B. Massumi, Trans.). University of Minnesota Press.
Foucault, M. (1986). Of Other Spaces. *Diacritics, 16*(1), 22–27.
Foucault, M. (1994). *The Order of Things: An Archaeology of the Human Sciences.* Vintage Books.
Hamera, J. (2012). *Parlor Ponds: The Cultural Work of the American Home Aquarium, 1850–1970.* University of Michigan Press.
Jackson, Z. I. (2020). *Becoming Human: Matter and Meaning in an Antiblack World.* New York University Press.
Kisling, Jr., V. N. (2001). Zoological Gardens of the United States. In V. N. Kisling, Jr. (Ed.), *Zoo and Aquarium History: Ancient Animal Collections to Zoological Gardens* (pp. 147–180). CRC Press.
Kurlansky, M. (2002). *Salt: A World History.* Walker and Company.
Micael, J., Sigurjónsdóttir, S., & Gíslason, S. (2023). Marine Aquarium Trade: An Open Door for Invasions in Iceland. *Journal of Coastal Conservation, 27*(57), 1–9. https://doi.org/10.1007/s11852-023-00986-4
Muka, S. (2018). Conservation Constellations: Aquariums in Aquatic Conservation Networks. In B. A. Minteer, J. Maienschein, & J. P. Collins (Eds.), *The Ark and Beyond: The Evolution of Zoo and Aquarium Conservation* (pp. 90–104). University of Chicago Press.
Muka, S. (2023). *Oceans Under Glass: Tank Craft and the Sciences of the Sea.* University of Chicago Press.
Tsing, A. L. (2015). *The Mushroom at the End of the World: On the Possibility of Life in Capitalist Ruins.* Princeton University Press.
Tsing, A. L., Mathews, A. S., & Bubandt, N. (2019). Patchy Anthropocene: Landscape Structure, Multispecies History, and the Retooling of Anthropology. *Current Anthropology, 60*(20), 186–197.

van Dooren, T. (2019). *The Wake of Crows: Living and Dying in Shared Worlds*. Columbia University Press.

Watson, G. J., Kohler, S., Collins, J. J., Richir, J., Arduini, D., Calabrese, C., & Schaefer, M. (2023). Can the Global Marine Aquarium Trade (MAT) Be a Model for Sustainable Coral Reef Fisheries? *Science Advances*, *9*(49).

Primary Sources

"Aquarium, 1980." [Report by S. E. Hamlet mailed to T. H. Reed]. Record Unit 365, Box 34, Folder 4. Smithsonian Institution Archives, Washington, DC. (Abbreviated RU #, Box #, Folder # below. And SIA will be implicit with these citations.)

Baker, F. (1899). [Letter to S. Langley dated November 17, 1899]. RU 74, 55, 8.

Baker, F. (1900). [Letter to Messrs. E. E. Saunders & Co. dated March 14, 1900]. RU 74, 55, 8.

Baker, F. (1900). [Letter to D. C. Turner dated April 14, 1900]. RU 74, 55, 8.

Baker, F. (1900). [Letter to Tagliabue dated July 28, 1900]. RU 74, 55, 9.

Baker, F. (1900). [Letter to R. Rathbun dated July 30, 1900]. RU 74, 55, 9.

Baker, F. (1900). [Letter to D. C. Turner dated September 15, 1900]. RU 74, 55, 10.

Baker, F. (1900). [Letter to Messrs. A. Kerr, Bro. & Co dated December 27, 1900]. RU 74, 55, 10.

Baker, F. (1901). [Letter to D. C. Turner dated April 16, 1901]. RU 74, 55, 11.

Baker, F. (1901). [Letter to R. Rathbun dated April 1, 1901]. RU 74, 55, 11.

Baker, F. (1901). [Letter to R. Rathbun dated April 17, 1901]. RU 74, 55, 11.

Baker, F. (1901). [Letter to J. L. Young dated September 30, 1901]. RU 74, Box 55, Folder 12.

Baker, F. (1901). [Letter to Messrs. Hurst & Fass dated December 18, 1901]. RU 74, 55, 12.

Baker, F. (1901). [Letter to Messrs. Odell Brothers dated December 18, 1901]. RU 74, 55, 12.

Baker, F. (1902). [Letter to Norfolk Sand Company dated June 5, 1902]. RU 74, 55, 15.

Baker, F. (1902). [Letter to R. H. Hegan dated September 9, 1902]. RU 74, 56, 3.

Baker, F. (1902). [Letter to Messrs. Lord & Burnham Co dated October 27, 1902]. RU 74, 56, 1.

Baker, F. (1902). [Letter to R. H. Hegan dated November 1, 1902]. RU 74, 56, 2.

Baker, F. (1902). [Letter to S. Langley dated November 1, 1902]. RU 74, 56, 2.

Baker, F. (1902). [Letter to W. U. Simons dated November 1, 1902]. RU 74, 56, 2.

Baker, F. (1902). [Letter to G. W. Warner dated November 1, 1902]. RU 74, 56, 2.

Baker, F. (1902). [Letter to Allen dated November 6, 1902]. RU 74, 56, 2.

Baker, F. (1902). [Letter to J. T. Boyhan dated November 8, 1902]. RU 74, 55, 15.

Baker, F. (1902). [Letter to J. T. Boyhan dated November 8, 1902]. RU 74, 55, 15.

Baker, F. (1903). [Letter to R. H. Hegan dated January 6, 1903]. RU 74, 56, 3.

Baker, F. (1903). [Letter to D. C. Turner dated January 6, 1903]. RU 74, 56, 3.

Baker, F. (1903). [Letter to Washington Granite Monumental dated January 16, 1903]. RU 74, 56, 4.

Baker, F. (1903). [Letter to C. G. Corliss dated January 19, 1903]. RU 74, 56, 4.

Baker, F. (1903). [Letter to D. C. Turner dated January 19, 1903]. RU 74, 56, 4.

Baker, F. (1903). [Letter to G. M. Bowers dated January 22, 1903]. RU 74, 56, 4.

Baker, F. (1903). [Letter to J. T. Boyhan dated March 23, 1903]. RU 74, 56, 5.

Baker, F. (1903). [Letter to W. Cuddy dated April 10, 1903]. RU 74, 56, 6.

Baker, F. (1903). [Letter to J. C. Robinson dated May 15, 1903]. RU 74, 56, 6.

Baker, F. (1903). [Letter to J. C. Robinson dated May 21, 1903]. RU 74, 56, 6.

Baker, F. (1904). [Letter to Messrs. R. H. Richardson & Son dated October 10, 1904]. RU 74, 56, 7.

Baker, F. (1905). [Letter to W. A. Watson dated May 13, 1905]. RU 74, 56, 8.

Baker, F. (1905). [Letter to C. H. Townsend dated November 2, 1905]. RU 74, 56, 10.

Baker, F. (1906). [Letter to W. Googhegan dated June 1906, with an illegible date]. RU 74, 56, 11.

Baker, F. (1907). [Letter to C. W. Parker dated April 13, 1907]. RU 74, 56, 12.

Baker, F. (1911). [Letter to G. H. Selkirk dated December 20, 1911]. RU 74, 56, 14.

Bashford, D. (1896). Public Aquariums in Europe. *Popular Science Monthly 50* (November), RU 74, 101, 9.

Bishop, H. (1903). [Letter to F. Baker dated May 11, 1903]. RU 74, 56, 6.

"Boston Aquarial + Zoological Gardens." (1861). [Broadside dated May 27, 1861]. "Broadsides Small Transport Box." Massachusetts Historical Society, Boston, MA.

Boyhan, J. T. (1902). [Letter to F. Baker dated November 7, 1902]. RU 74, 55, 15.

"Gosse's Aquarium." (1854). *Littell's Living Age*, July 29, 1854.

Hamlet, S. E. [Unpublished manuscript.] RU 365, 37, 1.

Hegan, R. (1903). [Letter to F. Baker dated January 14, 1903]. RU 74, 56, 3.

J. W. Buchanan & Sons. (1901). [Letter to F. Baker dated December 28, 1901]. RU 74, 55, 12.

Langley, S. (1899). [Letter to W. R. Emerson dated March 15, 1899]. RU 74, 55, 8.

Langley, S. (1902). [Letter to F. Baker dated October 14, 1902]. RU 74, 56, 2.

"Memorandum as to Requirements for a New Aquarium Building." (1899). [Memorandum dated March 14, 1899]. RU 74, 55, 8.

Rathbun, R. (1900). [Letter to A. Dohrn dated March 8, 1900]. RU 74, 55, 8.

Rathbun, R. (1908). [Letter to J. E. Jones dated July 30, 1900]. RU 74, 55, 9.

Robinson, J. C. (1903). [Letter to F. Baker dated May 12, 1903]. RU 74, 56, 6.

"Sea Water for the Zoo." (1902) [Newspaper clipping from *The Washington Times*]. "Scrapbook 1896–1907." RU 74, 286, 3.

Smith, J. A. (1903). [Letter to F. Baker dated January 15, 1903]. RU 74, 56, 4.

The Smithsonian Aquarium at Washington. (1857). *Scientific American, 13*(15), 113.

Traylor, J. G. (1901). [Letter to F. Baker dated September 23, 1901]. RU 74, 55, 12.

Townsend, C. H. (1908). [Letter to F. Baker dated April 15, 1908]. RU 74, 56, 13.

Turner, D. C. (1900). [Letter to F. Baker dated May 31, 1900]. RU 74, 55, 9.

Turner, D. C. (1902). [Letter to F. Baker dated July 19, 1902]. RU 74, 55, 15.

Warner, G. W. (1903). [Letter to F. Baker dated January 12, 1903]. RU 74, 56, 3.

White, J. (1902). [Letter to F. Baker dated November 5, 1902]. RU 74, 56, 2.

White, J. (1903). [Letter to F. Baker dated January 13, 1903]. RU 74, 56, 4.

Young, J. L. (1901). [Letter to F. Baker dated October 5, 1901]. RU 74, 55, 12.

Chapter 9

Beyond the Farm – Towards Multispecies Anarcho-communities

Amina Grunewald

Introduction

When exploring opportunities for modes of egalitarian and solidaric multispecies co-existence and animal emancipation in the Anthropocene,[1] there are pathways for envisioning local speculative geographies of animal freedom as nodes of agency[2] and resistance, set against a Western exploitative speciesist capitalist system. In this chapter, I focus on the fable *Anarchist Farm*[3] (1996) by Jane Doe[4] to investigate an animal rebellion informed by an anarcho-geographical stronghold (Springer, 2013) at the intersection of species and spatiality. Doe's political novel in the form of a fable provides the ultimate contrast to a human-controlled farm regime: a multispecies community that functions as a zooheterotopian alternative. The farm-turned-anarcho-community is inhabited by anthropomorphised free-living animal subjects that have taken over the farm after their human owner's natural death and transformed it into a commune of freedom. Thus, *Anarchist Farm* points to posthumanist futures beyond human-regulated farm businesses (Allen, 2015 in Cudworth & Hobden, 2018a, p. 36).

Anarchism as a political philosophy has always been executed from a terrain of resistance. According to Springer, Anarchist geographies "question the spatiality upon which governance is premised and argue for an unstructured 'field of action,' where individuals may voluntarily and/or collectively decide their own direction, free from the presence and pressures of any higher or ultimate authority" (2016, pp. 61–62). They are a fertile ground for civil disobedience (Springer, 2013) such as practised in Doe's fable by various animal agents to save their lives from commodification and erasure in an autonomous anti-capitalist safe haven: Circle A(narchist) farm.[5] Miller's concept of zooheterotopia complements this framework and explicitly extends it to non-human animals. As Miller explains, through his concept of heterotopia,

> Foucault signals a spatial imaginary as a counter hegemonic force; to be out of the way is to be located at a remove from normative ideology in a space that allows us to reimagine the world and, potentially, our relations with other creatures.
>
> (2015, p. 164)

DOI: 10.4324/9781003366706-13

Foucault's work (1977, 1986) on acts of resistance against authoritative systemic orders and performed normative conventions prepares the ground for discussing biopolitical oppression, as exemplified in Doe's fable, when envisioning alternative worldviews at a distance from an all-controlling power centre. In Doe's narrative, animals are treated by the corporation/big business and its complicit state as objectified consumable bare lives rather than as valuable individual sentient creatures. Hence Springer, Miller, and Foucault are combined here in analytical force to unfence the deviant beastly Other, who is continuously commodified as mass food in the agricultural farm sector.[6]

The value of Doe's fable lies in the depiction of a rebellious multispecies community as imagined from multivocal animal perspectives with special emphasis on the potential of anarchism as political praxis that incorporates animal freedom in terms of equality, solidarity, and cooperation. As *Anarchist Farm* functions as a fable, all non-human animals are anthropomorphised. Their subjectification highlights commonalities between humans and non-humans, thereby facilitating identification and empathetic engagement. Scholars have highlighted that depictions of animal sentience (Bremhorst et al., 2023; Webster, 2022), agency (Cudworth & Hobden, 2018b; Roscher, 2016; Wirth et al., 2016; McFarland & Hediger, 2009; McHugh, 2009; Latour, 2007), (dis)abledness (Nocella, 2012, 2016), partnership, and mutual support (Kropotkin, 1998 [1902]) enable emotional engagement and are key components of multispecies interfaces, triggering recognition of entangled human and non-human animal commonalities despite supposedly ontological differences.[7]

I argue that Doe subversively and powerfully employs the medium of the traditional fable as a lens through which to reimagine egalitarian posthumanist multispecies relations. Hence, Doe's work offers thought food for alternative creaturely futures. To demonstrate the power of that liberatory vision and the fable's didactic purpose for analysis, I will investigate Doe's counter-narrative as antispeciest intervention that criticises animal de-subjectification via anthropomorphic individualisation as a communicative strategy (Taylor, 2011; Barcz, 2017) to draw attention to real-life warfare against animals. *Anarchist Farm* offers political guidance and direct action motivation to transform a hegemonic speciesist status quo by educating readerships extradiegetically, that is, in the wider world beyond the story (including classrooms), on a zooheterotopic vision of animal freedom, equality, and solidarity in the form of an autonomous anarcho-community at grassroots level.

Doe leverages literary familiarity to draw readers into her depiction of an anarcho-community of animal emancipation. First, she plays with intertextual knowledge, basing her work on Orwell's well-known *Animal Farm*. In this literary-political context, Doe's animals are outspoken about their situation and ideas for radical change as in Orwell's original. Second, the familiar genre of Aesop's fable lets readers settle into its typical features: speaking animal characters radically flip conventional human–animal relations, with readers bearing witness to the deadly harms humans inflict on the animal kingdom. Consequently, the story works like a Trojan Horse, smuggling radical ideas (farm-turned-anarcho-community, systemic change) into readers' minds, inviting them to reflect on the deficits in human–animal hierarchies and be part of a growing posthumanist discourse. Fables such as Doe's offer a future vision of anarchism-informed animal emancipation from captive exploitation as represented in the fable via an egalitarian, solidaric, and cooperative geographic reorganisation of unequal species power relations (Springer, 2015). My analysis draws from and adds to existing literary studies scholarship in relation to anarchist geographies and pro-animal literature (Ortiz-Robles, 2016; Borgards, 2015; Harel, 2009), for anarchism, as theorised by Kropotkin (1901) and revitalised by Springer (2012), proves productive as antispeciesist liberatory philosophy and as a driving force for species thriving.

In the following sections, the fable is framed as a literary vessel to transport anarchist-posthumanist thinking and practice as antispeciesist methodology for progressing animal liberation. The fable's narrative power is twofold (Harel, 2009; McHugh, 2009; Borgards, 2015): First, it figuratively and literally zooms in on animal subjects' rights to life and bodily integrity (Stucki & Kurki, 2020, p. 1) and, second, it advances an animal liberatory understanding of anarchism in a fictionalised practice as antispeciesist intervention. Both are explored with reference to the fable's plot as summarised next.

Anarchist Farm – A Summary

Doe's 190-page fable is told from an omniscient narrator's perspective. Readers learn that farm owner Margaret Hansen has died, and her once co-dependent farm animals need to run the farm on a self-reliant leaderless basis. Meanwhile, the Forest Protectors, a guerilla group of free-living animals that also includes Sabo the black cat and Pancho the pig – characters that also appear in Orwell's *Animal farm* – are battling "the corporation." The corporation is a collective reference to big business whose only goal is to earn money for their shareholders by capitalising on animals and the forest as natural resources. Hence, the corporation plans to close Circle H farm and sell its non-human inhabitants as bare life commodities: "LAND, EQUIPMENT, LIFESTOCK . . . EVERTHING MUST GO" (p. 90). In a farm-forest collaborative effort of resistance based on consensus discussions, the corporation is ultimately defeated when The Great Stampede – a literal liberation run of all animals and human friends (soldiers, city dwellers, eco-activists) – advances from the beleaguered Circle A-farm to the city and the corporation's headquarters. Dreamed up by the Sacred Cow, the finale is stimulated by the animals' grief over their killed friends. The Great Stampede becomes unstoppable with all animals and human allies running together in a climactic sequence:

Every animal had his own reason for running, but the same goal. The raccoons ran for the forest and the shepherds for the kennel, but that did not matter. They ran for freedom. . . . They would always remember being part of the Great Stampede. . . . The raccoons still ran for the forest, but also for the dog kennel.

The dairy cows ran for the death of their Mother, but also for the horses' ranch. Pancho ran for Goldie and the farm, but also for the humans. . . . They were one unit now: one giant animal bent on justice for all.

(pp. 180–181)

The triumphant rally demonstrates a direct solidaric action and the limit of the corporate-governmental authoritative grip, since multispecies solidarity, in this fable, trumps big business/state power. Circle A Farm animals clarify that they do not wish to negotiate an agreement to restructure the current oppressive system but rather to abolish it thoroughly, as cow Rosy explains: "I do not accept this or any offer made by the evil corporation. I came here to destroy it, not to reform it." (p. 183). Denying the corporation, its "wicked stepsisters the bank and the government," and its "prisoner, the press" (p. 187) any opportunities for manipulation and continuing abuse by further luring the rebellious animals into their system, cow Rosie marches into the corporation's headquarters as iconic leader to the cause. With the animals behind her, they stampede into the boardroom: "Here they stood, face to face with their nemesis, the head of the corporation . . . small and powerless compared with the minds, hearts and courage of the animal revolution" (p. 184). Turkey Emma[8] discloses the corporate CEO's attempt to put the blame on the government, accusing him of lobbying the administration and reducing it to the CEO's puppet:

You own the government. You pay for the election campaigns, you own the newspapers that determine the elections, you appoint the judges, and hire the police. Maybe getting rid of you won't fix everything, but I say you're a very good place to start.

(p. 185)

Despite the CEO's statement in front of a summoned jury of animal survivors that everything the corporation did was legal, sow Martha cuts him short, claiming not to uphold the law but to uphold justice. Consequently, the urban corporation's CEO is found guilty and punished to "clean up the mess" by planting trees in the destroyed forest (p. 187). The confrontation ends with the abolishment of the corporation, the government, and the bank, with all animals returning to their natural state of freedom. The happy ending presents Circle A farm flourishing as an autonomous multispecies anarcho-community.

The Power of the Fable as Anarchist Counter-narrative

Ortiz-Robles suggests, "[I]n order to make animals count politically, we must first reconfigure the distribution of the perceptible within literature" (2016, p. 144). In traditional fables, animals exist but are perceived primarily as substitutes for human existence (Harel, 2009). However, drawing on Harel, Borgards suggests that the central aim of cultural literary animal studies is "to study these kinds of real animal and animal representation interrelations in their full complexity" (2015, p. 156). In Borgards' and Harel's understanding, fables do not always exclude animals as subjects of their own lives. Some describe "animal behavior or enable critical view[s] regarding human treatment towards other animals" (Harel, 2009, p. 10). Harel encourages a fable analysis towards a broader literal reading in favour of real complex animal realities. Animals in their readings are intentionally anthropomorphised to comprehend their individual lives, and Harel perceives this method as an opportunity to

> use anthropomorphism in order to discuss unfamiliar subjects, such as the nonverbal consciousness, in familiar terms . . . when a human mistreats an ass in a fable, we may allegorize it, but we may also judge this treatment literally.
>
> (2009, pp. 13–14)

In this way, non-human animals perform as metonymical representatives for real animals (literal level), while also standing for political ideas such as anarchism (allegorical level). Doe's politicised fable recycles the genre as a lens to shed light on animal justice issues literature. Its subversive power comes from letting animals voice and act out a possible future vision for their emancipation by overturning authoritarian anthropocentric corporate rule. Bolton Hall's *Monkey Shines; Little Stories for Children* (1904) follows a similar model. Hall's collection offers moral messages with the twofold intention to entertain and to teach, and his didactic writing is linked to animal plights, such as when critically assessing human control over and misrepresentations of animals as trained property in "A Bark that Bit" or as "vermin" in "The Flying Mouse." There are many examples of animal-centred stories that offer two levels of interpretation. Just some of these include *Black Beauty* (Sewell, 1877), which propagates animal welfare from a horse's perspective, and *Charlotte's Web* (White, 1952) – an emphatic appeal for farm animals' lives that depicts animal cooperation as agency to serve the cause of animal rights, albeit still accepting farm confinement. Adams' *Watership Down* (1972) describes a dystopic world in contrast to a peaceful, utopian existence of free-living rabbits. In *Fox 8*, Saunders (2018) criticises "yuman" cruelty against free-living foxes, and Henry Hoke's fable *Open Throat* (2023) describes multispecies inequalities from the perspective of a mountain lion. These and many other fable narratives entertain, teach, and tap into entangled human and animal lives while extending their agenda to political critique and a call for transformation, offering a framework for literary discursive and politicised contextualisation in which *Anarchist Farm* can be situated.

The common ground of these fables lies in their use of animal protagonists as lenses to witness, teach, and deplore specific facets of a *conditio humana*. More progressive contemporary works in tune with Animal Studies and Critical Animal Studies (such as Doe, Hoke, Saunders) focus on the testimonies, survival, and rights claims of centralised animals who "speak up" for their own cause. In sum, the power of Doe's and other anarchic stories lies in their embedding of a political stance and systemic critique in favour of multispecies anarcho-networks in a traditional format. With its additional intertextual connection with Orwell's classic *Animal Farm*, Doe's *Anarchist Farm* aptly unites various anarcho-spatial strands into an animal justice narrative. These fables can arouse reader interest to empathically and cognitively engage with the war raged against animals (Gruen, 2013; Wadiwel, 2015; Cudworth & Hobden, 2015).

Some scholars highlight the fable's conventional fault of reducing animals to mere metaphors (Simons, 2002; Derrida, 2002; Fudge, 2002). However, I agree with Harel (2009), Schönbeck (2019), and Fudge (2020) that fables can be employed to refer to the literal animal alongside the fictional narrative. Another potential risk of the animal fable lies in its tradition of anthropomorphising the animal Other for human behaviour and morals. As such, anthropomorphisation, from a Critical Animal Studies perspective, could be regarded as inherently counterproductive to animal plights in real life, for a non-human animal as sentient conscious subject of agency and individual being is already capable of performing intentional actions as an

> agent who is in possession of capacities and intentions, who is able to self-identify. . . . The holder of this capacity is a self-conscious subject, in possession of the designs of one's will and able to act upon them.
>
> (Zaharijević, 2021, p. 22)

Anthropomorphism risks a counter-performative backlash and return to ventriloquism, robbing animals of their inherent subject agency and individuality. An uncritical anthropomorphism risks speciesist misrepresentations, trivialised romanticised images, and hyberbolic expectations of real animals (see also Taylor, 2011; Barcz, 2017), which certainly is a potential flaw of the genre in general. A way out of that impasse in pedagogic settings is addressing such a reduction from a sensitive posthumanist position of critical self-awareness Anthropomorphism is arguably unavoidable as humans interpret the natural world (Taylor, 2011; Karlsson, 2012; Barcz, 2017). However, Barcz and Taylor suggest that critical anthropomorphism is a communicative instrumental strategy firmly situated within a critical anthropomorphist discourse that critiques human-non-human animal analogies as difficult literary tradition.

The critical, anarchist elements involved in these fables suggest a look at studies by Springer (2012, 2016) and Rouhani (2012), who investigate the history, spaces, and prospects for anarchist pedagogies in praxis. Rouhani (2012) states that anarchism as a pool of theory and practices has a long history of engagement with radical pedagogical experimentation. Fictional representations of anarchy in practice can serve as non-coercive, practical learning material, and potential inspiration for readerships. I suggest an anarcho-pedagogical approach to building alternative learning models is necessary for educational change – one that advances animal rights and liberation through storytelling, and locates future generations in an imagined Post-Anthropocene.

Kinna's *Anarchism: A Beginner's Guide* (2005) conceives political fiction as anarchism's "genre in arms" and their authors as political activists. She cites novels promulgating change, including Henry James's *Princess Casamassima* and Paul Auster's *Leviathan* (2005, p. 172). I would add Edward Abbey's classic *The Monkey Wrench Gang*, which Doe intertextualises when the animals attack the logger camp by monkey-wrenching various equipment (1996, pp. 33–38). *Anarchist Farm* draws on various direct action and civil disobedience strategies (guerilla tactics, property-destroying,[9] insurrections, mass rallies, occupation, and blockades) – all practised on a grassroots level.[10] I agree with Kinna's assertion that anarchism in writing reflects direct commitment: "Actions are about cultural subversion, building solidarity, debate and the destruction of corporate power. Whatever kind of actions militants take – fluffy or spiky – they reflect ideological commitment" (2005, p. 158).

The term "anarchism" itself is defined by the online dictionary Merriam-Webster as "a political theory holding all forms of governmental authority to be unnecessary and undesirable and advocating a society based on voluntary cooperation and free association of individuals and groups."[11] However, anarchism is by no means a fixed ideology but rather an evolving idea in practice, continuously unfolding in writing as well as in performative acts (Ramnath, 2011, p. 37). One extradiegetic performative act might be writing a fictionalised account of an anarchist community in praxis (see Kinna earlier in the chapter) to serve the intradiegetic remodelling of a Western anthropocentric worldview in favour of collective animal liberation from an anarchist bulwark (Circle A farm) – this time encompassing non-human animals. This zooheterotopian space is fictionally set apart but is at the same time informed by external real-life twenty-first-century animal justice issues which resonate with animals' priorities regarding bodily integrity and the right to lead one's own life. Springer's anarcho-geographies can be used to frame Circle A farm as a network of animal care "where alternatives to the state flourish" (Araujo, 2016, pp. 94–95). Abolishing political authority and a monitoring normative order in favour of independent nodes of "free and equal members based on a harmony of interests and the voluntary participation of everybody in carrying out social responsibilities" (Kinna, 2005, p. 68) seems to be a promising experiment. Animals, as presented in *Anarchist Farm*, would arguably benefit from losing the shackles that tie them to confinements legalised by state regulations, which restrict their autonomy, physical integrity, individual personality, potential, and reciprocal actions, in collective combination (Kinna, 2005, p. 76).

Anarchists have always been critical thinkers outside normative mindframes and, according to Cornell (2016), Anarchist experimenters and innovators in the realms of theory, art, and tactics proved trailblazers for liberation and novel forms of education, the creation of intentional communities of living "otherwise," and cultural transformations. Anarchist principles of decentralisation and antiauthoritarian self-governance (Cornell, 2016) inform this fable's participatory democracy, environmentalism, and animal freedom. Indeed, Pellow considers anarchism to present one of the most fertile grounds for planting the seeds of a politics of equality that recognises and includes species membership – thus expanding society beyond the human (2014, pp. 93–94, 99). Likewise, Dominick includes animal freedom as an outcome of the anarchist critique of human domination that extends to the animal kingdom: "A practicable theory opposed to causing suffering to any sentient animals, favouring widespread personal behavioral changes and system-targeted social activism to curb and eventually eliminate abuses" (2015, p. 39). Dominick's call to action reaches out to individual engagement in real life – a call echoed by Doe, who invites readers to step into animal characters' hooves, paws, and claws to experience their lives. Fables can appeal to readers' moral conscience about the treatment of animals whose lived experiences, emotions, and mind-frames are made accessible through anthropomorphised characterisation which builds empathetic engagement. Gruen's (2013) entangled empathy hereby proves an important strategy to inform and influence readers, going beyond feeling pity for suffering species to engage cognitively, emotionally, and sensorially with the animal characters' lives and precarious living conditions.

So how does Doe represent the formation of a posthuman anarchist stronghold using fictional animal characters? The following analysis describes various characters of *Anarchist Farm* in direct action mode as they implement an autonomous anarcho-community against the corporatist neoliberalist farm. I explore how the fable motivates a change by guiding its readers through a fictional praxis of how to apply anarchist thinking in a rebellion for independence from the State's governing principles of control and punishment directed against the bestial Other.

Analysis: Anarchist Farm and Its Liberatory Vision

In *Anarchist Farm*, the animal farm, and by extension the industrial meat plant, together the ultimate expression of animal oppression, ironically serve as a zooheterotopian stronghold of liberty and resistance. It is at Circle A farm, a multispecies grassroots community, where animals reclaim their physical integrity as individual beings. With reference to the potential of farm animal fables as part of animal literature, McHugh observes:

> Neither simply sacrificed nor gaining significance as "speaking meat," . . . farm animals emerge as irreducibly social creatures in the interstices of . . . representational media like novels that for so long have been seen as crafted only to give voice to human subject-forms. Literary narratives include a still broader range of possibilities, including that meat itself can serve as an agent of protest . . . to sketch community formation, reformation, and revolt.
>
> (2009, p. 32)

Doe's represented multispecies community consists of various individual animals, with every central character having a name, but who are collectively categorised as anonymous livestock from the corporation's perspective. Resisting de-subjectification and commodification, Doe individualises the animals' capabilities, emotions, and social ties, offering an alternate multispecies community in which animals can live their lives to the fullest. In addition, in Doe's fable, two characters from Orwell's *Animal Farm* (previously unnamed black cat Sabo and smart pig Pancho, formerly Snowball) realise their full revolutionary potential in search of what animal liberation might look like on an abandoned farm yard.

Anarchist Farm is depicted as a species-diverse community centred at and around Circle A Farm and comprising forest dwellings, which are threatened by a logging company, as well as other animal confinements including bird cages, a dog kennel/breeding factory, horse ranch, and zoo. Doe groups the non-human community into free-living animals (raccoons, squirrels, owl, raven, deer, bears), environmental activists, farmed animals (pigs, sows, cows, etc.), so-called companion animals (dogs), tested animals, and all kinds of animals abused for entertainment. Various human and non-human agents of resistance, including the human environmentalists Redwood, River, and Rainbow, and free-living forest animals, become members of the farm community. All animals follow a vegetarian/vegan diet (even the carnivores) feasting on acorn soup, vegetables, and fruit, echoing Reclus' 1901 work uniting vegetarianism, anarchism, and animal freedom as well as North American nineteenth-century Transcendental farm experiments (e.g., Fruitlands and Brook Farm).

Doe uses the fable's didactic tradition to instruct readers about anarchism's potential in defensive community-building against the war against animals. For example, the dead farm owner's companion dog Goldie explains to the newly arrived pig Pancho the anarchist idea of leaderless self-organisation:

> "Leader? "Goldie thought a bit. "I guess Margaret was sort of our leader, but she died," . . ., "the leader thing just never came up." Pancho raised his eyebrows. "But it's so organized here . . . who?" "It just hasn't been a problem. I guess we all organize ourselves." Goldie was tired of explaining.
>
> (p. 51)

After Margaret's death the animals take over the farm's garden and its homegrown food, having liberated themselves, as Goldie explains, from human husbandry by referring to their own abledness and natural skills:

> Margaret always kept her kitchen garden fenced to keep us animals out, but she spent a lot of time fighting the bugs, snails and slugs for ownership of the vegetables. Well, after she got sick . . ., the ducks made a good suggestion. They liked to eat snails and they would be careful not to step on the little plants . . . now the ducks are mostly responsible for that garden patch. . . . With so many different abilities from all the animals and so many of us, we hardly have to work at all!
>
> (p. 50)

Former Circle H Farm as well as the exotic birds in cages (p. 126) thus serve as an example for Michel Foucault's notion of shepherd-flock co-dependency suggesting complete dependence of the flock on the human shepherd as ruling authority, in permanent submission to his will and law (Foucault, 1979, p. 237).

As the narrative's impetus derives from the threat of Circle A farm being discovered by external neoliberalist interests, this farm community primarily functions as an anarcho-alternative stronghold that is care-ful and happily out of joint with the capitalist market activities operating in parallel beyond the farm's boundaries. These speciesist settings are shown through the eyes of animals as witness and testimony. During farm turkey Emma's discovery flight across the topography of human terror, and in dialogue with the animals she meets on her journey, she describes with horror humankind's exploitation of the environment and the enslavement of its other-than-human inhabitants, providing readers with a map (the interior book cover) of her flight itinerary through these dystopian spaces:

> This was the way to truly understand the horror of clearcuts. . . . From the sky Emma could see the ravaged hillsides, brown and dead, of the corporate lands to the north. Beyond them lay the concrete monoliths of the city.
>
> (p. 119)

Emma's map depicts the novel's hegemonic topography with the subversive zooheterotopian farm at the very bottom on the right from which her journey starts. Between Circle A farm and its spatial antagonist at the top left – the city and its toxic surroundings (dump, factories, clearcuts) – are situated various animal confinements such as horse ranch, dairy, zoo, and bird garden.

During her journey, Emma collects animal partisans for the rebellion against the urban corporate power centre (p. 119). The horse ranch, in the map's lower left corner, is a space of confinement signalling a history of animal abuse for human entertainment. However, horse Edsel who is isolated as a rebellious dangerous Other for displaying resistance (by kicking humans) joins the resistance movement as an agent of his own fate. Emma subsequently crosses the zoo prison complex where "she saw legs chained to posts and walls barring escape, in the monkey house, animals peered through the bars dreaming of freedom. The tigers paced back and forth in frustration" (p. 129). Emma moves on to the city dump/animal slum for discarded animals, returning finally with important knowledge of the landscape, recruited animals, and a mapped pathway to the city that only an eco-abled avian can achieve for her community.

Their potential commodification and consumption as industrialised products prove to be the greatest threat to Circle A farm's animals and the motor of the fable's most radical act of resistance by its animal inhabitants against a planned auction for the sale of land, equipment, and "livestock." The auction leaflet reads: "Every*thing* must go." (p. 90, emphasis added). Its objectifying language of de-individualisation evokes the a-relational anthropocentric mind-set of the auctioneers whose speciesism is characteristic of the present animal treatment via their treacherous language.

To resist the selling of the farm and its resident animals, a consensus-directed[12] community meeting is organised where all participate in decision-making, which leads to a plan to fight the corporation together with the kennel dogs, ranch horses, forest animals, and human environmentalists – a model to empower every individual group member (pp. 98–100). Physical resistance, storing food supplies, and building barricades are part of the plan as well as distributing leaflets among neighbours, a strike, and the ultimate public confrontation in the form of the Great Stampede.

Consensus-based grassroots democracy and communal solidarity in defence against the growing exterior threat combines with teach-ins for both the animal inhabitants of Doe's fable and its readers on a meta-level. For example, with the aim of activating Sabo's courage for the final farm fight and the Great Stampede, the elderly mother cow a.k.a. the Sacred Cow tells her a fable of a cat, a tiger, and a crippled fox as an allegory of emancipation from a former shepherd-flock stewardship and subordinate mind-set (Foucault, 1979). Sabo learns that she is not a passive crippled fox waiting to be fed but the hunting tiger that can nurture not only himself but also support others in mutual aid. Sabo's first step of self-empowerment is to revenge her friend Goldie's killing: As the golden retriever is being shot dead by a soldier, Sabo turns into a miniature tiger and attacks the military (pp. 179–180). As the State's war against Circle A farm escalates, other acts of multispecies solidarity occur in response to military attacks, such as "human forest defenders [tearing] their clothes into strips to bandage the injured dogs" (p. 178).

Despite passages of comical relief, the narrative leaves no doubt that forest and farm stand firstly for human versus animal war zones, especially when contextualised in an animal industrial complex discourse (Taylor, 2013). Freedom comes with human and non-human deaths, thereby shining a light on the dangers of direct actions carried out by the anti-deforestation activists (raccoons, deer, bears, human environmentalists) as during the attack on the logger company's equipment:

> To explode the big truck they needed extra dynamite. . . . The huge explosion caused metal pieces to fly in a much wider circle than they expected. . . . Many animals were injured and one, John Deer, was killed by a piece of the engine. . . . The flames leapt to the tent where the drugged guard lay sleeping. . . . he never woke up.
>
> (pp. 62–63)

Violence in this specific war zone seems unavoidable, but it is also deplored by body counting both the non-human and human victims. The fable is by no means a call to kill in order to overcome human-controlled animal exploitation. It rather sets an example of how a zooheterotopia can thrive as a safe haven for animals and their allies – until it becomes a threat to corporate profits – and highlights how violence can be state-driven and initiated by corporate lobbyists. Hence, animal resistance that started off as civil disobedience against the destruction of their environment and lives can be viewed as a necessary reaction to authoritarian state terror. Ironically, the animals' defence to protect their lives and homes is framed as terrorism when actually rebelling against exploitation and oppression as "strategically and specifically targeted direct action against corporate interests" (Kinna, 2005, p. 161).

Conclusion: Creating Liveable Spaces and Valuable Lives – Motivation and Guidance for Change?

This analysis has examined the fictionalisation of a farm's anarcho-communal transformation, and subsequent successful revolt over the corporate anthropolis, as offering motivation and guidance via a fable. A future of creaturely posthumanist co-existence can be negotiated in fictional zooheterotopian spaces (Miller, 2015). Drawing on Cornell (2016), who states that local prefigurative strategies provide oases of freedom and solidarity in a world marked by oppression, I argue that Circle A farm, and the fable as performative text, represent such oases. Doe's narrative attempts to convert and to delegitimise an urban establishment by modelling anarchist values in action within a specific local geography. Anarchists have often built prefigurative counter-institutions, such as edition houses, bookshops, activists groups, alternative schools, animal liberation initiatives, and DIY initiatives, from which to practice anarchism on a grassroots level in small anarchic local spots for smaller scale projects in participatory democratic formations (Cornell, 2016, p. 282, 284).

Fictional works such as Doe's fable can be discussed in classrooms and serve as small-scale interventions, providing empathetic and cognitive inspiration and guidance for motivating change in terms of animal liberation from injustices (Gruen, 2013; Nocella, 2012, 2015). Hence, the fable featuring anthropomorphised "messengers" can have signal function and serve as food for thought by offering a visionary platform for speculative thought experiments beyond literature. It invites critical thinking by creating re-imagined zooheterotopian spaces of relational multispecies network and mutuality. Cultural activities have been vital to anarchism, both in the sense of daily practices with shared significance and in the production of representations and meaningful "texts" (Cornell, 2016) to challenge society's conventional assumptions on personal freedom, equality, domination, and oppression of the Other. Thus, narrative space-making as cultural evidence production to show alternative counter-performances clearly references the fable genre's potential as activist platform for speculations (Harel, 2009; Schönbeck, 2019; Fudge, 2020) by employing a critical but instrumental anthropomorphism as communicative strategy (Taylor, 2011; Karlsson, 2012; Barcz, 2017). Fable narratives that draw from critical anthropomorphism can highlight shared characteristics across human and non-human animals, which "attract recognition beyond difference, and where commonalities and identification are resources for understanding" (Eisen, 2020, p. 154).

Doe's fable speculates on the contribution of direct grassroots action to systemic change in anarchic geographies (Springer, 2013, 2016). "Just" reforming the system within a protective paternalistic welfare culture is not sufficient according to *Anarchist Farm*. *Anarchist Farm* speaks to an imperative for active resistance anchored in civil disobedience and direct action on a micro-level. Consequently, this fable in particular functions as a political fiction-to-action call for animal justice. It may reflect back to the real world and inspire readerships to inform themselves on animal justice causes. Doe's fable as literary exemplification, motivation, and guidance adds itself to a discourse of antispeciesist animal narratives set against an orthodox anthropocentrism and unrestrained capital maximisation. Capitalism cannot infinitely continue unchecked. Change is only possible by resetting individual mindsets within a posthumanist political systemic reload informed by innovative alternative thinking (traditionally a strong point of anarchism). In the context of Cornell's (2016) emphasis on cultural activities as means of reimagination and deconstruction, this fable and the introduction of more "animal literature" into classrooms and academic curricula (Harel, 2009; McHugh, 2009; Borgards, 2015; Ortiz-Robles, 2016) can propagate a reassessment of cultural categorisations of animals. This aligns also with Rouhani's demand for educational activities to disperse knowledge on animal sentience and freedom from a posthumanist point of view (2012).

Autonomous, real-world micro-geographies in an anarchistic sense of "collectivist, non-capitalist and anti-normative forms of solidarity and affinity" (Springer, 2013, p. 53) can be motivated by animal literature through an affective embrace of readers. Such models, for example Rewilding Europe, attempt to leave animals alone by rewilding once agriculture-d land, and re-introducing free-living animals. On a micro-level, implemented by small-scale communities as lighthouse projects, such efforts could be a first step to animal independence, albeit acknowledging their multiple challenges.[13] Perhaps the key question is what animals themselves seem to value most (Eisen, 2019, p. 152). Should we humans recreate spaces where animals could flourish and develop their own communities in the truest sense of animal and earth liberation? (Cudworth & Hobden, 2018b, p. 92). Anarchism as a philosophy and methodology continues to be innovative, positively deviant, and instrumental in resisting normative regulations. It includes thinking animals outside human domination (Rouhani, 2012; Springer, 2013; Ortiz-Robles, 2016) and using antispeciesist literature to envision zooheterotopian anarcho-geographies that can plant the seeds of animal liberation into the minds of future generations.

Notes

1 The idea of a Post-Anthropocene, or Chthulucene (an epoch in which humans and non-humans would be linked in tentacular network practices, Haraway, 2016), suggests a future era of co-emerging communities, in which humans cannot escape their responsibility to the more than human world, and must acknowledge their duty to renegotiate human exceptionalism as measured against the reality to be one species among multiple species in interrelated eco-systems.

2 Agency: a model with an agent as autonomous entity that interacts with others and can influence a situation and possesses desires, intentions, and a will (Latour, 2007).

3 *Anarchist Farm* is a spin-off fable of George Orwell's Stalinist dystopia *Animal Farm*.

4 Undisclosed US-American author – Jane Doe is a pseudonym.

5 The upper two legs of the former Circle 'H' farm (denoting farmer Margaret Hansen's ownership of the land, animals, and property) have been hammered into an A-shape by monkey Bonkers. Circle A farm's letter A is encircled by a metal frame and fixed to the farm's entrance gate. The emblematic letter is turned into the iconic abbreviation of anarchism.

6 See standardized cartographies of anonymous de-individualized animals divided into meat body parts (Arcari, 2019, p. 177).

7 The terminology used here is based on Western-European scholarly perspectives and traditions. It does not encompass alternative non-Western thought systems, for example Indigenous conceptualisations of human and other-than-human relationalities.

8 The turkey's name is an allusion to anarchist-feminist Emma Goldman.

9 See forerunners Foreman (1985), Abbey (1976), and Powell (1971), who justified sabotage, notably damage to machines to preserve ecosystems.

10 Nocella II (2016) merges Disabilities Studies and Critical Animal Studies into a productive approach that celebrates collaboration and difference to challenge the human-non-human animal divide.

11 Merriam Webster Online Dictionary (2024, March 4). *Anarchism*. www.merriam-webster. com/dictionary/anarchism

12 Consensus finding is a procedure to exercise collective power and a deeper form of democracy (Pellow, 2014, pp. 104–105.)

13 Anarchist movements have continuously been marginalized phenomena with small monetary resources due to their state-critical decentralized antiauthoritarian principles (Cornell, 2016, p. 288) and independent grassroots character. This suggests shortcomings in a broad acceptance of anarchist ideas within societies which often equate anarchism with chaos.

References

Abbey, E. (1976). *The Monkey Wrench Gang*. Avon.

Araujo, E. (2016). What Do We Resist When We Resist the State? In M. L. de Souza, R. J. White, & S. Springer (Eds.), *Theories of Resistance* (pp. 79–100). Rowman & Littlefield.

Arcari, P. (2019). The Ethical Masquerade: (Un)masking Mechanisms of Power Behind 'Ethical' Meat. In M. Phillipov & K. Kirkwood (Eds.), *Alternative Food Politics – From the Margins to the Mainstream* (pp. 169–190). Routledge.

Barcz, A. (2017). *Animal Narratives and Culture: Vulnerable Realism*. Cambridge Scholars Publishing.

Borgards, R. (2015). Introduction: Cultural and Literary Animal Studies. *Journal of Literary Theory*, *9*(2), 155–160.

Bremhorst, A., Caeiro, C., Zamansky, A., & Karl, S. (2023). Animal Sentience, Animal Emotions, and Human-Animal Interactions. In A. H. Fine, M. K. Mueller, & N. Y. Zenithson (Eds.), *The Routledge International Handbook of Human-Animal Interactions and Anthrozoology* (pp. 169–182). Routledge.

Cornell, A. (2016). *Unruly Equality: U.S. Anarchism in the Twentieth Century*. University of California Press.

Cudworth, E., & Hobden, S. (2015). The Posthuman Way of War. *Security Dialogue*, *46*(6), 513–529.

Cudworth, E., & Hobden, S. (2018a). *The Emancipatory Project of Posthumanism*. Taylor & Francis Group.

Cudworth, E., & Hobden, S. (2018b). Anarchism's Posthuman Future. *Anarchist Studies*, *26*(1), 79–104.

Derrida, J. (2002). The Animal that Therefore I Am (More to Follow). *Critical Inquiry*, *28*(2), 369–418.

de Souza, M. L., White, R. J., & Springer, S. (Eds.). (2016). *Theories of Resistance: Anarchism, Geography, and the Spirit of Revolt*. Rowman & Littlefield.

Doe, J. (1996). *Anarchist Farm*. III Publishing.

Dominick, B. (2015). Anarcho-Veganism Revisited. In A. J. Nocella II, R. J. White, & E. Cudworth (Eds.), *Anarchism and Animal Liberation: Essays on Complementary Elements of Total Liberation* (pp. 23–39). McFarland & Company.

Eisen, J. (2020). Down on the Farm: Status, Exploitation, and Agricultural Exceptionalism. In C. Blattner, K. Coulter, & W. Kymlicka (Eds.), *Animal Labour: A New Frontier of Interspecies Justice?* (pp. 140–159). Oxford University Press.

Foreman, D. (Ed.). (1985). *Ecodefense: A Field Guide to Monkeywrenching*. Earth First! Books.

Foucault, M. (1977). *Discipline and Punish: Birth of the Prison*. Vintage.

Foucault, M. (1979). Omnes et Singulatim: Towards a Criticism in 'Political Reason'. *The Tanner Lectures on Human Values*, Stanford University, October 10 & 16, pp. 225–254. https://tannerlectures.utah.edu/_resources/documents/a-to-z/f/foucault81.pdf

Foucault, M. (1986). Of Other Spaces (J. Miskowiec, Trans.). *Diacritics*, *16*(1), 22–27.

Fudge, E. (2002). *Perceiving Animals: Humans and Beasts in Early Modern English Culture*. University of Illinois Press.

Fudge, E. (2020). What Can Beast Fables Do in Literary Animal Studies? Ben Jonson's Volpone and the Prehumanist Human. In S. McHugh, R. McKay, & J. Miller. (Eds.), *The Palgrave Handbook of Animals and Literature* (pp. 195–208). Palgrave Macmillan.

Goldman, E. (1934). *Living My Life*. Knopf.

Gruen, L. (2013). Entangled Empathy: An Alternative Approach to Animal Ethics. In R. Corbey & A. Lanjouw (Eds.), *The Politics of Species. Reshaping our Relationships with Other Animals* (pp. 223–240). Cambridge University Press.

Haraway, D. (2016). *Staying with the Trouble: Making Kin in the Chthulucene*. Duke University Press.

Harel, N. (2009). The Animal Voice Behind the Animal Fable. *Journal for Critical Animal Studies*, *VII*(2), 1–20.

Karlsson, F. (2012). Critical Anthropomorphism and Animal Ethics. *Journal of Agricultural and Environmental Ethics*, *25*(5), 707–720.

Kinna, R. (2005). *Anarchism a Beginner's Guide*. Oneworld.

Kropotkin, P. (1998[1902]). *Mutual Aid*. Freedom Press.

Latour, B., & Roßler, G. (2007). *Eine neue Soziologie für eine neue Gesellschaft: Einführung in die Akteur-Netzwerk-Theorie*. Suhrkamp.

Lawson, V. (2007). Geographies of Care and Responsibility. *Annals of the Association of American Geographers*, *97*(1), 1–11.

McFarland, S. E., & Hediger, R. (Eds.). (2009). *Animals and Agency: An Interdisciplinary Exploration*. Brill.

McHugh, S. (2009). Literary Animal Agents. *PMLA*, *124*(2), 487–495.

Miller, J. (2015). Zooheterotopias. In M. Palladino, & J. Miller (Eds.), *The Globalization of Space: Foucault and Heterotopia* (pp. 149–164). Pickering & Chatto.

Nocella II, A. J. (2016). Disability, Animals, and Earth Liberation: Eco-Ability and Ableism in the Animal Advocacy Movement. In J. Castricano & L. Corman (Eds.), *Animal Subjects 2.0: An Ethical Reader in a Posthuman World* (pp. 325–345). Wilfrid Laurier Press.

Nocella II, A. J., Bentley, J., & Duncan, J. M. (Eds.). (2012). *Earth, Animal, and Disability Liberation: The Rise of the Eco-Ability Movement*. Peter Lang.

Nocella II, A. J., White, R. J., & Cudworth, E. (Eds.). (2015). *Anarchism and Animal Liberation. Essays on Complementary Elements of Total Liberation*. McFarland & Company.

Online Merriam Webster Dictionary. (2024). *Anarchism*, March 4. www.merriam-webster.com/dictionary/anarchism

Ortiz-Robles, M. (2016). *Literature and Animals Studies*. Routledge.

Pellow, D. N. (2014). *Total Liberation. The Power and Promise of Animal Rights and the Radical Earth Movement*. The University of Minnesota Press.

Powell, W. (1971). *The Anarchist Cookbook*. Stuart.

Ramnath, M. (2011). *Decolonizing Anarchism: An Antiauthoritarian History of India's Liberation Struggle*. AK Press.

Reclus, E. (1901). *On Vegetarianism*. https://theanarchistlibrary.org/library/elisee-reclus-on-vegetarianism.

Rouhani, F. (2012). Practice What You Teach: Facilitating Anarchism In and Out of the Classroom. *Antipode*, *44*, 1726–1741.

Roscher, M. (2016). Zwischen Wirkungsmacht und Handlungsmacht: Sozialgeschichtliche Perspektiven auf tierliche Agency. In S. Wirth, A. Laue, & M. Kurth (Eds.), *Das Handeln der Tiere: Tierliche Agency im Fokus der Human-Animal Studies* (pp. 43–66). transcript Verlag.

Schönbeck, S. (2019). Return to the Fable. Rethinking a Genre Neglected in Animal Studies and Ecocriticism. In F. Middelhoff, S. Schönbeck, & R. Borgards (Eds.), *Texts, Animals, Environments: Zoopoetics and Ecopoetics* (pp. 111–125). Rombach.

Simons, J. (2002). *Animal Rights and the Politics of Literary Representation*. Palgrave.

Springer, S. (2013). Anarchism and Geography: A Brief Genealogy of Anarchist Geographies. *Geography Compass*, *7*(1), 46–60.

Stucki, S., & Kurki, V. (2020). Animal Rights. In M. Sellers & S. Kirste (Eds.), *Encyclopedia of the Philosophy of Law and Social Philosophy*. Springer.

Taylor, C. (2013). Foucault and Critical Animal Studies: Genealogies of Agricultural Power. *Philosophy Compass*, *8*(6), 539–551.

Taylor, N. (2011). Anthropomorphism and the Animal Subject. In R. Boddice (Ed.), *Anthropocentrism* (pp. 265–280). Brill.

Wadiwel, D. J. (2015). *The War Against Animals*. Brill.

Webster, J. (2022). *Animal Welfare: Understanding Sentient Minds and Why It Matters*. Wiley Blackwell.

Wirth, S., Laue, A., Kurth, M., Dornenzweig, K., Bossert, L., & Balgar, K. (Eds.). (2016). *Das Handeln der Tiere: Tierliche Agency im Fokus der Human-Animal Studies*. transcript Verlag.

Zaharijević, A. (2021). On Butler's Theory of Agency. In A. Halsema, K. Kwastek, & R. van den Oever (Eds.), *Bodies that Still Matter. Resonances of the Work of Judith Butler* (pp. 21–30). Amsterdam University Press.

Chapter 10

The Post-human Ontology of Gothic Enviro-toons

Defying Anthropo-denial in *Watership Down*, *The Plague Dogs* and *Padak*

Sutirtho Roy

Introduction

Since the mid-twentieth century, a certain category of animated films called the enviro-toon (James Weinman, 2004) has become an increasingly dominant mode for expressing concerns and generating awareness about animal welfare. Characterised by strong ecological subtexts, these stories elicit meta-narrative empathy for non-human suffering by creating heterotopic spaces where a counter-gaze to anthropocentric negligence and oppression is envisaged through speculative means. Heterotopias[1] in enviro-toons take many forms, ranging from the ocean and aquarium of *Finding Nemo* (2003) to the rainforests of *Rio* (2011) and *Ferngully: The Last Rainforest* (1992), and the laboratory in *The Plague Dogs* (1982) to the rural landscape of *Watership Down* (1976). Nor are such portrayals limited to Western enviro-toons. The seafood restaurant in the Korean indie animation *Padak* (2012), the magical locales envisaged in the Japanese film *Princess Mononoke* (1997), and the towns in the Indian animation *Delhi Safari* (2012) also become fictional heterotopias which emerge as sites that celebrate non-human resistance and liberation by subverting the anthropocentric gaze.

This chapter explores the potential for a certain category of enviro-toon, which I term the Gothic enviro-toon, to create heterotopic spaces which discursively challenge anthropocentric ways of perceiving the world. Drawing on three films, namely *Watership Down*, *The Plague Dogs*, and *Padak*, a combination of textual analysis, critical readings, and audience reception (using IMDb) is used to gauge these films' potential to generate awareness about non-human animal pain under anthropocentric oppression and cruelty, and to thereby counter anthropo-denial. In doing so, the chapter assesses the extent to which shock tactics and jeremiadic modes of storytelling can conceptualise non-human justice that highlights human oppression of other animals, non-human agency, and humankind's responsibility towards non-humans. Before addressing each film in more detail, I describe common features of enviro-toons and the factors that distinguish Gothic enviro-toons, while highlighting the themes and theories that inform the subsequent analysis.

DOI: 10.4324/9781003366706-14

Anthropomorphic Non-humans in Gothic Heterotopias

An effective enviro-toon uses the animated medium's penchant for modifying and anthropomorphising non-human bodies on screen (Heise, 2014, p. 304) to maximise "their cuteness, their resemblance to human children" (Barrier, 2008, p. 150). As such, enviro-toons can use anthropomorphism in two ways: 1. to play upon the subconscious appeal of infantile features in the minds of perceivers and 2. to place the non-human animals in a liminal space between the Cartesian categories of 'human' and 'animal' through the superimposition of "humanized faces, dialects, languages, emotions and postures on non-human creatures, enabling human beings to empathize with the[ir] fictional plight" (Roy, 2022, p. 231). The animated bodies become important semiotic devices that portray non-human animals as sentient and expressive subjects possessing complex affective states as opposed to blank slates that are simply "good to think with" (Lévi-Strauss et al., 1969).[2]

The enviro-toon thus becomes the most obvious visual adaptation of the literary animal autobiography (Herman, 2016, pp. 1–17) while addressing Anthropocenic human/non-human conflicts. Here, anthropomorphism becomes a useful tool to create an empathic human approximation of an animal's perspective and skills (Wells, 2009, p. 53), even if the same is done for different ends through the anthropocentric gaze. Of the inherent paradox in humans representing the non-human, biologist Marc Bekoff writes:

> As humans studying other animals we cannot totally lose our anthropocentric perspective. But we can try as hard as possible to combine the animals' viewpoints to the ways in which we study, describe, interpret and explain their behaviour.
>
> (2004, p. 74)

However, the use of weak anthropomorphic[3] features can humanise non-human pain and affective states without erasing their non-human alterity or Otherness. For instance, the use of weak anthropomorphism ensures that viewers' sympathies remain with *Bambi's* woodland creatures as they "hear gunshots that shatter the forest calm and . . . experience the loss of a mother along with Bambi" (Murray & Heumann, 2011, p. 43). By inviting audiences to experience the world through an orphaned fawn's eyes in a space where *Homo sapiens* emerge as monsters, the forest is transformed into a deeply unsettling heterotopia.

Similarly unsettling heterotopic spaces are established in *Finding Nemo*, where young clownfish Nemo is captured by a gigantic diver (magnified on screen to portray the terrifying size difference), and in the opening scenes of *Rio*, where, after falling onto the forest floor, Blu, a young Spix's macaw, experiences the metallic bars of a cage clamp down around him with a resounding *clang*. This genre of enviro-toons, situated against a backdrop of Anthropocenic biodiversity loss and the climate crisis, lends itself to Gothic fictional narratives (Smith & Hughes, 2006) that lurk beneath the comedic lightheartedness, and can create tangible impacts for the depicted (anthropomorphised) non-human protagonists from whose perspectives the story is told.[4] However, most of these films still embrace the idea of a happy ending, with the journeys of Bambi, Nemo, and Blu taking the form of a hero's journey culminating in a triumph over enemies and reunion with loved ones. Hence, despite the ecocritical significance of these enviro-toons (see for example Preiser et al., 2017), Rebecca Rose Stanton argues, "representations of animals in animation are romanticized for the sole benefit of humans" and are thus anthropocentric (Stanton, 2020, n.p.).

In contrast, a smaller number of enviro-toons abandon lightheartedness and/ or romanticisation altogether, choosing instead to establish almost entirely Gothic heterotopias.[5] While this study analyses different animated narratives which typically lack the conventionally happy ending and are characterised by more character deaths and portrayals of blood and violence than the typical animated fare, what I term the 'Gothic enviro-toon' is not a homogeneous category. The distinction between a Gothic and non-Gothic enviro-toon is not always clear. For example, both of Martin Rosen's animated films *Watership Down* and *The Plague Dogs*, adapted from Richard Adams' novels of the same titles, narrate stories of non-human resilience. *Watership Down* tells the story of a group of rabbits guided by a prophetic vision to find a new home after anthropogenic activities destroy their warren. These rabbits are plagued by dangers, with threats looming around every corner. Similarly, in *The Plague Dogs*, the titular canines, Rowf and Snitter, escape from a government animal testing facility and remain on the run from military forces while searching for a place to live.

However, while *Watership Down* builds an idyllic picture of nature and non-humans and features a happy ending that includes a dignified death for the main non-human protagonist Hazel, the heterotopic space established in *The Plague Dogs* leaves Rowf's and Snitter's fate darkly ambiguous as they swim off into the mist while a barrage of bullets rains down on them. This chapter explores these semiotic aspects of *Watership Down* and *The Plague Dogs* because, despite being borne of the same authorial and directorial minds, the stories envisage different degrees of Gothic-ness on screen and have achieved differing degrees of box-office success. By reading into the portrayal of non-human animals as well as human/non-human interactions, the chapter also gauges the scope of non-human liberation envisioned in these films, as suggested by their impact on viewers.

The aforementioned aspects of Gothic storytelling are not limited to Western enviro-toons, as illustrated by *Padak*, also known as *Swimming to Sea*. Narrating the titular mackerel fish's ill-fated attempts to escape from an aquarium in a Korean seafood restaurant, the story (and the film's brightly coloured poster depicting the titular fish leaping out of the water in ecstasy) bears similarities to *Finding Nemo*. However, the story subverts such similarities through its realistic and macabre ending in the protagonist's gruesome demise. The narrative has three major characters: upbeat mackerel protagonist *Padak*, innocent greenling Spotty, and grumpy flatfish Master, leader of their shared aquarium tank. Unlike *Finding*

Nemo or even Rosen's films, *Padak* has received little to no critical scholarship. This, and the limited scholarship on Gothic enviro-toons, make it difficult to gauge the genre's effectiveness in using the portrayal of non-human pain, shock, and/ or trauma to unsettle audiences' normative anthropocentric modes of perceiving human/non-human relationships.

Regarding the use of such pain and trauma in non-human advocacy, Fernandez highlights "the extensive possibilities of moral shock to break with the silencing of speciesism and nonhuman animals' voices by reaching people's emotions" (2020, p. 74). Wrenn also notes the potential of graphic depictions of violence to generate empathy for non-human animals (2013, pp. 390–392). However, both scholars contend that such shock tactics are effective only in limited frameworks and contexts. In contrast, Slovic argues the shock effect of jeremiadic or elegiac storytelling in ecological narratives can lead to immediate, short-lived awakenings among audiences, although for long-lasting impacts, a celebratory tone is often needed that offers hope towards the salvation of non-humans and the environment (1996, p. 84).

Despite the arguable ephemerality of their impact and limited academic reception, the move towards more Gothic narrative approaches to enviro-toons has been more prevalent in the twenty-first century. All three films have high IMDb scores with *The Plague Dogs* at 7.7 (101 reviews), Watership Down at 7.6 (240 reviews) and *Padak* at 7.4 (10 reviews), attesting to their popularity among these audiences,[6] even if those audiences might constitute a more cultic fanbase.[7] As such, the following sections analyse the shock effect of these films as well as the lament for non-human life they envision. In doing so, the chapter assesses the degree to which a jeremiadic mode of storytelling including moral shocks can generate awareness towards non-human justice via the Gothic enviro-toon.

Watership Down

Watership Down[8] uses weak anthropomorphism to portray non-human animals in a way that attempts to preserve interspecies differences while conveying an idea of non-human "intentionality, subjectivity [and] communicativity" (Plumwood, 2002, p. 56). As such, the film portrays rabbits as per the original author's intention, as resourceful, brave creatures desperately struggling to survive and thrive in a world fraught with dangers, as opposed to vehicles of cartoonish whimsy such as Brer Rabbit and Peter Rabbit (Adams, cited in Meyer, 1994; Jordan, 1983). This makes the ecological theme of their journey more poignant, as the small animals face insurmountable odds (including wild predators, dogs, gunmen, and rival warrens) while fleeing their home for a new place to live. With rabbits portrayed as complex beings akin to humans, the narrative of *Watership Down* questions the ethics of expanding upon non-human spaces and habitations without consent, conceiving the destruction of the protagonists' home as a form of colonisation. This emphasises the notion of anthropocentrism as emerging from a colonial speciesist mindset (Huggan & Tiffin, 2010).

The titular location of *Watership Down* is the only place where the rabbits can live largely free from the impact of human activities. The film visually depicts Adams' original idyll:

> Few places are far from human noise – cars, buses, motorcycles, tractors, lorries. . . . But here, on Watership Down, there floated up only faint traces of the daylight noise below.
>
> (Adams, 1972, p. 139)

Interspecies contact zones (Haraway, 2007, p. 216) between humans and rabbits in *Watership Down* become Gothic heterotopias where the non-human animals become refugees on the run, struggling to escape the seemingly omnipotent and all-encompassing might of humanity.

During the rabbits' journey to Watership Down, they are chased by a domestic dog, doe Violet is picked off by a hawk, and the group barely survives a run-in with a fox. At one point, Hazel is shot and although he survives, the sound of the bullet, his pained expression, and near-death experience create a Gothic heterotopic space where viewers see through his eyes, realising the inescapable ruthlessness of human actions. Humankind is depicted as

> an incomprehensible and corrupting force of cosmic horror . . . selfish and disinterested gods who disfigure environments and ecologies, bringing death, destruction, and madness to the rest of the world for the sake of a few pelts, a cabbage patch, and apartment complex.
>
> (Botts & Sipos, 2022, n.p.)

This is made evident in an exchange between the terrified rabbits after their journey:

Blackberry:	Men have always hated us.
Captain Holly:	No. They just destroyed the warren because we were in their way.
Fiver:	They will never rest until they've spoilt the earth. (Rosen, 1978)

The film adaptation of *Watership Down* incorporates not only Adams' knowledge of the work of British naturalist Ronald Lockley on rabbit behaviour (Pennington, 1991; Meyer, 1994) but also his experiences as an animal rights activist (Mullins, 2016).

For example, Captain Holly's[9] narration of the rabbits' deaths in Sandleford Warren begins with his flashback from inside the burrows as a construction crew starts filling them in, coupled with the sound and sight of falling rubble which gradually obliterates the rabbits' vision. This horrific scene shifts to ghastly whispers of buried rabbits rising to a haunting din as they slowly suffocate – their red, desperate eyes set against a pale ghostly blue. A giant rake scores the ground with bright red tracks, symbolising the bloody scourge of the earth and its non-human inhabitants. These visuals are paired with the Captain's agonised narration in short, traumatised sentences:

Men came. Filled in the burrows. Couldn't get out. There was a strange sound. Hissing. The air turned bad. Runs blocked with dead bodies. I couldn't get out. Everything turned mad. Warren, herbs, roots, grass, all pushed into the earth.

(Rosen, 1978)

Humans are distinguished from other animals in Captain Holly's frustrated words: "Animals don't behave like men. If they have to fight, they fight. If they have to kill, they kill. But they don't set their wits to devise ways of hurting other creatures" (Rosen, 1978). While this statement seemingly reinforces the human/animal binary even if the same is done to villainise human exceptionalism, it nevertheless paints an authentic picture of the incomprehensibility of human selfishness and greed as perceived by these rabbits.

Such depictions of violence featuring non-human animals have led *Watership Down* to be embroiled in controversy owing to its impact on young viewers (Stanton, 2021). Critic/film reviewer Margaret Hinxman believes she might have enjoyed the film more had it been Disney-esque (Hinxman, 1978), whereas fellow critic Walter Chaw argues "the picture may be *more* appropriate for young children than a legion of condescendingly sugar coated Disney fare" (2002, emphasis added). These contradictory viewpoints highlight the on-screen prevalence of the Disney Syndrome, or disnification.[10] While originating from Disney animated features, the Disney Syndrome or disnification now refers to a stylistic choice across the majority of big-budget animated features to portray cute animals with different degrees of anthropomorphism. This choice often aligns with a directorial desire to create a family-friendly film and validates the common consensus about animated films being primarily geared for younger audiences.

Despite feelings of discomfort and shock that *Watership Down* might elicit among younger audiences, it is also acknowledged that the film "encourage[s] empathy for [animals'] wellbeing" (adrian_rawling, 2022), and "asks us to spend time with those elements of existence that we will always find most troubling (and haunting and moving), and that we so rarely allow our children's culture or our own entertainment to dwell on" (Jones, 2015). The importance of negotiating with troubling elements of existence bears even greater relevance in the Anthropocene with its massive loss of non-human life. The horror experienced by the rabbits as they interact with humans or their machinery creates a heterotopic space where human negligence casts them as a disturbing and malevolent force before the anthropomorphic rabbits with whom the viewers empathise.

Based on 240 viewer reviews, *Watership Down* has been simultaneously praised for its elevated storytelling style and raw character moments, and critiqued for its realism and depiction of blood and gore. However, with a score of 7.6 out of 10, an overwhelming number of reviews are positive and the film was not only a box office success but its heterotopic spaces achieved a tangible impact for non-human advocacy, as emphasised by viewer ed321:

> Here is a movie I saw as a child that has stayed with me all my life and even gave me some direction in life (I am a huge animal rights welfare supporter).
>
> (ed321, 2018)

The Plague Dogs

Despite being "resurrected recently as a cult favourite by audiences who appreciate it as a unique and original vision in animated film history" (Kozak, 2022) and often spoken of in the same breath as *Watership Down*, *The Plague Dogs* bombed at the box office. Partially attributed to receiving no distributors for its initial release and negative reviews, this failure could also be attributed to the absence of a happy ending.

Reviews of *The Plague Dogs* on IMDb often compare it with *Watership Down*, mostly reaffirming the former's bleakness – indeed, "so relentlessly bleak that it makes *Watership Down* look like *The Rescuers Down Under*" (zetes, 2018). Another user, who rates the film 10/10, remarks that neither the book nor the film version received the same levels of critical attention or success as *Watership Down*, attributing this to an intentional lack of subtlety:

> While *Watership Down* hid violence and severe socio-political criticism behind a disguise of a children's tale, *The Plague Dogs* is much more in-your-face, much less subtle, and makes no attempts to hide itself behind pretty words. *The Plague Dogs* is a tragic tale that is mercilessly critical toward modern society, taking a strict stand on the subject of cruelty to animals.
>
> (itsmarscomix, 2005)

The same reviewer suggests it might be difficult to conceive that an animated film could be solely for adults and concedes that the realistic depiction of animal cruelty is unsuitable for children.

Most viewers connect this animated feature with real-world issues about animal cruelty and experimentation. One viewer remarks that the tribulations of the fictional dogs searching for a loving home will cause one to develop a greater appreciation of their own dogs; to "hug [their] pets tight and be glad they wound up as [their] pets" (a_fox_tabatha, 2005). This highlights the potency of *The Plague Dogs*, and the Gothic enviro-toon in general, as a storytelling medium that elicits empathy for non-humans without disnifying nonhuman bodies (Mullan & Marvin, 1987). Anna Hoing points to the successful use of the Gothic medium in *The Plague Dogs* which:

> harnesses the narrative power inherent to the Gothic to its animal rights agenda, using the Gothic's ability to unsettle its audience in order to forcefully draw to

the reader's attention that the animal that really matters is the real animal, and that humans averted their gaze from this animal for too long.

(Hoing, 2021, p. 71)

From the beginning, the *Plague Dogs* sets the tone for an anti-anthropocentric narrative, depicting Rowf, a large black dog, being drowned for supposedly experimental reasons in a tub of water. Rowf and fox terrier Snitter soon escape, and their journey is represented in black, white, and grey colour palettes, aligning with Gothic aesthetics that paint the dogs' world as bleak and unforgiving.

The Gothic recurs in a scene of Snitter's seizures, hallucinations and guilt-induced trauma after being responsible for the deaths of the two human characters who had shown compassion for the dogs. Gothic elements manifest in eerie silences and sudden noises, and in close-ups of haunted monkeys and rabbits in the laboratory. The Gothic returns more violently in a graphic, bloody scene where Snitter mistakenly causes a gun to go off in a farmer's face, and in the death of a man who trips off a cliff. An aerial shot pans over the man's mangled corpse, hinting that the starving dogs have feasted on him. To the humans on screen, the dogs represent the monstrous Other of a Gothic space who carry the plague and eat human flesh. However, the film portrays humans as the true monsters, whose overarching power reaches out in negligently cruel manners, either intentionally or as the result of normalised and invisibilised violence. As such, the dog's journey is marked by Gothic heterotopias that shock audience sensibilities and unsettle the anthropocentric gaze. Even if non-human animals are not liberated in these spaces, the portrayal of their pain and lack of agency act as a clarion call for the liberation of non-humans in the real world. As such, Anna Hoing explains, the film "invite[s] reflections on the processes that make us demonise an 'other'" (2021, p. 71).

Elements of Gothic storytelling are also etched into the animated physiques of the film's characters – Rowf, with his shaggy fur as a result of starvation and constant drowning, and Snitter with his disfigured, scarred head as a result of vivisection. Rather than adhering to anthropocentric notions of what companion animals should be, Rowf and Snitter are portrayed as feral. Rowf is further alienated because of his blackness, playing upon subconscious and racially charged notions of the dark 'other.'[11] His black fur camouflages him against the sombre and dark night-time background, and renders his facial expressions impossible to see (Thomas and Johnston, 1981, p. 268), making him difficult to connect with and subverting the tendency towards disnification.

This consciously anti-anthropomorphic tendency helps Rosen blur the line between conventional usages of light and dark colour palettes in animated films to assign moral categories of good and evil (Hoing & Husemann, 2016, p. 106). For example, one scene depicts Rowf approaching imaginary monsters in the distance. The camera focuses on the pitch-black dog walking towards these monsters with his teeth bared and closes in on his demonising red eyes (Rosen, 1982). However, the sympathies of the author, director, narrator, and audience are with Rowf, and the struggles he has endured, establishing a Gothic heterotopia where the normative order is upended. Instead of villainising the black-furred dog, traditionally associated with evil and bad luck in European folklore (Naish, 2022), he is depicted as a resilient and heroic survivor. Through these interactions, the film critiques dominant anthropocentric understandings of dog's innate duties and "the famous 'mutual bond' between human and dog" (Hoing, 2016, p. 70).

Before escaping from the laboratory, Rowf utters the words, "I'm not a bad dog, Snitter," as if seeking to justify his escape and cope with the guilt of prioritising himself over the cruel actions of government officials constructed as leaders of his pack. To this, Snitter replies, "They're not masters. I had a master once, and I know whatever the white coats are, they are not masters." In Snitter's traumatised mind, the status of the master is afforded to only his own previous master, who, in contrast to the 'whitecoats,' is characterised as benevolent. While the dogs attempt to gain autonomy over their lives, the limits to their agency, their capacity to change their fates, and the redundancy of the 'companion animal' bond are made painfully clear. The heterotopic spaces established in the film continually offer audiences hope in the dogs' liberation before snatching it away, thereby turning a mirror to the real world, forcing audiences to acknowledge the non-humans who suffer from similar pain, and to imagine an alternative future where other animals are free of anthropocentric oppression.

Being a 'good dog' is still shown as being important to the canine protagonists, but what Rosen depicts as the dogs' sense of duty that drives them to befriend human beings constantly leads to misfortune. Furthermore, the dogs' physical attempts to escape humans are mirrored in their subconscious efforts to undo their species' domestication via a return to the 'wild' lifestyle of their ancestors. The narrative shows the dogs progressing through this de-domestication to become distinctly un-doglike (or un-companion species-like), as they kill (rather than protect) sheep and mistakenly kill a human and apparently consume their flesh, thereby committing taboo and heinous acts beyond what even a 'bad' dog would do. In one scene, Rowf bites a man's hand while trying to save Snitter, visually and metaphorically signalling his transformation into a wild, undomesticated animal that can bite the hand that once fed it. The scars inflicted on the canine characters extend beyond the physical to include Rowf's emotional trauma associated with repeated drowning and Snitter's hallucinations – the result of brain surgery – which cause him to blur the line between the real and the imaginary.[12]

Snitter is constantly fearful of his surroundings and cannot correctly process sensory information. The fox-terrier's trauma-riddled, blurry perceptions cause him to interpret his visions in ways that defy real incidents. His unreliable narration is thereby used to emphasise his Otherness, and to offer a discursive, non-human counter-gaze to anthropocentric oppression (Hoing, 2017, p. 11). The realism of Snitter's visions and their closeness to real-world animal experimentations makes the film's final scene more potent, enhancing its Gothic nature.

Snitter sees an island in the middle of the sea and convinces Rowf to swim with him towards it. Deviating from the novel's happy ending where the dogs are rescued from drowning before being reunited with Snitter's master, the film has them disappear into the mist with the implication that they drown and find peace in death. While their death can be poignant, Snitter and Rowf die because their attempt to escape humankind fails, as opposed to the heroic passing-away of loyal and dutiful canine companions.

Regarding the seeming failure of *The Plague Dogs* and its comparatively limited reception when compared to less Gothic enviro-toons or stories featuring animals, Hoing and Husemann observe:

> Audiences may not want to witness images of realistic, suffering non/human animals aimlessly wandering in a bleak world in desperate need of human aid that will never come. Viewers . . . demand their ideological foundations to be affirmed, not attacked.
>
> (2016, p. 113)

As such, while making use of moral shock through the imagery of graphic suffering like *Watership Down* (albeit to a greater degree), *The Plague Dogs* does not offer any form of hope. As such, its limited success may be because audiences found neither a happy ending to indulge in, nor any sense of optimism pointing to an alternative where 'all is well.' Hoing and Husemann posit there might be a risk in attacking viewers' expectations through moral shock, especially in the absence of an alternative and more optimistic possibility to celebrate.

Padak

Contrary to *The Plague Dogs*, it is being taken *out* of the water – specifically a tank located in a Korean restaurant – that leads to the demise of *Padak's* central character, filleted alive as she gasps for breath. *Padak*, which translates as 'Swimming to the Sea,' was a little-known film to even avid animation fans before it was given prominence by YouTube content creators.[13] Even now, it has garnered only 10 reviews on IMDb[14] with many identifying its similarity to *Finding Nemo*. Indeed, the movie's poster features *Padak* nose-to-nose with a clown fish, suggesting its deliberate positioning as an un-disnified counter-gaze to Pixar's family-friendly tale.

The film follows three fishes who inhabit an aquarium tank in a sushi restaurant, from where they view the goings-on and see their fellow fish being served up, not knowing when they might be next. The story is narrated from the fish's perspective, whereby the tank is transformed into a Gothic heterotopia with humans as anthropocentric monsters whose whimsies and taste buds dictate the fishes' fates. Padak arrives in the tank with hope of liberation and urges them all to escape with him to the sea. However, like *The Plague Dogs*, this hope is repeatedly dashed.

The Gothic-ness of the aquarium is accentuated by muted palettes and a sense of hopelessness brought on by the fish's realisation of their impending demise. Unlike the strongly anthropomorphised animals of *Finding Nemo* and other Disney films, who espouse strong humanistic moral values and whose optimism and resilience are rewarded, *Padak* paints a bleak portrait of a world where the fish have little to no agency, and no rewards for being hopeful. Consigned to the murky waters of the aquarium, they are unable to reach the open waters located a few meters away.[15] Instead of the brightly coloured exotic members of *Finding Nemo's* Tank Gang, *Padak's* fish are depicted in dull, muted tones, and derive their sustenance from dead and dying bodies dropped into their tank. The lead characters, Padak, Spotty, and Master, serve respectively as subversive echoes of *Finding Nemo's* Dory, Nemo, and Gill the Tank Leader. Indeed, Gill's statement to the other fish that they are not meant to be in a box because "it does things to you" is truly realised in *Padak*.

Padak's flatfish Master is portrayed contrastingly as a liar who concocts false stories about the sea and hides away in the most inaccessible area while ordering his crew to play dead and even engage in cannibalism. In one horrifying scene, the crew sets upon a halibut who is still alive and tears her/him apart. Padak's moral code prevents her from joining in. However, this decision comes to haunt her as her hunger forces her to devour a tank of small clownfish. Padak, right down to her blue skin tone, ability to talk to different creatures and optimistic attitude recalls Dory – if Dory ate Nemo. This decision to use clownfish as victims explicitly orients the Gothic heterotopias of *Padak* as stark countersites to *Finding Nemo's* bright locales. Indeed, there is no liberation here, including for Padak who is killed and eaten after this encounter. The 'sea' in the English title of *Padak* becomes an out-of-reach Arcadia, like Snitter's hallucinatory island. Neither Snitter, Rowf, nor Padak make it to their promised land.

The potential power of Padak for nonhuman advocacy lies in its shock value. However, its limited reception might be traced to how, like *The Plague Dogs*, it fails to showcase an optimistic alternative in which audiences can find solace, or a better future they can celebrate. One viewer reports they were "not expecting to be this affected" (Avwillfan89, 2020) while another observes that the scenes of marine life being eaten "would make anybody want to join PETA or go vegan" (ironhorse_iv, 2020).

Conclusion: Jeremiadic vs Celebratory Storytelling in Gothic Enviro-toons

Watership Down showcases a happy ending where rabbits reach their promised land, free from anthropocentric oppression. In contrast, the hapless victims of experimentation and culinary desire in *The Plague Dogs* and *Padak*, like their real-world counterparts, never achieve salvation. As such, both critical and audience reception repeatedly refer to the darker narrative undertones of these Gothic enviro-toons and the disruptive heterotopic spaces of human/non-human conflict they create. In fact, Walter Chaw (2002) praises *Watership Down* as forerunner of what American animation could have become, and what anime is currently: "a mature medium for artistic expression of serious issues."

Critical and audience reviews of these films indicate the underutilised potential of dark, Gothic enviro-toons in generating awareness about the plight of non-human animals. By emotionally appealing to and engaging with viewers, evoking shock, guilt, and grief on behalf of the on-screen characters and their real-life counterparts, they can make a tangible contribution to animal advocacy (see Fukano et al., 2020). This potential is perhaps starting to be realised with the renewed popularity of these films, especially *Watership Down*, which is being re-told in a Netflix mini-series. Animation is no longer considered to be solely the domain of young children or simple storytelling tropes. All three films create a post-human[16] ontology where viewer's empathies are split between non-human characters who are humanised on screen, and who struggle to survive under the direct or indirect onslaught of human oppression, and humans who are alienated from their non-human victims by their cruelty, but with whom viewers identify and can thereby recognise their role and responsibility in this oppression.

However, only *Watership Down* provides a happy ending. The story of rabbits overcoming extreme mental, physical and emotional trauma to finally reach their utopia is more appealing to audiences seeking light at the end of the tunnel, not only for the nonhuman characters in which they are invested but their real-world counterparts who face similar ordeals. As such, while the absence of a happy ending is not to suggest that *The Plague Dogs* and *Padak* are filmmaking failures (they have proven their place in advocating for non-human justice), *Watership Down* achieves a balance of optimism and grief – shocking while also offering a glimmer of hope.

Watership Down successfully juxtaposes jeremiadic and celebratory modes of storytelling. It creates Gothic heterotopic spaces of human/animal conflict which disturb anthropocentric ways of perceiving the world, holding up a counter-gaze to human negligence while also offering an alternative vision where rabbits retire to a wilderness untainted by human expansions. Hence, though the film advocates for the necessity of wild spaces devoid of anthropocentric interference, it also reflects the dual-sided narrative of conservation whereby we celebrate the saving of some species while lamenting the loss of others (Dean & Wilson, 2023; Hance, 2016). The jeremiadic element functions as an essential reminder. It is, then, using a mixture of hope and shock that Gothic enviro-toons might unlock their potential as vessels of change, generating awareness about and advocating for non-human animals through fictional means.

Notes

1 A heterotopia, in the Foucauldian sense, refers to a discursive or physical space that is disturbing or somehow 'other' (Foucault, 1984).

2 Such affective states are often denied to non-human animals, before using their supposed lack of these qualities as a pretext for interspecies exploitation – a cultural mindset termed anthropo-denial (Waal, 258)

3 In contrast to strong anthropomorphism, such as clothing an animal and making it drive a car (like in *Zootopia*), which seeks to deny the "animal's own differences and excellences" (Plumwood, 2002, p. 59), weak anthropomorphism as used in *Watership Down*, *Rio*, and *Happy Feet* portrays non-human animals in ways that preserve said differences while still allowing for viewer empathy (ibid., p. 56).

4 The emotional impact of the death of Bambi's mother in the 1942 Walt Disney film is thought to have influenced US attitudes towards hunting (Fukano et al., 2020). Furthermore, a study of the conservation potential of animated media found that animals depicted in the show *Kemono Friends* "attracted more financial supporters from citizens after the broadcast as compared to the animals that were not" (Fukano et al., 2020, p. 7).

5 A heterotopia, already a space of Otherness that unsettles normative orders, becomes Gothic through the use of darker narrative styles and colour palettes, and a looming sense of horror. Some enviro-toons employ the tendency of Gothic storytelling to unsettle, shock and disturb the status quo.

6 Scores surpass those of *Rio* and *Happy Feet* – enviro-toons with conventionally happy endings, produced by larger animation studios, and with a considerably larger reach.

7 This does not negate that conventional animated features have been reviewed by a larger group of viewers than Gothic enviro-toons. Hence, despite the high performance of the Gothic toons, this is borne of reviews from a more niche group of dedicated viewers.

8 *Watership Down* refers to Rosen's film except when referring specifically to Adams' novel.

9 Head of the warren guards

10 Baker and Adams describe disnification as the process of rendering non-human animals in visual form in a manner that "seems somehow to incline towards the stereotypical and the stupid, to float free from the requirements of consistency or of the greater rigour that might apply in other non-visual contexts" (2006, p. 176).

11 The associations of darkness or blackness with evil in Western epistemology is known to be a contributing factor to the racial alienation of certain groups of people (Hall, 2018, p. 4)

12 Snitter's ability to visualize aspects of the world not readily before him can be used to contrast him with Fiver from *Watership Down*, whose story is also marked with seeming hallucinations. However, while the narrative treats Fiver's abilities as prophetic visions destined to guide the rabbits to safety, *The Plague Dogs* offers a grimmer take on Snitter's hallucinations.

13 YouTubers (Saberspark, 2020; Steve Reviews, 2020), well-known among fans of animated features, have made extensive analyses of the film.

14 While IMDb is not the only platform for viewers to review movies, it remains the largest and most comprehensive database of films. For consistency and coherence, this chapter refers to IMDb for all reviews.

15 Observing such fishes in inhumane conditions in the real world so moved the director Dae-Hee Lee that he decided to voice dissent against the cruel practices of the Korean fish market through *Padak* (Eun-San, 2012)

16 The idea of post-humanism is anti-anthropocentric, as per Francesca Ferrando's use of the term, which critiques the use of the word "human" as a privileged marker used by a dominant culture to alienate and justify the oppression of non-human animals and othered humans (Ferrando, 2020, p. 80).

References

Adams, R. (1972). *Watership Down*. Penguin Books Canada, Limited.

Adams, R. (2016). *The Plague Dogs: A Novel*. Vintage Books.

adrian_rawling. (2022). A Beautiful, Poetic Film. Review of *Watership Down*. IMDb. www.imdb.com/title/tt0078480/reviews?sort=curated&dir=desc&ratingFilter=0

Advani, N. (Director). (2012). *Delhi Safari* [Film.] Krayon Pictures.

a_fox_tabtha. (2005). I Hugged My Puppy. Review of *The Plague Dogs*. IMDb. www.imdb.com/title/tt0084509/reviews/?ref_=tt_ov_rt

Aitken, R., Moran, E., Pugh, B., & Varney, J. (Executive Producers). (2018). *Watership Down* [TV series]. Biscuit Filmworks; BBC; Netflix.

Anderson, D. K., & Reher, K. (2002). *A Bug's Life* [DVD]. Sandy City, CA; Distributed by Kit Parker Films.

andrewchristianjr. (2021). WTH. Review of *Watership Down*. IMDb. www.imdb.com/title/tt0078480/reviews?sort=curated&dir=desc&ratingFilter=0

Avwillfan89. (2020). Was Not Expecting to Be this Affected. IMDb. Review of *Padak*. https://m.imdb.com/title/tt2164058/reviews/?ref_=tt_ov_rt

Baker, S., & Adams, C. J. (2006). *Picturing the Beast: Animals, Identity and Representation*. University of Illinois Press.

Barcz, A. (2017). *Animal Narratives and Culture: Vulnerable Realism*. Cambridge Scholars Publishing.

Barrier, M. (2008). *The Animated Man: A Life of Walt Disney*. University of California Press.

Bekoff, M. (2004). Wild Justice and Fair Play: Cooperation, Forgiveness, and Morality in Animals. *Biology & Philosophy*, *19*(4), 489–520. https://doi.org/10.1007/sbiph-004-0539-x

Belcourt, B.-R. (2014). Animal Bodies, Colonial Subjects: (Re)locating Animality in Decolonial Thought. *Societies*, *5*(1), 1–11. https://doi.org/10.3390/soc5010001

Bencich, S., Friedman, R. J., & Mauldin, N. (2006). *Open Season* [Film]. India; Sony Pictures Animation.

Bölinger, L. (2022). Animal Subjectivities and Anthropocentrism in Richard Adams' Watership Down. *Göttinger Schriften Zur Englischen Philologie*, *15*, 95–180. https://doi.org/10.17875/gup2022-1919

Borgi, M., & Cirulli, F. (2016). Pet Face: Mechanisms Underlying Human-Animal Relationships. *Frontiers in Psychology*, *7*, 1–11. https://doi.org/10.3389/fpsyg.2016.00298

Botts, E., & Sipos, M. (2022). *Watership Down & the Incomprehensible Power of Humanity*. The Other Folk, July 20. Retrieved March 31, 2023, from www.theotherfolk.blog/dissections/watership-down

Chapman, E. L. (1978). The Shaman as Hero and Spiritual Leader: Richard Adams' Mythmaking in 'Watership Down' and 'Shardik'. *Mythlore*, *5*(2), 7–12.

Chaw, W. (2002). Watership Down (1978) [The Criterion Collection] – Blu-Ray Disc. *Film Freak Central*, May 8. Retrieved March 31, 2023, from www.filmfreakcentral.net/ffc/2015/02/watership-down-1978-the-criterion-collection-blu-ray-disc.html

Cruelty Free International. (2021). Richard Adams – a Man of Stories and Animals. *Cruelty Free International*, December 17. Retrieved March 30, 2023, from https://crueltyfreeinternational.org/latest-news-and-updates/richard-adams-man-stories-and-animals

Dean, A. J., & Wilson, K. A. (2023). Relationships Between Hope, Optimism, and Conservation Engagement. *Conservation Biology*, *37*(2), 3–12. https://doi.org/10.1111/cobi.14020.

Derrida, J., & Wills, D. (2002). The Animal that Therefore I Am (More to Follow). *Critical Inquiry*, *28*(2), 369–418. https://doi.org/10.1086/449046

de Waal, F. B. M. (1999). Anthropomorphism and Anthropodenial: Consistency in Our Thinking About Humans and Other Animals. *Philosophical Topics*, *27*(1), 255–280. www.jstor.org/stable/43154308.

Disney, W., & Hand, D. (2011). *Bambi* [Film]. Walt Disney Studios Home Entertainment.

Donkin, J. C., & Anderson, B. (Director). (2011). *Rio* [Film]. Blue Sky Studios; Twentieth Century Fox Home Entertainment.

ed321. (2018). Amazing, Just Amazing Film. *IMDb*. www.imdb.com/title/tt0078480/reviews?sort=curated&dir=desc&ratingFilter=0

Eun-San, L. (2012). Padak, Sign of a Boom in Korean Animation. *Korean Film Biz Zone*, August 6. Retrieved April 1, 2023, from www.koreanfilm.or.kr/eng/news/interview.jsp?mode=INTERVIEW_VIEW&pageRowSize=10&seq=16

Fernandez, L. (2020). The Emotional Politics of Images: Moral Shock, Explicit Violence and Strategic Visual Communication in the Animal Liberation Movement. *Journal for Critical Animal Studies, 17*(4), 53–80.

Ferrando, F. (2020). *Philosophical Posthumanism*. Bloomsbury Academic.

Firstmagnitude. (2019). Paradise Alley, Magic, Midnight Express, Watership Down, Comes a Horseman, 1978. *Siskel and Ebert Movie Reviews*, November 10. Retrieved March 31, 2023, from https://siskelebert.org/?p=8038

Fischer, N. (1987). Peace and Animal Liberation Themes in Martin Rosen's Film *The Plague Dogs*. *Peace & Change, 12*(1–2), 45–50. https://doi.org/10.1111/j.1468-0130.1987.tb00092.x

Foucault, M. (1984). *Of Other Spaces: Utopias and Heterotopias*. https://foucault.info/documents/heterotopia/foucault.heteroTopia.en/.

Fukano, Y., Tanaka, Y., & Soga, M. (2020). Zoos and Animated Animals Increase Public Interest in and Support for Threatened Animals. *Science of the Total Environment, 704*, 1–8. https://doi.org/10.1016/j.scitotenv.2019.135352

Gigliotti, C. (2015). The Struggle for Compassion and Justice Through Critical Animal Studies. In L. Kalof (Ed.), *The Oxford Handbook of Animal Studies* (pp. 188–207). Oxford University Press.

Gipson, F., & Tunberg, W. (Director) (2002). *Old Yeller* [Film]. Walt Disney.

Grandi, R. (2021). 'Animals Don't Behave Like Men: They Have Dignity and Animality.' Richard Adams's Watership Down and Interspecies Relationships in the Anthropocene. *Textus, English Studies in Italy, 34*(3), 159–186. https://doi.org/10.7370/102411

Haley, A. (2019). *Kemono Friends*. Yen Press.

Hall, K. F. (2018). *Things of Darkness: Economies of Race and Gender in Early Modern England*. Cornell University Press.

Hance, J. (2011). Could Blockbuster Animated Movies Help Save Life on Earth? *Mongabay Environmental News*, September 26. Retrieved March 26, 2023, from https://news.mongabay.com/2011/09/could-blockbuster-animated-movies-help-save-life-on-earth/

Hance, J. (2016). Why Don't We Grieve for Extinct Species? *The Guardian*, November 19. www.theguardian.com/environment/radical-conservation/2016/nov/19/extinction-remembrance-day-theatre-ritual-thylacine-grief

Haraway, D. J. (2007). *When Species Meet*. University of Minnesota Press.

Haraway, D. J. (2020). *The Companion Species Manifesto: Dogs, People and Significant Otherness*. MTM.

Hastings, A. W. (1996). Bambi and the Hunting Ethos. *Journal of Popular Film and Television, 24*(2), 53–59. https://doi.org/10.1080/01956051.1996.9943714

Heise, U. K. (2014). Plasmatic Nature: Environmentalism and Animated Film. *Public Culture, 26*(2), 301–318. https://doi.org/10.1215/08992363-2392075

Henson, R. (2019). A Prairie Dog's Linguistic Abilities. *Discover Magazine*, November 12. Retrieved March 30, 2023, from www.discovermagazine.com/mind/a-prairie-dogs-linguistic-abilities

Herman, D. (2016). Animal Autobiography; or, Narration Beyond the Human. *Humanities, 5*(4), 1–17. https://doi.org/10.3390/h5040082

Hinxman, M. (1978). What a Beastly Affair! *Daily Mail*, pp. 32–33.

Hoing, A. (2016). 'Snit's a Good Dog!' Dogs' Innate Duty in Richard Adams's the Plague Dogs. In C. Blanco & B. Deering (Eds.), *Who's Talking Now? Multispecies Relations from Human and Animals' Points of View* (pp. 69–76). Oxford Interdisciplinary Press.

Höing, A. (2017). Unreliability and the Animal Narrator in Richard Adams's the Plague Dogs. *Humanities, 6*(1), 1–13. https://doi.org/10.3390/h6010006

Hoing, A. (2021). Devouring the Animal Within: Uncanny Otherness in Richard Adams's the Plague Dogs. In R. Heholt & M. Edmundson (Eds.), *Gothic Animals: Uncanny Otherness and the Animal With-Out* (pp. 57–74). Palgrave Macmillan.

Höing, A., & Husemann, H. (2016). The Vicious Cycle of Disnification and Audience Demands: Representations of the Non-Human in Martin Rosen's Watership Down (1978) and The Plague Dogs (1982). In A. E. George & J. L. Schatz (Eds.), *Screening the Nonhuman: Representations of Animal Others in the Media* (pp. 101–115). Lexington Books.

Huggan, G., & Tiffin, H. (2010). *Postcolonial Ecocriticism*. Routledge.

IMDb.com. (n.d.). The Plague Dogs. *IMDb*. Retrieved April 2, 2023, from https://m.imdb.com/title/tt0084509/trivia/?ref_=tt_ql_trv

ironhorse_iv. (2020). Holy Mackerel! This Computer Animation Motion Picture Is Really Disturbing!" *IMDb*. Review of *Padak*. https://m.imdb.com/title/tt2164058/reviews/?ref_=tt_ov_rt

itsmarscomix. (2005). A Tragic and Moving Tale. *IMDb*. Review of *The Plague Dogs*. www.imdb.com/title/tt0084509/reviews/?ref_=tt_ov_rt

Jones, G. (2015). Watership Down: 'Take Me with You, Stream, on Your Dark Journey'. *The Criterion Collection*, February 26. Retrieved March 31, 2023, from www.criterion.com/current/posts/3475-watership-down-take-me-with-you-stream-on-your-dark-journey

Kelly, S. (2023). Princess Mononoke: The Masterpiece that Flummoxed the US. *BBC News*, June 8. www.bbc.com/culture/article/20220713-princess-mononoke-the-masterpiece-that-flummoxed-the-us.

KlangSmithToo. (2018). Finding Nemo Meets Oldboy. *IMDb*. Review of *Padak*. https://m.imdb.com/title/tt2164058/reviews/?ref_=tt_ov_rt

Kozak, O. E. (2022). An Appreciation for the Plague Dogs. *Paste Magazine*, July 27. Retrieved April 2, 2023, from www.pastemagazine.com/movies/plague-dogs/an-appreciation-for-the-plague-dogs

Kroyer, B., & Faiman, P. (Director). (2005). *Ferngully: The Last Rainforest* [Film]. Twentieth Century Fox Film Corporation; Youngheart Productions; FAI Films.

Kynaston, E. (2020). *A Symphonic Discussion of the Animal in Richard Adams' Watership Down*. BA Thesis, Lund University.

Lee, D.-H. (Director). (2012). *Padak* [Film]. CJ Entertainment.

Lester, C. (2022). Watership Down: Family-Friendly BBC Version Risks Losing the Power of Epic Original. *The Conversation*, September 13. Retrieved March 31, 2023, from https://theconversation.com/watership-down-family-friendly-bbc-version-risks-losing-the-power-of-epic-original-108699

Lévi-Strauss, C., Needham, R., & Poole, R. C. (1969). *Totemism*. Penguin Books.

Livni, E. (2019). What the Study of Animal Emotions Shows Us About Being Human. *Quartz*, March 9. Retrieved March 29, 2023, from https://qz.com/1568701/what-the-study-of-animal-emotions-shows-us-about-being-human

Lockley, R. M. (1985). *The Private Life of the Rabbit: An Account of the Life History and Social Behaviour of the Wild Rabbit*. Boydell.

Marstal, T. (2022). Recentering the Rabbits. *Leviathan: Interdisciplinary Journal in English*, 8, 109–122. https://doi.org/10.7146/lev82022132077

Meyer, C. A. (1994). The Power of Myth and Rabbit Survival in Richard Adams' 'Watership Down'. *Journal of the Fantastic in the Arts*, 3(3/4), 139–150.

Miller, G., Collee, J., Morris, J., & Coleman, W. (2006). *Happy Feet* [Film]. India; Warner Home Video.

Miyazaki, H. (1997). *Princess Mononoke* [Film]. Studio Ghibli.

Moore, R., Howard, B., & Bush, J. (2016). *Zootopia*. Walt Disney Studios Motion Pictures.

Mullan, B., & Marvin, G. (1987). *Zoo Culture*. Weidenfeld & Nicolson.

Mullins, A. (2016). 'Watership Down' Author Was a Staunch Animal Rights Advocate. *PETA*, December 27. Retrieved March 31, 2023, from www.peta.org/blog/richard-adams-watership-down-author/

Murray, R. L., & Heumann, J. K. (2011). *That's All Folks? Ecocritical Readings of American Animated Features*. University of Nebraska Press.

Naish, D. (2022). Legend of the Black Dog. *Tetrapod Zoology*, August 16. https://tetzoo.com/blog/2022/8/13/legend-of-the-black-dog.

O'Brien, D. H. (2016). Other(ed) Rabbits: Using Otherness as a Frame to Teach Critical Approaches to Human-Animal Relations in Japan. *Otherness: Essays and Studies: Animal Alterity*, 5(2), 47–78.

Pennington, J. (1991). From Peter Rabbit to Watership Down: There and Back Again to the Arcadian Ideal. *Journal of the Fantastic in the Arts*, 3(2), 66–80.

Plumwood, V. (2002). *Environmental Culture: The Ecological Crisis of Reason*. Taylor and Francis.

Preiser, R., Pereira, L. M., & Biggs, R. (O). (2017). Navigating Alternative Framings of Human-Environment Interactions: Variations on the Theme of 'Finding Nemo.' *Anthropocene*, 20, 83–87. https://doi.org/10.1016/j.ancene.2017.10.003

Rawls, W. (2018). *Where the Red Fern Grows*. Delacorte Press.

Rodriguez, F. C. (1988). Beyond Satire: Richard Adams's The Plague Dogs. *The International Fiction Review*, 15(1), 51–53.

Rosen, M. (Director). (1978). Watership Down [Film]. Cinema International Corporation; Nepenthe Films.

Rosen, M. (Director). (1982). The Plague Dogs [Film]. United Artists; Nepenthe Productions; Goldcrest Films.

Roy, S. (2022). Can the Non-Human Subaltern Speak? Addressing Injustice Through Parakeets, Penguins and Blue Macaws. *Studia Universitatis Babeş-Bolyai Philologia*, 67(2), 227–250. https://doi.org/10.24193/subbphilo.2022.2.13

Saberspark. (2020). "What the Hell Is Padak?" *YouTube*, July 3. www.youtube.com/watch?v=Gk4YMoHX-wA.

Silk, M. J., Crowley, S. L., Woodhead, A. J., & Nuno, A. (2018). Considering Connections Between Hollywood and Biodiversity Conservation. *Conservation Biology*, 32(3), 597–606. https://doi.org/10.1111/cobi.13030

Slovic, S. (1996). Epistemology and Politics in American Nature Writing: Embedded Rhetoric and Discrete Rhetoric. In C. G. Herndl & S. C. Brown's (Eds.), *Green Culture: Environmental Rhetoric in Contemporary America* (pp. 82–110). University of Wisconsin Press.

Smith, A., & Hughes, W. (2016). *Ecogothic*. Manchester University Press.

Stanton, A., Peterson, B., & Reynolds, D. (2003) *Finding Nemo* [Film]. Pixar Animation Studios.

Stanton, R. R. (2020). Animals, Animation and Academia. *Animationstudies 2.0*, June 15. Retrieved March 30, 2023, from https://blog.animationstudies.org/?p=3587

Stanton, R. R. (2021). Controversial or Comfortable: What is the Secret to Acceptable Violence in Animation. *Animationstudies 2.0*, October 26. Retrieved March 30, 2023, from https://blog.animationstudies.org/?p=4241

Steve Reviews. (2020). Steve Reviews: Padak. *YouTube*, May 24. www.youtube.com/watch?v=MOP6vz5i5XM.

Sung-yoon, O. (Director). (2018). *Underdog (A Dog's Courage)* [Film]. Odoltogi.

Thomas, F., & Johnston, O. (1995). *The Illusion of Life: Disney Animation*. Hyperion.

Weinman, J. (2004). Something Old, Nothing New Thoughts on Popular Culture and Unpopular Culture. *A Good Enviro-Toon*, September 6. Retrieved March 26, 2023, from http://zvbxrpl.blogspot.com/

Wells, P. (2009). *The Animated Bestiary Animals, Cartoons, and Culture*. Rutgers University Press.

Whitley, D. (2016). *The Idea of Nature in Disney Animation: From Snow White to Wall-E*. Routledge.

Williams, Fr., I. (2021). Richard Adams and Animal Welfare. *Father Irenaeus Williams*, September 5. Retrieved March 31, 2023, from https://fatherirenaeuswilliams.com/2020/05/17/richard-adams-and-animal-welfare/

Wrenn, C. L. (2013). Resonance of Moral Shocks in Abolitionist Animal Rights Advocacy: Overcoming Contextual Constraints. *Society & Animals*, *21*(4), 379–394.

Zanette, L. Y., Hobbs, E. C., Witterick, L. E., MacDougall-Shackleton, S. A., & Clinchy, M. (2019). Predator-Induced Fear Causes PTSD-Like Changes in the Brains and Behaviour of Wild Animals. *Nature News*, August 7. Retrieved April 2, 2023, from www.nature.com/articles/s41598-019-47684-6

zetes. (2007). Depressing but Amazing. *IMDb*. Review of *The Plague Dogs*. www.imdb.com/title/tt0084509/reviews/?ref_=tt_ov_rt

Chapter 11

Laugh to Liberate
Futurabilities of Posthumanist Comedy

Katya Krylova

Introduction

"The worst part of my Monday is hearing you complain about yours," says one of Grumpy Cat's[1] memes circulating across various social media platforms. She passed away, but 'her' misanthropic voice still sounds liberating. By lifting the taboo on negative thoughts and emotions, one cute cat with an angry face and the collective creativity behind her memes challenge regimes of toxic positivity. Grumpy Cat reminds us that being alive means having ups and downs: from time to time, we might want to suspend our obligatory productive and optimistic selves to be annoyed, inconvenient, and antisocial. Her questioning of idealised online identities made me think about my corporeal selfhood, which I tend to neglect, edging towards the deadline set for another exciting project.

In contrast with my everyday routines, subdued to ambitions and plans, my cat practices unwavering self-care, which demonstrates a healthy sense of proportion. We know, if cats are otherwise well, they sleep, eat and groom themselves regularly. They never neglect responsibilities to their own bodies for maintaining energy balance, while their humans time and again choose to postpone rest for the sake of productivity or entertainment. The more I study human–animal relationships, the more obvious it becomes that if someone or something is to save humans from the Anthropocene, it will be animals and their capacity to take care of themselves and enjoy their time before anything else. However, I wonder if they will also be able to save themselves from us and all the anthropogenic risks escalating too quickly. In what follows, I argue that Grumpy Cat and other cute animals delivering ironic messages through mainstream social media can benefit nonhuman species if the content producers managing their profiles channel the affective powers of animal images and voices to challenge anthropocentric attitudes.

Claire Colebrook notes that satire is "*immanently* historical . . . produced from the flow of life" (2004, p. 139); the same is said about cuteness, described by Simon May (2019) as being "attuned to the zeitgeist . . . a playful, often trivial, but accurate expression of the times" (2019, pp. 9, 143). The existing scholarship on humour (Colebrook, 2004; Critchley, 2004; Eagleton, 2019) shows that some of its functions overlap with those of cuteness when it comes to specific processes of

DOI: 10.4324/9781003366706-15

mental and political ecology, such as stress relief and struggle against oppression (Colebrook, 2004; Critchley, 2004; Eagleton, 2019). Cute animals 'pretending' to hate Mondays remain one of the most popular types of content on social media. Most of them 'post' light-hearted jokes that ease our anxieties without questioning obsolete cultural constructs, such as the belief in human superiority and authority, affecting these animals' lives and wellbeing. However, as I will show, a small number of content producers based in Europe, United States, Japan, South Korea, and other places are negating conventional representations of animals as 'objects' for pleasure. Instead, they indicate that animals pursue their own agendas and in most cases are better off without humans. Some species do not even acknowledge our existence, and almost all face worse life challenges than any human.

One such author is Brooke Barker, the US artist behind a popular Instagram profile called @sadanimalfacts. Brooke is fascinated with nonhuman ways of living and, by finding a delicate balance between cuteness and irony, represents animal stories in messages we can relate to. "Can we all just hug?" asks the mouse from one of Brooke's Instagram comics, pointing to the fact that mice are empaths who can sense sadness in each other and feel it. Her drawings typically show animals ironically reflecting on their species traits – such as colour blindness or having 15,000 teeth – that make their structuring existential experiences different from ours or, on the contrary, surprisingly similar. The posthuman turn in political theory and the development of critical art practice concerned with animals' rights to autonomy lead me to speculate about the liberating potential of such content and ask myself if this albeit small range of hybrid, human-animal-algorithmic comedians, challenging anthropocentric attitudes to nonhuman animals on Instagram and other social media with potentially high impact, have the capacity to be vessels for the mobilisation of more-than-human ethical futures.

Thus, in what follows, I build on these rare examples of posthumanist humour, emerging in popular culture and contemporary art and as user-generated content for mainstream participatory platforms, to explore the possibility of responsible speech on behalf of nonhuman others. Applying Gilles Deleuze's concept of superior irony unfolding towards multiple points of view beyond anthropocentric logic (1994, p. 182) and Elisa Aaltola's notion of simulative empathy (2018, p. 28),[2] I imagine AI-assisted production of humorous narratives 'voiced' by animal characters. In my preferred version of the technology-enabled future envisioned in this chapter, bits of posthumanist comedy spread across social networks by content distribution algorithms evolve into a systemic practice leading towards an alternative state of participatory media where the responsible representation[3] of nonhuman species and individuals is adopted as a principle of new digital ecology.

The Ambivalence of Humour, and Its Transformative Power

As Simon Critchley points out, humour reveals the "depth" of what we (as human beings) share "not through the clumsiness of a theoretical description, but more quietly, practically and discreetly" (2004, p. 18). This *delicate* agency of humour interests me as a productive pathway for revealing obsolete cultural constructs behind anthropocentric values. The emerging sites for animal-mediated satire make me believe that humour can challenge conventional perceptions and representations of animals as pests or playthings across media – in mainstream fiction and movies, or on social networking sites, where the majority of content on behalf of nonhuman others is still detached from critical discourses of posthumanism.

Countless memes related to animals exercise so-called *superior humour*, that is, the "powerful laughing at powerless" or distribute *reactionary* jokes that reinforce stereotypes and pre-existing speciesist attitudes (Critchley, 2004, p. 12). Many animal profiles on Instagram also encourage *laughing at the expense of others*. Gradually displaced from open communication for being discriminatory towards different groups of people, such humour finds refuge inside simulacra of talking animals who invite their audiences to laugh at obese or deformed bodies and 'dumb' behaviours. Such jokes highlight the ambivalent nature of humour: similar to cute aesthetics, it can be utilised for "defence or affirmation, subversion or celebration, solidarity or critique" (Eagleton, 2019, p. 42). Therefore, if we look carefully, alongside the hundreds of animals depicted as laughing at themselves to provoke the release of our "pent-up nervous energy" (Critchley, 2004, p. 3), we can notice a few others who make us face "a sort of truth about people . . . about what hurts and terrifies [us], about what is hard, above all, about what [we] *want*" (Griffiths, cited in Critchley, 2004, p. 10, emphasis in original).

Similar to cuteness, the subversive potential of humour lies in its ability to highlight the abuse of power without taking the position of superiority in relation to those who hold this power or suffer from its misuse. As pointed out by Colebrook, this aspect allows Deleuze to draw a line between irony and "radical humour" (or "superior irony"), which, in contrast to the former, operates horizontally, "dissolves high-low distinctions," and "insists on multiplicity" (2004, pp. 130, 133). This means that radical humour can surpass the inertia of human supremacy and unfold towards multiple points of view beyond the anthropocentric perspective. Since radical laughter defamiliarises the mundane, it can make animal suffering seem extraordinary and surreal, invite us to laugh at power and, by doing so, realise the invalidity of the "oppressive" social order that seems "fixed" (Critchley, 2004, pp. 10–11). Existing scholarship on humour indicates many potentially subversive strategies, such as bathos, black humour, out-of-placeness, or hyperbole (Eagleton, 2019, p. 88). Each of them, along with new styles of writing that negate subject/ object logic, can be mobilised to fight animal suffering.

In the same way as it can reveal non-obvious manifestations of racism, sexism, or ageism, radical humour can light up obscure symptoms of the misrepresentation of animals. In this regard, Mark McGurl distinguishes the body of literary texts constituting *posthuman comedy*, defined as "critical fiction . . . in which scientific knowledge of the spatiotemporal vastness and numerousness of the nonhuman world becomes visible as a formal, representational, and finally existential problem" (2012, p. 537). Today, posthumanist jokes on social media 'narrated' by animals appear to fit the same purpose – to resist a "long history of ignoring, forgetting, and erasing" animal suffering (ibid., p. 538).[4]

Not the Self, Nor the Other: Who Are Voiced Animals Laughing At?

It is not a coincidence that images of cute animals fit so well with humorous captions. Jokes provide a "brief vacation from the mild oppressiveness of everyday meaning" (Eagleton, 2019, p. 15), and so does cuteness. In response to the Covid-19 pandemic and later wars in Ukraine and Palestine, the therapeutic effects of cute and funny posts featuring capybaras and quokkas became even more evident. As Eagleton points out, "humour allows us to relax our mental muscles" (ibid., p. 16).

As cost of living continues to increase together with precarious employment, environmental anxieties, and political instability, stand-up comedy and cute animal videos are acknowledged as accessible ways to manage the stress overload associated with material and psychological pressures. Engagement with this content can be seen as a form of "desublimation" (Eagleton, 2019, p. 22). As Ian Buchanan notes, referring to Herbert Marcuse's concept of desublimation, "banal and powerless" art diminishes our impulses to resist the ordinary and to strive beyond the conventional (2010, n.p.). Striving beyond convention demands "psychological strain, not having to do so can be a gratifying sensation" (Eagleton, 2019, p. 14). Eagleton's arguments explain why humour is so widespread in times of emotionally and intellectually intense labour. However, stress-relieving jokes narrated by cute animals are not always banal and powerless. Plenty involve self-mockery and invite reflection.

For example, in 'his' jokes about Mondays, another cat-influencer Eric (Instagram profile: @eric_the_c.at) is depicted expressing solidarity with humans (self) consumed by work. Like memes of puppies and kittens falling asleep wherever they can, unable to resist fatigue, Eric's posts facilitate the acknowledgement and acceptance of our shared vulnerability to strenuous labour regimes. Performing the role of an office worker or entrepreneur in his posts, Eric creates a situation in which "the object of laughter is the subject who laughs" (Critchley, 2004, p. 14). In her jokes addressed to 9-to-∞ workers, Eric's guardian "give[s] voice to what others shy away from" (Eagleton, 2019, p. 142). Using Eric's profile as an outlet for self-irony, she finds a way to publicly talk about her own and neoliberal humanity's weaknesses and frustrations.

Similar examples emerge in popular culture, where companion animals reflect and mock the lifestyles of their guardians by engaging in human experiences, such as psychoanalysis. In the 2021–2022 adult animated comedy series *HouseBroken*, a white poodle named Honey leads a group of neighbourhood animals through amateur therapy sessions but struggles to succeed as a therapist, being caught up in her own worries. Liberated zoo animals from the post-apocalyptic animated TV series *Wild Life* (2020) also reflect neoliberal subjects in their desires to become other or 'better' than they are. For instance, pacifist cheetah Glenn tries to suppress his predatory impulses with the help of an imagined therapist; dolphin Marnie dreams of becoming a terrestrial animal and ignores giant blisters popping up on her skin in reaction to the sun; sloth Viv drinks energy tonics; koala Darby leads an unnaturally active life – driving a car, shopping, building companion robots for his zoo neighbours. It is not made clear if these animals have suppressed their nature in captivity or lost it due to mutation. However, as a result of this essential disconnect, they become unbearably bored. Day after day, they try to entertain themselves with sports, parties, clubs, and other ways to socialise and experience emotions constituted as especially human, including negative ones (jealousy, envy, disappointment, and self-loathing). Their lives in the ruins of post-scarcity appear like a dysfunctional anarcho-communist utopia, in which joy becomes a new, but no less tedious job.

Although humorous narratives that systematically challenge anthropocentric perspectives on social media are rather rare, as I have indicated, there are several profiles that invite us to imagine a better future in which our capacity for laughter changes individual and collective destinies of nonhuman others. With @sadanimalfacts, Brooke Barker has found a way to mobilise liberating agencies of cuteness and humour to talk about animal experiences while avoiding celebrification of particular breeds or species. Her jokes on behalf of animals (Figure 11.1) highlight the gaps in our knowledge about them and, at the same time, do not involve exploiting animal individuals as a source of cute images – all nonhuman narrators appear in the form of depersonified cartoons.

For example, "Being awake is the worst," says Barker's cat, referring to the caption stating that, "cats sleep 70% of their lives." This kind of 'truth' might have smashed any fantasy of unconditional love felt by 'my' cat towards me (if I held such a fantasy). However, it would have done this in a way that, rather than hurting my feelings, invites me through laughter to change my optics and actions. This joke made me think about what it is like to be a cat – a nocturnal animal expected to synchronise with their diurnal companion, a creature appreciating silence but living in a noisy urban neighbourhood, inhabiting a space designed for humans without any cat-specific hideouts, a living creature vulnerable to being bothered by humans during their sleep as a result of their irresistible cuteness. In turn, as Critchley points out, the displacement resulting from the attempt to see the world through the eyes of others, brought me closer to a laughable and simultaneously tragic "enigma" of my ordinary life (2004, p. 22).

My cat is having his best time while he sleeps. From this radical idea, marked as a sad cat fact, my thoughts turn to the self – to my stress-induced insomnia, disturbed biorhythms accompanying my precarious existence, my reluctance to wake at 6:30 am to attend language classes so I can reside in a foreign country, and wander further, to the problem of mass sleep deprivation.[5] Brooke's radical humour in @sadanimalfacts invites me to recognise and "examin[e] the follies of others with a full perception of [my] own weakness" (Colebrook, 2004, p. 143). Her jokes unfold in their social significance by making me face the "folly" of my own world in order to imagine a better one in its place and act upon it (Critchley, 2004, p. 17).

Figure 11.1 Collection of comic illustrations created by Brooke Barker for her Instagram account @sadanimalfacts. Courtesy of the artist.

I believe that this "messianic power" of humour can work in favour of both animals and humans (ibid., p. 16).

Inspired by rare examples of subversive jokes narrated by voiced animals, I imagine in what circumstances such content can become dangerous for the politics of using animals as affective labourers and treating them as 'objects' for manipulation and amusement. In what follows, I engage in a speculative design of a radical utopia in which posthumanist humour enabled by fictional characters brings about a new future without animal influencers serving as avatars for promoting further commodification of cute nonhuman bodies.

My speculative utopia originates in the emergence of *Naked Derrida (ND)*,[6] an imagined movement focused on promoting animal otherness, human–animal differences, and ways of undermining anthropocentric practices through comedy. The conceptual core of this scenario combines Deleuze's notion of superior irony and Aaltola's thoughts on simulative empathy as a form of other-directed narrative, enabling the adoption of posthumanist ethics as a set of active values guiding everyday choices. This thought experiment demonstrates that other-directed comedy can facilitate radical perceptual shifts and re-evaluations of culturally inherited attitudes to both nonhuman animals and other humans.[7]

Imagining an Event That Will Turn Everyone Into a *Naked Derrida*

In my imagined future, *Naked Derrida* (ND) is a think tank and content production network popularising posthumanist humour. It represents all kinds of nonhuman animals – charismatic and, according to conventional taste, abominable, young and aged, captive and free-living, loved and despised, overabundant and endangered. The core product of the movement is an animated comedy series, *Naked Derrida*, featuring diverse groups of animals whose stories introduce multiple other-than-human points of view to facilitate a shift in perceptions of animal species and individuals.

Each ND guest, be they a companion, pest, exotic animal from the zoo, or stray, must have observed humans in some way. This is the only criterion potential guests have to meet. Their opinions on humanity are the motif of the show. Targeting a mass audience, ND producers must be honest with each other: humans are primarily interested in themselves and will listen to a fly if only she has a funny, human-centred story to tell.

The show's pilot episode presents a thought-provoking discussion between a pet otter, the ghost of a white cockroach, a chevrotain (mouse-deer), and a feral black cat with a tipped ear. The roach, known as Survivor, speaks of human curiosity. Once, he met a human during moulting, one of the most vulnerable moments of his relatively long seven-month life. Terrified and unable to move, he amazed the human predator with his body colour, which resembled her favourite white chocolate. This was enough to make her put down the slipper. The chevrotain, known as Saber-toothed Mouse, tells of being terrified of the overexcited humans he used to see at the zoo, and describes being in awe of Survivor and his species' resourcefulness. Survivor explains that the secret to perseverance has nothing to do with his panoramic vision (cockroaches' eyes are wrapped around their heads) but his learning to wait out moments of danger on the surface of a very expensive painting and squeezing into a bin of organic waste at night. Keanu the feral cat, on the contrary, despised his formerly domesticated apartment life. He could not bear the smell of jealousy emitted by his guardian. Back then, his ear was intact,[8] and his human was envious of his great morning looks. The stench of this envy remained in the flat for hours on weekdays after she left for work and permeated weekends. Keanu was sick of licking off her kisses and eventually lost the appetite for tuna flakes. "Once she was upset for a whole year," he recalls, "she thought she turned 30, but all this time she was still 29. On the eve of her actual 30th birthday, the mistake came out, and she was furious." Aiko, the smiley otter with a mosquito voice, one of nine 'pets' in the busy household of an Instagram influencer, agreed: "Patriarchy tampers with their brains."

Appealing content eligible for a prime time slot has to be, or at least masquerade as, human-oriented. Nevertheless, ND creators manage to put the show into service of nonhuman animals and create the glimmer of a future where all creatures are appreciated for their intrinsic value. This effect requires a combination of comedy and simulative empathy. Dialogues written for *Naked Derrida* offer humans a "way out of self-directedness" by making them face their existential crises through the eyes of the many other species they live alongside (Aaltola, 2018, p. 35). To imagine and simulate the experiences of animals, ND producers rely on scientific research, critical texts, and empathetic fictional narratives. An interdisciplinary perspective helps them contextualise and politicise animals. Thus, while having a great laugh, ND fans are "diving into what it may be like to inhabit the body-mind and context of another animal" (ibid., p. 7).

In their attempt to bring these "simulative takes" (ibid) into perspectives of animals closer to a mass audience, ND's think tank initiates collaborations with artists and comedians managing successful profiles on social media. After all, the ND movement was originally inspired by zoomorphic characters featured in artistic videos appearing on Instagram, such as the *2 Lizards* series[9] created by New Yorkers Meriem Bennani and Orian Barki in 2020, during the first months of the Covid-19 lockdown (Joseph, 2021, n.p.).

In this series, Benanni and Barki used animal characters as avatars of humans. Their videos feature lizards and other animals talking to each other and reflecting upon their pandemic experiences. Some scenes are funny, others romantic, both emerging from the disruptions of corporate routine and forced aloneness. While humans around the world were curating their bookshelves to have smart titles appear in the background of Zoom conferences, Bennani and Barki's *humanimals* dared to laugh at the crisis of the ordinary out loud, making self-ironic confessions, voicing thoughts that usually remain unspoken. "I feel like most people who've been working from home have been working without their pants on," says one of the lizards in the last episode of the series (Bennani & Barki, 2020). At the same time, companion animal influencers across Instagram were discussing the insulting behaviour of their guardians not washing their leggings for months. None of these profiles pursued animal rights agendas per se, nevertheless, they managed to inspire alternative ways of perceiving animals by imagining their thoughts and feelings in new circumstances.

In Bennani and Barki's series, animals are infatuated by all the things they put aside before the pandemic. Lizards enjoy empty streets and the opportunity to spend time with themselves, while some of their friends (mostly represented by noncharismatic species) explore their sexuality, protest against systemic racism, or finally focus on work, which presumably was not so easy before social distancing. Something about these 3D lizards, dragonflies, and racoons attracted thousands of viewers. One reason could be that they did not sound judgmental of humans. After all, they were designed as humans in animal bodies, however, somewhat more honest – not afraid to appear vulnerable and deprived of control over their individual futures, bringing to mind the struggles of the city 'pests' from another animated comedy series, *Animals*,[10] in which rats, pigeons, and cats focus on the brutally mundane, not-so-promising present.

The ND team decides to extend these humorists' praxis of ironic reflection by prioritising the animal gaze. The task is challenging. While exposing patriarchy through high-quality comedy, they must think about how to resist the celebrification of animals appearing on the show. Also, they must find ways to make viewers quit the habit of ranking other species by their charisma. Finally, they must decide how to turn human-centred discussions into a driver for the appreciation of animal otherness. At this point, ND producers decide to develop an AI algorithm to match species and personalities for each conversation (episode) to ensure none of the guests are disadvantaged in relation to the most 'charismatic.'

All characters must score highly in terms of attractiveness to humans but not high enough to boost impulse buying or adoption. None of the show's characters are corporeal. To distance the viewer from the appeal of realistic bodies, *Naked Derrida* features only CGI-animated animals. Similar to millennial monsters manufactured by Pixar, all characters appearing on the show are designed to look cute. The decision to sweeten the appearance of repulsive bodies is problematic, but in order for animals to be heard, ND creators agree to compromise and balance the comforting and challenging effects of the cute in favour of their mission. To avoid celebrification, ND never shows the same character twice. Audiences are left to wonder what species and personalities will feature in the next episodes. There are no hosts. A maximum of four guests meet on their own in a Zoom-like teleconference that allows all participants to be equally represented. Their bodies are merged with minimalist backgrounds, each animal depicted as the centre of her or his own universe.

After each premier, ND fans cut the four-part frame into separate memes. These circulate on social media for months, exploiting the free labour of platform users, and serving as existential ruptures in the anthropocentric worldview. Millions look up ND guests' social media profiles and subscribe. Within a week or two of each release, every ND character becomes a mega influencer. To avoid homogenisation and automation, content is constantly reinvented by the ND team. The only thing that does not change is their motto, a quote from Mary Douglas: "A joke is a play upon form . . . [that] affords opportunity for realising that an accepted pattern has no necessity" (Douglas, 1999, p. 150).

Finally, when the audience's attention shifts to narratives, the ND creators let go of cute animal faces and initiate a second season. This shows the world through the eyes of fictional nonhuman characters as they go about their everyday lives, inviting viewers to acknowledge how rarely animals look at humans. However, to keep audiences invested, the animals continue to discuss what they think it is like to be human nowadays, how it happened that the people they meet do not belong to themselves, become lost in things that exhaust them, and forget how to enjoy life. A bat asks, "Can you imagine, instead of sleeping, my human was scrolling through social media accounts of sleeping animals from 2 to 4 am." She is happy that she could get her morning dinner before her guardian went to sleep. The bat's cat neighbour also felt blessed with his human servant becoming nocturnal, losing their sense of time and switching to an automatic pet feeder, which is more disciplined and predictable.

The second season of ND does well in translating animals' umwelten to humans and affirming their otherness. Posthumanist humour becomes the new black. Those who still objectify animals on social media lose followers and advertising contracts. ND episodes are rotating everywhere, facilitating spontaneous collective experiences of laughter in the subway, in waiting areas of banks, coffee shops, clinics, and airports, and in school canteens, prisons, and hospices. By the time the sixth episode of the first season is released, the show and social media profiles of its characters are being studied at universities as a form of political technology. Despite the success, the show's authors remain unknown to avoid the consequences of fame and prevent the shift of audiences' attention from the characters to their creators.

Self-Directed Humour as a Path Towards Other-Directedness

In essence, the *Naked Derrida* fantasy actualises the possibility that a few scattered intuitive attempts of using voiced animals to produce media content that is appreciative of nonhuman otherness can inspire a series of purposeful, carefully orchestrated intellectual interventions that challenge anthropocentric representations of animals and affirm multiple nonhuman perspectives as a new standard of content production.

While imagining this "futurability" (Berardi, 2017, p. 3),[11] I came across several challenges. How can animal voices be stress-relieving and subversive at the same time? What are the specific circumstances in which anthropocentric perspectives can be reversed without detracting from the positive implications of animal presence across media? And most importantly, what qualities does animal content have to possess to facilitate the adoption of posthumanist ethics as a set of active values guiding everyday choices?

Self-deprecating humour[12] as a type of content that is amusing and, at the same time, intrinsically antagonistic to the politics of treating animals as inferior to humans, seems promising in terms of bringing posthumanist ideas closer to the public through the same channels that are currently utilised to commodify animals. It is true that in this scenario, animal characters featured in ND and circulating in memes across social media still perform the role of affective labourers. However, unfolding through subversive strategies of radical humour, such as incongruity, 'their' efforts finally benefit them and not only the platform owners and content producers. The fusion of two "incompatible codes" (Weeks, 2020, p. 4) – cute appearance and acute voice – allows them to engage audiences in the paradoxically enjoyable practice of redemption or, in Critchley's words, a "shared prayer" for more attentive and compassionate coexistence based on the acknowledgement and appreciation of difference (2004, pp. 16–17).

It is worth mentioning that any intellectual interventions pursuing the transformation of social norms within the existing framework of profit-oriented media technologies that encourage competition, acquisition, and surplus labour, can cause certain side effects, such as the "commodification of otherness" (Abidin, 2018, p. 40). Therefore, the process of content production itself has to be transgressive, outpacing simplification and automation. To unpack the idea of "countercultural" components of comedy, those that "fundamentally subvert conventional social codes," Mark Weeks refers to Deleuze's *Nomad Thought* (1977) (2020, p. 8). For Deleuze, it is essential that the affect produced by the incongruity-driven humour (or superior irony) of Kafka or Beckett's prose is "nomadic," by which he means that it provokes an endless interpretative process, the very shift "from reactive being toward active becoming" (ibid., p. 10). The question is, what does it take to induce this "nomadic existential orientation towards difference" on a mass scale (ibid., p. 11)?

To prevail over most probable futures, marginal tendencies pursuing paradigm change require a disruptive event. Sometimes, this takes the form of innovative technology and, at others, philosophical argument. Defining effective intellectual interventions against political 'common sense,' David Graeber and Andrej Grubačić argue that such events "reveal aspects of reality that had been largely invisible but, once revealed, seem so entirely obvious that they can never be unseen" (2021, p. 18). To expose anthropocentric perspectives on animals as harmful for both humans and nonhumans, I imagine ND as a movement that supports consistent demystification of human–animal relationships by presenting a series of viewpoints on their interactions based on what is known about various lifeforms, their cognition, behavioural traits and history of adaptation to anthropogenic environments. In this sense, ND represents a specific case of what Franco Berardi calls a "conscious force intended to dismantle and reprogram the world according to the concrete usefulness of knowledge" (2017, p. 7).

To make the alternative future attainable, the disruption of the current code has to cause media furore, preferably based on positive affects (sensations, emotions, sentiments) (Mavrodieva et al., 2019; Roberts, 2021; Berger & Milkman, 2013). In contrast, exposure to disturbing images of animal suffering can evoke feelings of powerlessness and 'compassion fatigue' (Aaltola, 2014, p. 19) manifesting as "desensitisation and emotional burnout . . . associated with pervasive communication about social problems" (Kinnick et al., 1996, p. 687). Referring to pervasive communications of ecological harms that invoke individual culpability, Tim Jensen similarly describes the resulting environmental guilt registering "as a flicker of pursed lips when tossing a to-go coffee cup in the garbage," scrolling past too-familiar stories of endangered species, and general feelings of helplessness (2019, pp. 1–3). He suggests that this form of political impotence is one reason why "cognitive and affective awareness of one's participation in massively distributed acts of violence" rarely translates into action against it (ibid., p. 135).

My belief in the transformative potential of posthumanist comedy comes from questioning the capacity of negative emotions, such as shame and guilt, to facilitate ethical commitment, honouring the intrinsic value of nonhuman lives. As Jack Snyder points out, referring to human rights advocacy movements, "shaming is likely to produce anger, resistance . . . denial and evasion" (2020, p. 109). Furthermore, when weighed against each other, the not uncommon desire to ensure the wellbeing of animals, and the options actually available to achieve it, appear utterly disproportionate (Jensen, 2019, p. 135). It might sound unfair to animals suffering in silence all over the world, but the work of @sadanimalfacts and other content producers demonstrates that effective animal advocacy can be uplifting and liberating of guilt, drawing on the fact that every individual, human and nonhuman, faces countless frustrations and tries to cope with them. That is why, imagining ND as an alternative to the mainstream activist rhetoric of shame, I chose to avoid images of animal suffering and focus on comedy as an alternative way to expose the culturally inherited patterns of systemic (self)oppression structuring human values and actions that directly and indirectly affect animals.

In the context of participatory media, thousands of nameless victims are less powerful than an individual whose personality is contextualised and made complex. In the same way, one cute pug posting images every day can be more influential than news of a pod of whales beaching themselves to stay with a wounded member stranded on the shore. Aaltola rightfully argues in favour of narratives inviting "familiarity" with animals, a "recognition that individualises them from the faceless, distant, generic mass of 'animality'" (2018, p. 38). Sharing her desire to see posthumanist ethics in the core of mass action, I paraphrase her question – what should animal advocates do to make heterogenic openness practically relevant as a set of values structuring quotidian life? (ibid., p. 7). In response, I would say that, along with stories centred around animal individuality, we need more other-directed narratives based on positive affect (Roberts, 2021, p. 819). Non-anthropocentric content supplied in an enjoyable form can operate within the framework of mass entertainment like a Trojan horse, posing as an enjoyable time-filler and an alternative to doomscrolling while acting as a disruptive agent gradually rectifying biased perceptions of animals and reattuning social media users to the ideas of interspecies respect and cooperation.

To render widespread anthropocentric patterns of animal representation obsolete, the exposure to alternative ways of portraying them should be consistent. Sensational stories, both devastating and delightful, are easily forgotten unless their shelf life is long enough to make them a part of everyday life (Goode & Nachman, 1994, p. 155).[13] Therefore, the course of events interrupting the objectification and commodification of animals on social media and television has to be systematic. This is one reason why, unlike some activists I have encountered during my research, I do not argue for a ban on animal influencers on Instagram, Facebook, and other mainstream platforms. On the contrary, I offer to admit their efficacy as a format based on serial narratives.

Along with literary fiction, Aaltola argues that films and video games created to evoke simulative empathy can facilitate responsible representation and treatment of animals, functioning as a "bridge between theory and practice, between abstraction and our lived experiences" (2018, p. 19). On mainstream social media, where the simplification of content and reinforcement of stereotypes can beget ethical inertia, other-directed narratives risk losing the nonhuman other as a "reference point" (ibid., p. 37). This can lead to the transformation of content to become "one-sided, misleading and colonising" (ibid.). However, this risk is not a reason to avoid practising heterogenic openness through animal influencers and neglect the opportunity to push alternative voices to the point where they will be prioritised by recommendation algorithms. Social media can be effective in distributing other-directed narratives in terms of consistency of exposure and influencers' strong emotional connection with their audiences.

@sadanimalfacts, relying on informed imagination, successfully translates "dissimilarity into something we can more fluently comprehend without replacing it with similarity" (ibid., p. 36). It invites familiarity with 'noncharismatic' species by making them look cute and, at the same time, articulate. This praxis of imagining the experiences of others, informed by scientific studies of animal anatomy and behaviour and expressed in the form of superior irony, invites social media users to rethink the ethics of human interactions with other animals and search for better answers to these essential questions: (1) what efforts might improve the wellbeing of particular species and individuals? (2) when is it necessary to participate in their lives? and (3) when is it vital to let them be?

Conclusion

Studying representations of animals on social media and in mainstream entertainment (television series, video games, stand-up comedy) offers a valuable opportunity to identify the anthropocentric values behind conventional perceptions of animals, discover effective strategies to resist these attitudes, and create conditions that foster their transformation. Listening to some of the voiced animals proliferating on social media and in popular culture can teach us how to "situate lives of others in their political context, offer these lives credible, tangible, realistic agency and introduce ways to take action" (Aaltola, 2018, p. 44). Narrative forms created on behalf of animals are to be encouraged if they help evoke empathy for nonhuman others and resist their exploitation and suffering.

If posthumanist humour were to saturate stand-up comedy and other popular content experiences bringing joy, it might play a significant part in animal liberation in all spheres of oppression, including farming, testing, breeding animals for appearance, and exotic pet keeping. Be it profiles of lifestyle influencers or K-pop idols, fantasy series, or action-adventure video games, there is always room to exercise the 'soft power' of cultural products (Mavrodieva et al., 2019, n.p) with a delicate message challenging anthropocentric attitudes towards animals. As Berardi asserts in *Futurability*, the "future is not prescribed but inscribed" (2017,

p. 235). The self-deployment of optimistic *possibilities* into actualities requires the invention of concepts and aesthetic forms that can be shared and applied in the present as criteria for responsible content production, use of technologies, and practices of care (ibid., p. 236).

The simulacra of voiced animals dispersed across media contain multiple possibilities that can be defined and translated into ideas beneficial for all species. By mapping these alternative trajectories, we may identify new ways of reducing the adverse effects of neoliberal cultures mobilising affective powers of cute animal bodies primarily to intensify consumption. Instead, animal subjectivities can be put at the core of our resistance to "repressive desublimation" – a mechanism that "liquidat[es] the oppositional and transcending elements" from cultural narratives, with essential effect and function of these subjectivities to stand "in contradiction with social reality" and to highlight areas for improvement (Marcuse, 2007, pp. 59–60). The practice of amplifying more-than-human voices that call for pro-species plurality can become a way to "hit the system from without" (ibid., p. 261) and herald a future beyond convention.

Notes

1 The Grumpy Cat (2012–2019) was a cat with 'grumpy' facial expression caused by an underbite and feline dwarfism. She became an Internet celebrity in 2012 after her photo was posted on Reddit. Throughout the following years, her profiles on Facebook, Twitter, and Instagram attracted millions of subscribers.

2 An empathy evoked by other-directed projection – an "imaginative exchange of personas" with the other, "envisioning what it must be like" to be this other being (Aaltola, 2018, pp. 28, 31).

3 Here and further, using the phrase 'responsible representation,' I refer to a desirable state of content production within which the practice of portraying animals and animal characters for entertainment, advertising, and other forms of content with high impact is guided by the concern about animal wellbeing and the influence of particular representational patterns on nonhuman lives and ecosystems.

4 Although the act of 'voicing' real or fictional animals is a form of anthropomorphism, in the case of posthumanist comedy analysed in this chapter, the act of endowing animals with the ability to speak like humans serves as a critical instrument against speciesism. Here, ironic messages delivered by animals exemplify critical anthropomorphism, which aims to highlight differences between animals and humans by attributing exclusively human qualities to animals in an exaggerated way (Krylova, 2023, pp. 315–316).

5 The history of sleep reduction to approximately five hours per night in late capitalism is analysed in Jonathan Crary's book *24/7* (2014).

6 This title refers to the famous bathroom encounter between naked philosopher Jacques Derrida and his cat, described in *The Animal That Therefore I Am* (2008). The incident is represented as an event that encouraged Derrida to focus on the "insistent gaze of the animal . . . a seer" and his feeling of shame for being "naked in front of this cat" and for feeling "ashamed for being ashamed" (2008, p. 4).

7 While the artworks, texts, content creators, and scholars who inspired and shaped *Naked Derrida*, are real, the characters and events associated with this utopia are fictitious.

8 In New Zealand, France, Canada, and some other countries 'ear tipping' is used as a preferred method to mark spayed or neutered stray cats in order to identify them easily during further encounters.

9 A total of eight episodes of *2 Lizards* were published on Instagram by @meriembennani in March-July 2020.
10 Broadcast by HBO in 2016–2018.
11 Berardi coined the term "futurability" to describe the layer of possibility that lies in the "texture of present reality . . . and consciousness" and, depending on the number of factors, "may or may not develop into actuality" (2017, pp. 3, 17).
12 Here, I refer to the human 'self' perspective, since the content is created by humans trying to look at themselves through animals' eyes.
13 Goode and Nachman refer to the process of sensitisation through repeating "exaggerated, sensationalistic stories" as a way to escalate minor disturbances, which, in case of the negative news, leads to moral panic (1994, p. 155).

References

Aaltola, E. (2014). Animal Suffering: Representations and the Act of Looking. *Anthrozoös, 27*(1), 19–31. https://doi.org/10.2752/175303714X13837396326297.
Aaltola, E. (2018). *Varieties of Empathy: Moral Psychology and Animal Ethics*. Rowman & Littlefield.
Abidin, C. (2018). *Internet Celebrity: Understanding Fame Online*. Emerald Publishing Limited.
Animals. (2016–2018). Created by P. Matarese & M. Luciano. HBO Entertainment, Karen BBQ, Duplass Brothers Television, Starburns Industries, seasons 1–3.
Barker, B. (2022). Interview. Conducted by K. Krylova, October 7.
Bennani, M., & Barki, O. (2020). *2 Lizards*. https://www.instagram.com/meriembennani/?hl=en
Berardi, F. (2017). "Bifo". *Futurability: The Age of Impotence and the Horizon of Possibility*. Verso.
Berger, J., & Milkman, K. L. (2014). Emotion and Virality: What Makes Online Content Go Viral? *NIM Marketing Intelligence Review, 5*(1), 18–23. https://doi.org/10.2478/gfkmir-2014–0022.
Buchanan, I. (2010). *A Dictionary of Critical Theory*. Oxford University Press.
Colebrook, C. (2004). *Irony*. Routledge.
Crary, J. (2014). *24/7: Late Capitalism and the Ends of Sleep*. Verso Trade.
Critchley, S. (2004). *On Humour*. Routledge.
Deleuze, G. (1994). *Difference and Repetition* (P. Patton, Trans.). Columbia University Press.
Derrida, J. (2008). *The Animal that Therefore I Am* (M.-L. Mallet, Ed.). Fordham University Press.
Douglas, M. (1999). *Implicit Meanings: Selected Essays in Anthropology*. Taylor & Francis Routledge.
Eagleton, T. (2019). *Humour*. Yale University Press.
Goode, E., & Nachman, B.-Y. (1994). Moral Panics: Culture, Politics, and Social Construction. *Annual Review of Sociology, 20*, 149–171. https://doi.org/10.1146/annurev.so.20.080194.001053.
Graeber, D., & Grubačić, A. (2021). Introduction. In P. Kropotkin, *Mutual Aid: An Illuminated Factor of Evolution* (pp. 18–25). PM Press.
HouseBroken. (2021). Created by G. Allan, J. Crittenden, & C. DuVall. Kapital Entertainment, Bento Box Entertainment, Merman, AllenDen, Fox Entertainment, animated series, seasons 1–2.
Jensen, T. (2019). *Ecologies of Guilt in Environmental Rhetorics*. Palgrave Pivot.
Joseph, M. (2021). Crisis Ordinariness: Meriem Bennani and Orian Barki's *2 Lizards*. *Magazine*, February 8. www.moma.org/magazine/articles/505

Kinnick, K. N., Krugman, D. M., & Cameron, G. T. (1996). Compassion Fatigue: Communication and Burnout Toward Social Problems. *Journalism & Mass Communication Quarterly, 73*(3), 687–707. https://doi.org/10.1177/107769909607300314.

Krylova, K. (2023). *The Market of Convenient Animals.* New Literary Observer.

Marcuse, H. (2007). *One-Dimensional Man: Studies in the Ideology of Advanced Industrial Society.* Routledge.

Mavrodieva, A. V., Rachman, O. K., Harahap, V. B., & Shaw, R. (2019). Role of Social Media as a Soft Power Tool in Raising Public Awareness and Engagement in Addressing Climate Change. *Climate, 7*(10), 122. https://doi.org/10.3390/cli7100122.

May, S. (2019). *The Power of Cute.* Princeton University Press.

McGurl, M. (2012). The Posthuman Comedy. *Critical Inquiry, 38*(3), 533–553. https://doi.org/10.1086/664550.

Roberts, J. (2021). Empathy Cultivation through (Pro)Social Media: A Counter to Compassion Fatigue. *Journalism and Media, 2*(4), 819–829. https://doi.org/10.3390/journalmedia2040047.

Snyder, J. (2020). Backlash Against Human Rights Shaming: Emotions in Groups. *International Theory, 12*(1), 109–132. https://doi.org/10.1017/S1752971919000216.

Weeks, M. (2020). Affect Philosophy Meets Incongruity: About Transformative Potentials in Comic Laughter. *The European Journal of Humour Research, 8*(1), 1–13. https://doi.org/10.7592/EJHR2020.8.1.weeks.

Wild Life. (2020). Created by A. Davies. Nice Try, Octopie Network, Valparaiso Pictures, TZGZ Productions, animated series.

Personal Shifts and Transformations

DOI: 10.4324/9781003366706-16

Chapter 12

Love Beyond the Species Divide in Nizami Ganjavi's *Layla and Majnun*

Alexandra Isfahani-Hammond

I couldn't lie anymore, so I started to call my dog, 'God'

– Sant Tukaram Maharaj

I. Introduction

Might Islamic mysticism portend a remedy for humans' oppressive relationships with other creatures? I explore this question by contemplating transformative affective spaces. As mindsets alter, species boundaries erode, forging imaginal apertures to futures without animal exploitation. My point of origin is the intimate sphere I shared with my dog friend, Akbar, during his final months. I proceed to the adjacent poodle-human love story in André Alexis' *Fifteen Dogs* (2015). These sites lead ultimately to the wilderness of Nizami Ganjavi's classic Sufi text *Layla and Majnun* (1188), a visionary universe hinged on altered consciousness that reverberates through the centuries to prefigure post-anthropocentric kinship alliances.

Writing about Akbar began as the fulfillment of a need to honor him, going public about my love for the youthful dog who led me to veganism and critical animal studies (Isfahani-Hammond, 2013). Later, when he was elderly and ill, I wrote about my ministrations to him, describing the "world within a world" wherein we cared for one another (Isfahani-Hammond, 2021). Now I write to keep him with me. Finding my experience of love and loss reflected first in Majnoon, the poodle, then Majnun, the mystic, I offer this exploration as balm and inspiration for the bereft.

In December 2018, when Akbar returned home from the hospital after two major surgeries with a prognosis of days or weeks to live, I transformed our apartment in an effort to induce rest, eliminate stress, and give him the best possible chance of surviving. In "Akbar, My Heart" (2021), I describe the locale I came to think of as an incubator, including:

> my 24/7 presence, sound healing music, lavender oil diffuser, a nest by the fire and yoga mats lining the floors providing traction to keep his legs from splaying out. An obstacle course took up half the living room for days when he was up

DOI: 10.4324/9781003366706-17

to doing physical therapy intended to slow muscle degeneration. A pyramid of five dog beds functioned as stairs he could climb onto my bed, which was itself encircled by a foam moat to soften potential falls.

(pp. 17–18)

These ministrations embody what Katherine Gillespie defines as a multispecies death doula approach: "a political project of centering how animals' lives matter to themselves and others and how their deaths are ones to be mourned and acknowledged as significant losses" (2020, p. 3). Crucially, this practice has implications for broad, structural change:

This can begin with the animals with whom we live closely but this form of care at their end of life can model a radical shift in how it is possible to think about the lives and deaths of other animals beyond those closest to us.

(ibid., p. 4)

Adding to Gillespie, the locales wherein such epistemological transitions occur are what Michel Foucault defines as spaces "whose functions are different or even the opposite of others" (2002, p. 361) that "suspect, neutralize, or invert the set of relations that they happen to designate, mirror, or reflect" (1967, p. 3). The love-filled dwelling Akbar and I inhabited – not for days or weeks but for a full 19 months beyond what the vet predicted – was life-altering in more ways than one. In addition to the revamped function of our co-created space, I experienced our subject positions shifting into transformative juxtaposition:

The tapping of his claws against the floor behind me took on new meaning. He wanted to be near me, yet I wanted even more so to be near him. I stopped at every opportunity, kneeling down to hold his head in my hands: 'Are you following me? I'm following you, my love. Please know that. It is me who is following you.'

(2021, p. 17)

Our bond intensified, our identities were "confused," and I felt ever more assured that there was no separation between us. Still, I never ceased dreading Akbar's physical disappearance. I inhaled his scent in an effort to incorporate him, seeking a "solution" to his inevitable departure.

On Akbar's last morning in his body, I took him to a forest clearing to lay on a blanket together underneath a canopy of trees, their interlaced branches forming a protective dome against the robin's-egg-blue sky. Like the incubator of my apartment, the clearing was transitional. Though I'd thought of returning there for solace in his absence, I found that doing so was painful, so I turned to literature instead. Just as writing about Akbar keeps him close – I encourage him to accompany me, while he encourages and guides me – my grief-fueled study trajectory led to new and uncanny sites of communion with him. A few days after he passed away, I dreamt that he was a squirrel swimming in a stream, beckoning me to follow as the stream merged into a vast, dark ocean. I wasn't sure where it would lead but knew I had to jump in anyway.

I began with André Alexis' *Fifteen Dogs* (2015), a novel about a group of free-roaming Torontonian dogs wherein a central theme is the bond between the poodle, Majnoon, and the human, Nira. In "Dogs without Masters: Astray with Akbar and in André Alexis' *Fifteen Dogs*" (2022), I explore the "divine intimacy" (p. 144) that evolves between them, subverting the master/dog paradigm. Nira refuses to allow anyone to refer to Majnoon as "her" dog: "I'm as much his as he's mine," she says (p. 132), while Majnoon sees Nira as "a being who completed him, made him more than he would otherwise have been" (p. 142). As he puts it, "this human is not a master. I do not know what Nira is, but I am not afraid" (p. 71). As species binaries erode, Nira identifies with Majnoon to the extent that she "had the odd but fleeting sensation that Majnoon had entered her consciousness in some new way" (p. 125). This echoes Majnoon's experience of the divine when the god, Hermes, visits him in a dream: the deity, disguised as a poodle, "spoke no particular language. Its words were in Majnoon's mind, like a strange idea" (p. 123). Nira is initially destabilized and Majnoon feels "adrift between species" (p. 47), while "the distance between (them) narrowed until each could anticipate what the other wanted . . . By degrees, they had less use for words or English" (p. 132). Nira and Majnoon are ultimately so identified with one another that their sense of individuated selfhood dissipates:

One morning [Majnoon and Nira] discovered that they'd dreamed of the same field, the same clouds, the same house in the distance – wooden with a red-brick chimney. They had dreamed of the same squirrels and rabbits. They had drunk from the same clear stream. There was only one difference: when Nira, in her dream, looked into the water, she saw Majnoon's face reflected back at her, while Majnoon, in his, saw Nira's face where his should have been.

(p. 132)

Following the thread of Alexis' narrative, I proceeded to Nizami Ganjavi's *Layla and Majnun* (1188), with which *Fifteen Dogs* is in dialogue.[1] Therein, I beheld the unfolding of yet another heterotopia – one that both provides direction about how to carry on in Akbar's physical absence (Akbar is everywhere) as well as how to most skillfully use this incarnation. A zoocritical interpretation of Nizami reveals that altered consciousness – the condition of being "madly in love" – is a healing pathway to obliterating species dialectics, compelling humans to discard their arms.

Whereas animals in religious and literary texts are frequently deemed figurative,[2] I propose to remove the quotation marks around Majnun's animal friends, avowing their historicity and moral import, heeding Carol Adams' (1990) appeal to recompose animals absented through metaphor, and engaging in what Laura Wright and Emelie Quinn conceptualize as "reading through a vegan lens," expanding rather than foreclosing possibilities by attending to "the role of literary texts in forming, shaping, developing, exploring and challenging our relationship with nonhuman animals" (2022, p. 3). *Layla and Majnun* is by no means free of anthropocentrism nor other oppressive biases. It centers humans, whereas the animals are frequently figured as a shapeless mass lacking character individuation. Yet in Nizami's insistence upon kindling care-centered relationships with animals and his account of Majnun releasing his human façade, I identify the seed for dismantling the pretense of human exceptionalism. Such an analysis is indebted to Evan Maina Mwangi's (2019) consideration of postcolonial African fiction that, while not overtly engaged with animal liberation, exhibits a prefigurative "vegan unconscious" defined as "a latent possibility that the societies portrayed in the text will embrace a future where animals will not be killed to satisfy human needs" (p. 9).

Omid Safi (2018) distinguishes the Islamic mystical tradition of *Mazhab-e eshq* as a path of "radical love," discerning it from the flat, quotidian use of the word, "love," to denote something more "fiery, fierce and alchemical," a love that "constantly spills over again and again, overflowing whatever cup seeks to contain it" (p. xxi). With this reading, I venture the provocation that Nizami's treatment of human/animal relations surpasses metaphorical enclosures, advancing a conceptualization of fundamental tenderness that transgresses the species divide. I came late to an exploration of Persian literature, led by Akbar to the Sufi cosmology woven into my DNA, which I also like to think is an introduction to my grandfather, Saleh, the gentle mystic who provided shelter to neighborhood stray dogs and cats. My requirement of an English translation of *Layla and Majnun* inevitably generates oversights. For these, I'll request expert readers' generosity.

II. The Hunt Is Forgotten

"Layla and Majnun" is the best-known love story of the Muslim world. Of Arabic origin, it has been recounted in poems, songs, and epics from the Caucuses to the interior of Africa, from the Atlantic to the Indian Ocean (Gelpke, 1966, p. 200). I focus on the most famous version, by twelfth-century Persian poet Nizami Ganjavi, who was the first to utilize all the traditional versions, far-ranging and varied in detail, which he molded into a singular narrative poem (1188, p. 200). Specifically, I accompany Majnun's plight in the dog-eared copy of this legend which my mother gave to me as a teenager but which remained overlooked on a shelf until, grieving for Akbar, he guided me to crack open its yellowed cover.

Qays and Layla meet in their youth and fall deeply in love, but Layla's father deems Qays' enthusiasm excessive and improper, rejecting him as his daughter's suitor. Qays is so bereft that he comes to be called "Majnun" ("crazy"). His yearning inspires him to compose stunning quatrains that are repeated by word of mouth and eventually reach Layla's ears, while she also writes messages she scatters to the winds so that the birds carry their fragments to Majnun. Though his family and friends attempt to intervene, restoring Majnun's seemlier demeanor, he insists on delving ever more deeply into his longing for Layla.

Spatial translocation is an originary condition of Majnun's love journey. He is at different points described as living in the desert, on a mountain, in a gorge, a cave, steppes and, simply, the wilderness. When Nizami calls him "the wanderer through the world" (p. 185), this speaks to acceptance of the transitory nature of this plane of reality. Indeed, "[h]e who remains a stranger in this world and wanders, restless as the moon at night, will find peace" (p. 120). But alongside recognition of impermanence, Majnun's roaming enables access to a domain imbued with love. Fleeing from his parents' abode, he announces, "My home is my love; nowhere else am I at home" (p. 168). Angrily, Majnun's father endeavors to make him "accept this world as it is" and desist from "living in the wilderness like a beast among beasts" (p. 115), but Majnun quashes his efforts, declaring, "I am lost in my own wilderness! I have become a savage with wild beasts as my companions. Do not try to bring me back to the world of humans!" (p. 118). Together with his flight to an alternative terrain, the façade of Majnun's human identity dissipates. He is "a creature writhing on the stones like a snake" (p. 42) who "moved on all fours over the ground like an animal" (p. 113).

Leaving humans' turf to reside with animals, they become Majnun's companions:

He had come as a stranger into their realm, yet had not hunted them. He had crept into their caves without driving them out. Just as they, he was afraid and fled whenever men approached. Did Majnun, therefore, appear to the animals like an animal himself? Not entirely: they sensed that he was different. He possessed a strange power, unlike that of the lion, the panther or the wolf, because he did not catch and devour smaller animals. On the contrary, if he found one of them caught in a trap, he stroked its fur, talked until it had calmed down, and then released it. Why? What kind of a creature was he? Who could understand him? He fed on roots, grass and fruit – but even of these he ate sparingly – and showed no fear of the powerful four-legged beasts of prey which could so easily have torn him to pieces and devoured him. Yet they did not do so.
(pp. 126–127)

The heterotopology Majnun inhabits upsets normative anthropocentric relations of domination. "Creeping into their caves without driving them out," the wilderness is a space whose functions are the opposite of others. This alterity is reflected in Nizami's description of the region where he dwells as a foreign country: "He lived among all these creatures like an exiled ruler in a foreign country," a king who "never oppressed his own subjects, nor extorted taxes, nor sacrificed their blood to make war on other people" (pp. 127–128).

Majnun's boundless love for Layla causes him to see her reflected in the animals. He becomes their protector, engaging in counter-conduct to persuade hunters to discard their arms. At one point, Prince Nawfal is stalking some antelopes and wild donkeys who try to escape into the mountains. When Nawfal comes upon two or three of them huddling together in the semi-darkness, flanks trembling, at the entrance to a cave, he stops short. Behind one of the antelopes, he spots

Majnun, "a living being such as he had never encountered before" (p. 62). About this "creature," he wonders, "Was it an animal or a human being, a savage or one of the dead – maybe a demon?" (p. 62). Majnun recounts to Nawfal the tale of his thwarted love, with the result that "Nawfal listened attentively and his sympathy for Majnun grew with every word. The hunt was forgotten" (p. 63). At another point, Majnun comes upon two gazelles caught in a hunter's snare and demands that the hunter free them:

> Let these animals go free!. . . . Is there not room enough in this world for all creatures?. . . . Look how beautiful they are. Are their eyes not like those of the beloved?. . . . Let them go free, leave them in peace! These necks are too good for your steel, these breasts and thighs are not meant to be devoured, these backs, which have never carried any burden, are not destined for your fire!
>
> (pp. 85–86)

The hunter has never heard words like these and his mouth opens in astonishment. Majnun kisses the gazelles' eyes, singing "Dark as the night, like hers, your eyes!" (p. 86), then frees them from their fetters and watches them disappear toward the horizon.

The very next day, Majnun interrupts another hunter, telling him not to kill a stag:

> Torturer of the weak and the defenceless! Release this poor creature at once so that it may still enjoy its life for a short while. How will the hind feel tonight without her companion? What would she say if she could talk with a human tongue! She would exclaim: 'May he who has done this to us, suffer as we do; may he never see another day! . . .' Would you like that? Do you not fear the distress of those who suffer? Imagine yourself as the stag – the stag as the hunter and you as his victim!
>
> (p. 88)

Once the hunter has gone, Majnun approaches the animal gently, stroking and caressing him:

> Like myself, are you not also separated from your beloved? Quick-footed runner of the steppes, dweller of the mountains, how vividly you remind me of her! Go, hurry, search for her, your mate. Rest in her shadow – there is your place.
>
> (p. 89)

Requesting that, should he encounter Layla's tent, he repeat some verses describing his longing for her, Majnun removes the noose from the stag's legs and sets him free.

This is not the only time Majnun speaks with animals, recognizing their agency and identifying his sense of separation with theirs. For instance, he speaks aloud to a raven, asking whom the "Blackfrock" is mourning and requesting that they convey some lines of his poetry of yearning to Layla. The love goes both ways. The animals become used to Majnun's presence in the wilderness and are drawn to him: "Catching his scent from afar, they came flying, running, trotting, creeping, drawing narrowing circles around him" (p. 127). In the same way that he seeks to protect them, they follow suit. A lion watches over him, with others soon following, first a stag, a wolf, then a desert fox, every day there were more. A visitor to this wilderness observes that "gazelles give you their love, and you stroke lions as though they were house-cats" (p. 143). On the one hand, there are references to Majnun's lack, as in "I have nothing left but these few beasts" (p. 182). On the other, Nizami addresses the reader, "Was their love less rewarding than that of human beings? Do not believe it" (p. 128). This inconsistency evinces tension between the entrenched anthropocentric worldview and Nizami's breathtaking counter-map, while by no means diminishing the audacity of Majnun's defiance nor the reworlding potential of his transformation from man to "creature" (p. 62).[3]

Empathizing with animals' grief when humans kill their kin, Majnun desists from eating them, subsisting on roots, grasses, and fruit. Though there is self-abnegation in his veganism, for Sufis, this is a pathway to restoring a sense of interconnectedness. The artifice of alienation is healed by annihilating the ego and, with it, the perceived isolation from the Beloved, who is at one and the same time celestial and terrestrial; there is no antagonism between God and one's earthly loved ones. Majnun's mended relationships with animals nourish him: "Watching them stills my hunger too" (p. 163). His nonviolent, creaturely solidarity leads even carnivores to desist from hunting in his presence, nurturing those who would have been their prey: "Leaving the human world, he had come to the wilderness reconciling wild ones with wild ones" (p. 129) such that "the lion kept his claws off the wild ass, the lioness gave milk to the orphaned baby gazelle and the jackal buried his age-old feud with the hare" (p. 128). Humans journey to Majnun's wilderness to behold this unusual site, a "peaceful army" (p. 130), which appears like a "fairy tale, a saga from times gone by" (p. 129). Extinguishing carnivores' biological need for flesh is obviously unrealistic, yet Nizami's utopianism accentuates the potential for radical rupture envisaged in his post-oppression wild place.

III. Despairing Bodies Separate

Nizami is known for two principal "updates" to the tale of Layla and Majnun. The first is that he infuses it with the symbolism of the mystical journey back to God. The second is that he adds Majnun's love relationships with animals. How is the journey to God related to transspecies affect?

In Sufi cosmology, the pain of separation is the foundation of mystical life. On the one hand, death is the gate to the real world, to relief from separation, and to seeing things as they really are. For example, when Majnun meets Layla in a garden, he says, "What *there* is one, down *here* is forced apart./ Yet if despairing bodies separate,/ Souls freely wander and communicate" (p. 178). Yet one can die an ego death prior to physical death, thereby alleviating suffering while still alive. When Nizami advises his reader to "die your own death in this life" (p. 120), the death of the individuated self is the pathway to God and, therein, alleviation of suffering. Therefore, rather than "getting over her," Majnun's prayer is to be more and more in love, not cured of his love as his father wishes. His yearning and grief are themselves the pathway to connection with Layla. Majnun extinguishes his separate self so that only one being exists: "take what is left of my life and add it to Layla's" (p. 38). Majnun describes the annihilation of his ego:

> Who am I? I keep turning upon myself, asking 'What is your name? Are you in love? With whom? Or are you loved? By whom? . . .' A flame burns in my heart, a flame beyond measure, which has turned my being to ashes.
>
> (p. 118)

Majnun's cure is reflected in his declaration that, having delved so profoundly into his yearning, there is no separation between love, lover, and beloved: "I am one of those . . . who have eaten the eater within themselves" (p. 183), with the result that, "Love has moved in and adorned the house, my Self has tied its bundle and left. You imagine that you see me, but I no longer exist: what remains, is the beloved . . ." (p. 184). At the same time, paradox is at the heart of mystical consciousness. Majnun may have eaten the eater within himself, yet still he agonizes over Layla's death and his physical separation from her. He "fell down on her grave as if struck by lightning" (p. 193) and "[l]ike a serpent, twisting and turning over the treasure which it guards, he writhed in torment and his tongue was a flaming torch of lament" (p. 193). Majnun implores the maker, "Let me go where my love dwells" (p. 196) and, clutching her headstone with all his might, he dies (p. 197).

Throughout Nizami's narrative, renunciation of the individuated self is enmeshed with co-creaturely tenderness. Majnun is reduced to a skeleton with skin but still predators do not pick at him; rather, "he rested in the shade of the vultures' wings" (p. 127). He sheds his human form, so that humans cannot place his species identity, nor are the animals sure of it – they do not perceive him entirely as an animal, but "sensed that he was different" (p. 126) – a state of confusion Alexis echoes in his description of Nira, who perceives that Majnoon is more than a poodle, while Majnoon is not sure what Nira is: "I do not know what Nira is, but I am not afraid" (p. 146). As mentioned, like Nizami's Layla and Majnun, who "recognized in the mirror of each other's face their own fear, their own pain and love" (pp. 23–24), Alexis' Nira and Majnoon see each other in their reflections: "when Nira, in her dream, looked into the water, she saw Majnoon's face reflected back at her, while Majnoon, in his, saw Nira's face where his should have been" (p. 132).

This continuity between the breakdown of false binaries (other/self, beloved/ lover, animal/human) manifests poignantly in Majnun's attunement to animals as fellow lovers and to their yearning and grief for their slain mates. Crucially, attunement and transspecies empathy lead to action. When Nizami writes, "Break your cage, break free from yourself, free from humanity; learn that what you thought was real is not so in reality" (p. 171), he is telling us that the concepts of both the individuated self and "human" status are artifices that incarcerate us and keep us cut off. Breaking out of our human cage, we simultaneously break the cage that circumscribes "animals" to the zone of killability, as when Majnun frees innumerable animals from snares. The elimination of the species veil yields interconnectedness and the elimination of suffering. The dissolution of the hunter/hunted dyad is a balm for our yearning. We are cured by disrupting the hunt.

The entanglement of Layla's and Majnun's lives – Majnun's hope that the remainder of his life be added to Layla's and his declaration that "it is not ourselves who hold fate's thread in our hands" (p. 43) – is reprised in Alexis' apologue, wherein Nira and the free-roaming Majnoon are so intimately entwined that the fates err, snipping Nira's mortal string instead of Majnoon's. Majnoon refuses to leave the spot where he last saw Nira, accompanied by neighborhood dogs who hold vigil with him as a sign of respect. Likewise, Majnun's earthly form is surrounded by the animals who "stayed with him to the last, even longer" (p. 134), "unwilling to believe that he would never awaken again" (p. 198). When only his bones are left, the animals finally depart. Just as Majnun implores his maker to "let me go where my love dwells," Hermes eventually guides Majnoon the poodle to the other world and to Nira.

Majnun's confusion of self and other is a state of bewilderment that causes him to perceive the world differently from most. When one of Prince Nawfal's men observes, "If a gust of wind sweeps by, or a cloud sails past in the sky, he believes them to be greetings from [Layla] and he thinks he can inhale her scent" (p. 62), he insinuates that the messages Majnun receives are "all in his head." Majnun himself observes that his love is "a riddle without a solution, a code which none can decipher" (p. 35). Is Majnun seeing signs that aren't really there, or is his perception altered? One week after Akbar died, I lay down disoriented in the forest clearing where we'd spent his final morning. How to fathom being split apart?

IV. Madness

In Nizami's rendering, Majnun's "madness" – which is also the condition of being "madly in love" – brings him closer to God, to Layla, to ego dissolution and to animals, at one and the same time. For twelfth/thirteenth-century Muslim scholar Muyhiddin Ibn Arabi, bewilderment, or "the disabling of rational faculties," holds the promise of a more honest possibility of truth (Almond, 2002, p. 516). As Ian Almond puts it, "when we are confused we see things that we miss when we think we know what we are doing" (ibid., p. 535). Confusion conveys mingling together. Majnun's "confusion" dissolves speciesist conditioning, with its reification of separation. It allows him to see the world as it truly is; all beings are the Beloved, hence humans should put down their spears and disarm their snares. Majnun's longing for Layla is therefore a realm of ahuman, liberatory promise. The poetic echo of his bewilderment is magnified by the old English of bewilder, signifying to lure astray, into the wild, what we could think of as scrapping the human guise to revel in a collective creaturehood.

pattrice jones' (2007) account of the psychological origins of human exceptionalism also illuminates Majnun's mindset. jones ventures that traumatic events such as famine and natural disasters induced humans to attempt to "cut themselves out of the web of life and escape the vulnerability that comes with being alive" (p. 22). In so doing, they severed "their kinship ties with other creatures, setting themselves apart as isolated and embattled would-be rulers of the earth" (p. 21). Majnun's restoration of caring relationships with animals heals the traumatic and traumatizing illusion of separation that abetted the "breaking" and "domestication" of animals, leading us to the precipice of climate catastrophe. Wilding and bewilderment portend an alternative to the epistemic violence of the Anthropocene, allowing us to break with systems of knowledge that disavow animal suffering. Whereas speciesist society pathologizes animal liberationists – characterizing loving animals "too much" as mental illness (Gruen & Probyn-Rapsey, 2018) – Majnun's "madness" exposes the cognitive dissonance at the heart of human-inflicted terrorism that is overwhelmingly centered on rupturing animals' kinship bonds. His altered consciousness is the dystopic experience of attunement to animal suffering in a carnist world, whereby he perceives the normalized ethos of the hunting ground as dependent upon the suppression of fundamental empathy. Presaging a remedy for the illness of our age, with its extravagant violence and anthropogenic mass extinctions, his "bewilderment" is a condition of clarity that sees through and strives to dismantle the vicious cycle at whose center are damaged relationships. His lucidity is likewise a repudiation of gaslighting that recalls the declaration of seventeenth-century Hindu poet Sant Tukaram: "I couldn't lie anymore, so I started to call my dog God." Avowing his dog's godliness, Tukaram might as well have written, "I couldn't (go along with the) lie anymore."

A discussion of Majnun's bewilderment leads full circle back to personal reflection, and back to Akbar, whom I credit with pulling the wool from over my eyes both figuratively and literally, since loving him led to rejecting garments made from animals' bodies. In his canine-centered version of the story of Layla and Majnun, thirteenth-century Persian poet Jalal al-Din Rumi also addresses the distortion of speciesism. In response to an onlooker who objects to his petting, kissing, crooning over, and feeding rose syrup to a dog in Layli's neighborhood, Rumi's Majnun observes that the dog is a "talisman sealed by the Lord," whose blessed soulfulness the onlooker fails to perceive because his vision is limited to "form and size."[4] He further invites his critic to discard the frame through which he perceives separation rather than interconnectedness: "You just see form and size./ Come here and view beyond those through my eyes!" (p. 37) and

> If from mere outward form you can transcend,
> You'll reach such heavenly gardens, my good friend.
> When you've smashed your own form, then this will bring
> Knowledge of breaking forms of everything.

(p. 37)

The task is to lift the quotation marks holding Nizami's animals captive within the zone of the symbolic, delimiting and disempowering them, just as Majnun removes the fetters binding their feet, allowing them to go "flying, running, trotting [and] creeping" into a wilderness constituted as a concrete, rather than symbolic, ahuman future.

Much as I endeavor to remove the quotation marks from Nizami's animals, I have struggled to remove the lens mental health care professionals impose upon the animals who haunt me, rendering them symbolic. For instance, soon after Akbar passed away, I discovered a dead nursling squirrel, her eyes not yet open though she was covered in a soft layer of new fur. The encounter undid me, driving a knife into the raw ache of missing Akbar. When I conveyed this to my psychiatrist, she took a swift detour to hair-splitting questioning: "How do you know the squirrel fell from a nest? Maybe 'it' was sick." Of course, the point was not how the newborn got there, but the frame through which she perceived the world blinded her to the catastrophic import of a deceased infant, to whom she referred as an object: "it." Since she already believed I'd lost my grip on reality, there was no point recounting to her the dream in which Akbar had appeared as a squirrel swimming in a stream, nor, indeed, the consonance of Majnoon and Nira's shared dream – co-dreaming, if you will – about squirrels and rabbits. She would not have comprehended my conviction that the experience of following Akbar "out to sea" was a gift of guidelines for navigating grief. Had I at that point met Majnun, I might have expended less energy "making a case" for my haunting. Instead, channeling his resolve, I'd have met the psychiatrist's questions with my own: "How will the mother feel tonight without her nursling? Imagine yourself as her. Do you disregard the distress of those who suffer?"

I might also have presented her with another incident, years earlier, when my mother was slowly dying. I'd encountered a squirrel struck dead by a car along the edge of a busy street, the white fur on her chest clean and soft like the white stripe down the center of Akbar's chest; she'd only just been killed. She was the first squirrel for whom I would prepare a memorial in the forest, circling her with leaves and flowers on the earth's soft surface. When I related the episode to my psychotherapist, she observed that my actions reflected a preoccupation with closure driven by my mother's agonizing illness, inducing hypersensitivity to untended bodies. She reduced the animal to a metaphor for interpreting my human condition. In my distress, I may have "confused" the squirrel's body with that of both Akbar and my mother, but this perplexity enabled me to perceive what speciesist society strives to conceal. Echoing Majnun, I wish that I had asked, "Why can't I grieve both my mother and the squirrel? Look how beautiful she is. Are her eyes not like those of the Beloved?" Emboldened by Majnun's rebuke to the hunter, I could have pressed her further: "If I'm unhinged, tell me how I went swimming in a stream with a rodent Akbar?" She'd have deduced that the dream was wish fulfillment. But in Sufi cosmology, dreams can be portals.

I've long known that my depression and anxiety originate with my mother's history, but only more recently came to understand that the trauma is transspecies as well as transgenerational. One of the inaugural moments of my mother's "bewilderment" occurred when, as a young child, she witnessed a boy slitting the throat of a lamb adorned in garlands of flowers one Eid-e-Ghorban, a Muslim holy day featuring animal sacrifice in a theatrical feat of hostility masqueraded as commemoration of God's mercy. She was particularly affected by the perverse "kindness" of offering water to the infant at the moment of her execution and ceased, forever more, to eat meat. Perceiving animals as real beings rather than as metaphors, her reactions – crying when she smelled the fumes of animal flesh grilled on our neighbor's barbeque – were judged hypersensitive by a society that normalized torturing animals. This psychological warfare was poignantly highlighted for me when a family friend, Brie, asked my mother about the origin of her attunement to animal suffering. My mother recounted the memory of her own mother catching sparrows and forcing her to convey them to their beheadings. In a dazzling feat of epistemic violence, rather than absorb the story, Brie displaced its meaning. The problem was not the wicked beheading of birds, but making a small child participate; as she put it, "what a terrible thing to do to a child." There was a meeting of eyes, a duel wherein my mother's truth came up against Brie shutting it down: because my mother did not have a comeback, and I was too intimidated to speak up on her behalf, Brie's truth prevailed. My mother had indeed been tortured by the same woman who made her participate in killing sparrows. Domestic abusers frequently weaponize violence against animals to augment the suffering of the humans who care for them. But the sparrows were not mere props nor were they symbolic of my mother's anguish; clearly, my grandmother inflicted graver damage to the murdered sparrows than she had to her child, who survived in spite of her.

If pathologizing caring for animals is intrinsic to the war against animals (Wadiwel, 2015), Nizami's tale is nourishment for the resistance. I can't undo the wicked throat-slitting of lambs to which my mother bore witness, nor the beheading of sparrows whose fluttering in her hands sent terror rippling into her cells, yet this mystical love story is a balm for the aftershocks of fight or flight that arouse my own mental struggles. I am emboldened by the affirmation that distress about their murders signifies attunement to, rather than misapprehension of, reality. With my zoocritical reading, I hope to contribute to the undertaking of "becoming majnun," redressing the stigmatization of transspecies love to unveil a domain of post-anthropocentric consciousness. The remedy for Majnun's own suffering rests on dispelling the illusion of separation and engaging in praxis – counter-conduct not of a human acting *on behalf of* animals but, rather, creaturely co-care. Together with the curative text to which Akbar guided me, my despair lifts while tending to two dogs liberated from a biomedical laboratory. Walking Cosmo and Balsam, their engagement in sniffing and the breeze on their ears is infectious. More than anything, their vigorous immersion in the activity of forward movement is an imperative pulling me toward a future I'd lost faith in. In this wild ecology, I am filled with well-being and connection to Akbar who belongs both to the past and to the future, behind me and before me, a continuous loop of love. I am struck by the poetic resonance of Balsam's name, synonyms for which include balm, salve, soothing agent, and medicine. The illusion of separation dissolves, just as cohabiting with Akbar, I happily joined sides with him. Rather, lost in the flow of mutual care and play, I knew that there were no sides.

Notes

1 Though Alexis doesn't explicitly mention Nizami's renowned text, the relationship between Nira and Majnoon unmistakably parallels that between Layla and Majnun. Similarly, in a contemporary love story where a character named Juliet is prevented from being with her beloved due to a family feud, the reference to Shakespeare's famous tragedy, "Romeo and Juliet," would be self-evident.
2 Reluctance to recognize the anti-speciesist potential of Sufi texts is aptly demonstrated by Pasha M. Kahn's statement, in his analysis of Muhyiddin Ibn Arabi, that he does not want to make "any grand claims about how his ideas might be inserted into the modern debate over the rights of animals or ethics with regard to animals" (2008).
3 Alongside the many verses of Nizami's story describing Majnun's relationships of mutual protection and care with animals, others lament the deficiency of his animal kin, as in "his beasts surrounded him in dumb loyalty" (p. 194). Speciesiest breaches are not the only problem. Nizami's narrative also contains numerous anti-black slurs, including the "black devil" who bears the news of Layla's marriage (p. 107) and the association of blackness with impurity (p. 112) and shame (p. 165). Misogyny is likewise in abundance, such as when Layla's father says that, rather than marry Majnun, he would prefer she be killed, burned, or cut to pieces (p. 80) and even that he would rather behead and feed her to the dogs himself (p. 81). As Majnun's muse, Layla is a largely passive figure whose virginity Nizami repeatedly underscores with references to her "unpierced pearl" (pp. 174, 158, 159). On the other hand, Layla indicts gender inequity (pp. 145, 174) and the injustice of Majnun's freedom by contrast with her imprisonment: "*He* in his wilderness could be as free and as mad as he liked; *she* had always been a prisoner, first her father's, then her husband's" (p. 172), and exerts agency, such as when she hits her husband, Ibn Salam, preventing him from raping her (p. 104).
4 Layla and Layli are versions of the same name.

References

Adams, C. (1990). *The Sexual Politics of Meat: A Feminist-Vegetarian Critical Theory*. Continuum.

al-Din Rumi, J. (2012). *The Masnavi, Book Three* (J. Mojaddedi, Trans.). Oxford University Press.

Alexis, A. (2015). *Fifteen Dogs*. Coach House Books.

Almond, I. (2002). The Honesty of the Perplexed: Derrida and Ibn 'Arabi on 'Bewilderment.' *Journal of the American Academy of Religion, 7*(3), 515–537.

Foucault, M. (1984). Of Other Spaces: Utopias and Heterotopias. *Architecture/Mouvement/Continuite*, October, 1–9. No. 5. https://web.mit.edu/allanmc/www/foucault1.pdf

Foucault, M. (2002). Space, Knowledge, and Power. In J. D. Faubion (Ed.), *Power: The Essential Works of Michel Foucault 1954–1984* (Vol. 3, pp. 349–364). Penguin Books.

Ganjavi, N. (1978). *Layla and Majnun* (Dr. R. Gelpke, Trans.). Shambhala.

Gillespie, K. (2020). Provocation from the Field: A Multispecies Doula Approach to Death and Dying. *Animal Studies Journal, 9*(1), 1–31.

Gruen, L., & Probyn-Rapsey, F. (2018). Distillations. In L. Gruen & F. Probyn-Rapsey (Eds.), *Animaladies: Gender, Animals and Madness* (pp. 1–8). Bloomsbury.

Isfahani-Hammond, A. (2013). Akbar Stole My Heart: Coming Out as An Animalist. *e-misférica: performance and politics, 10*(1), n.p.

Isfahani-Hammond, A. (2021). Akbar, My Heart: Caregiving for a Dog During Covid-19. *Animal Studies Journal, 10*(1), 11–29.

Isfahani-Hammond, A. (2022). Dogs Without Masters: Astray with Akbar and in Andre Alexis' Fifteen Dogs. In E. Quinn & L. Wright (Eds.), *Edinburgh Companion to Vegan Literary Studies* (pp. 138–151). Edinburgh University Press.

jones, p. (2007). *After Shock: Confronting Trauma in a Violent World: A Guide for Activists and Their Allies*. Lantern.

Kahn, P. M. (2008). Nothing but Animals: The Hierarchy of Creatures in the *Ringstones of Wisdom. Journal of the Muhyiddin Ibn Arabi Society, 43*, n.p.

Maharaj, S. T. (n.d.). First He Looked Confused. In D. Ladinsky (Tr. & Ed.), *Love Poems from God: Twelve Sacred Voices from the East and West* (p. 333). Penguin.

Mwangi, E. M. (2019). *The Postcolonial Animal: African Literature and Posthuman Ethics*. University of Michigan.

Quinn, E., & Wright, L. (2022). Introduction. In E. Quinn & L. Wright (Eds.), *Edinburgh Companion to Vegan Literary Studies* (pp. 1–15). Edinburgh University Press.

Safi, O. (Ed.). (2018). *Radical Love: Teachings from the Islamic Mystical Tradition*. Yale University Press.

Shakespeare, W. (1597). *Romeo and Juliet*. John Danter.

Wadiwel, D. (2015). *The War Against Animals*. Brill.

Chapter 13

Choosing Snakes
Towards Unhampered Hospitality

Sue Hall Pyke

Introduction

A few years ago, a human neighbour picked me a handful of beans from her garden. She lived not far up the road from where I am writing this. It was a warm day and my first visit to her home. Her hands deep in the bean's thick green foliage, she chatted about the tiger snake who preferred that patch of ground at that time of day. I accepted the handful of beans with relief, awed at her courage. Now I wonder at her willingness to disturb the foraging of that snake. Since around about that time I have been consciously attending to snakes differently. I can now recognise the dazzling Daisy, a maturing female silvery tiger snake and a regal but rarely seen old male creamy tiger snake I know as His Majesty.[1] For a while last summer, a young female Eastern brown snake, Boldy, made herself very well known. Perhaps that boldness was what killed her. I found her dead underneath a wheelbarrow in autumn. I first began recognising snakes with a tiger snake I called Selena. I would think Daisy was her, had I not been told of Selena's death. I like to think of Daisy as Selena's daughter and imagine His Majesty had a hand in things as well.

The snakes sharing this house, this garden, these paddocks, are descendants of families who began living in this area an extremely long time before my family arrived. Their rights to this place are so much greater than any rights my invasive body might claim. My ignorance of snake rights is such that it has taken years to shuck off my decades of socialised antipathy. I was taught to hate snakes. Some of my human visitors maintain this mindset. They would prefer it if all snakes were removed from this place, by whatever means. I have no intention of killing Daisy or His Majesty, and I can't imagine them doing well if a 'snake catcher' relocated either of them to an unknown and therefore unsafe territory. The thought of either of these exquisite snakes secured in a plastic tub, held as a pet, as an educational prop or as a laboratory subject is abhorrent. Why shouldn't they live in this space? It's lovely, with water and plenty of food. It's always been that way.

Thankfully, most of my human visitors are delighted to see a snake 'in the wild,' unhampered in the choices that they make.[2] Much harm is done by mainstream white human assumptions that nonhumans exist only to fit around human lifeways

DOI: 10.4324/9781003366706-18

(Plumwood, 2000). Unhampering the two snakes who live around my house involves letting them live their own lives.

It's not easy. My determination to live better with snakes has the potential to be undone every time I get in the car and head down the gravel track to the highway. In making a choice to go where I want to go, I risk interrupting the 'community dynamics' of the snakes around here (Andrew & Gibbons, 2005, p. 772). Cars on roads can create isolation, they can bar a snake's shift in habitat. Southwest Victoria has a long way to go, in terms of providing ecopassages to suit snakes and other creatures who need to cross human roads (Baxter-Gilbert et al., 2015). It would be a good investment, not only here but elsewhere. Snakes become familiar with these under-road tunnels quite quickly (Dillon et al., 2022). Without them, the dead snakes on our local roads will continue to form part of the 'staggeringly high' number of non-human deaths from 'animal-vehicle collisions' across the globe (Blackwell et al., 2016, p. 246). Currently, it is quite legal for me to accidentally drive over a snake, even though if I were to do the same to a human I could be punished for negligence and recklessness. Sadly, travelling at speed it's hard to see a metre of potential death on a road streaked with rain or sun or shadow, and travelling slowly invites the danger of road rage. When I was younger, the trace of tyre tracks burning a death trail to a snake was common. This is less the case now, perhaps because *deliberately* killing a snake can earn a fine of $10,000 and two years in prison.[3] It might also be that more humans are working on getting better at co-existing with snakes.

In short, I am merely being a law-abiding citizen, not demonstrating any kind of moral generosity, by doing nothing when I see Daisy cruising from the hydrangeas to the daisy bushes, or His Majesty coming for a drink from the house drain. My next work is to ensure fuller lifeways for these snakes.

Freedom, Choice, and Justice

His Majesty is old enough to know it's best to only rarely come down to the house from his rise of rocks and fallen wood that I call Habitat Hill. This summer I saw Boldy a lot, Daisy three times, and nothing at all of His Majesty. I've seen him twice in 10 years. Daisy is also getting old enough to be more cautious. As for Boldy, I thought she might be learning my smell (Nagabaskaran et al., 2021). But now, having found her dead without apparent injury, I'm thinking perhaps she was missing something she needed to survive. She was way too bold.

As a child, the only snakes I observed for long enough to know them again were either dead or held captive in enclosures. The reprehensible reptile enclosure at the Melbourne Zoo gave me my first chance, as a primary school student, to gaze at a live snake for a period longer than it takes to yell *snake*.[4] At home, snakes were unprotected by the law. Killed snakes were slung over barbed wire fences to save the killers from killing them again. A dead snake can easily be mistaken for a live snake. People around here called this hanging an offering to the kookaburras, as if these snake-eating birds need humans to give them a taste for such flesh. Anthropologist Deborah Bird Rose might have described these murdered snakes as death trophies, a human dysfunction she illustrates with tree-hung dingoes (2011, p. 66). I've grown up a lot since then and so has the law.

Domains	Freedoms		Hierarchy of needs	
Nutrition	From hunger and thirst		Physiological	
Environment	From discomfort		Safety	
Health	From pain, injury and disease		Love and belonging	
Behaviour	To express normal behaviour		Esteem	
Mental state	From fear and distress		Self-actualization	
Key to shading	Existing	Living		Choosing

Table 13.1 Towards choice: domains, freedoms, and posthumanist needs.

My aim, to hinder my own choices to unhamper the choices available to Daisy and His Majesty, assumes their full personhood.[5] As nonhuman personhood becomes better understood, play, creativity, and spirituality are increasingly being explored for a range of species (Massumi, 2014; Gigliotti, 2022; Brooks Pribac, 2021).[6] Keeping the personhood of Daisy and His Majesty in mind, I feel it is necessary to extend the 'five domains,' that spectrum from bare existence to less precarious living.[7] There are limitations behind this category, as well with the more liberal 'five freedoms'. Even while they certainly offer provisions for nonhuman animal rights, they are not 'liberties' (McCausland, 2015, p. 653). The broader spectrum from existing (where bare needs are met), to the liberty to make choices, can be seen when the domains and freedoms are set against Maslow's humanist hierarchy of needs (Table 13.1).[8] As political scientist Erika Cudworth argues, when choice is differentially circumscribed by humans for nonhumans, this creates an 'anthroparchy' dominated by humanist logics of oppression (2005, p. 8). Ways of living fully and creatively to the potential of one's being must surely be expressed in snake life.

When I apply these frames to the lives of Daisy and His Majesty, thinking about snake lifeways from bare existence to scant living to rich choice, there seems to be no reason why they should not be afforded the darker shading towards the bottom of these three frameworks. A fair ask for humans is a fair ask for other animals. Daisy and His Majesty may well enjoy friendship, self-esteem and self-actualisation, based on studies of snakes' sentience and cognition.[9] However, as Table 13.1 demonstrates, such richer living can only be possible when one's life involves more than the absence of harm. I have watched Daisy halt her slow steady emergence from a hole in the broken concrete of the path to the washing line, shifting to fast reverse, as smoothly and quickly as cooked spaghetti sliding off a spoon. Scared by me watching. His Majesty has always seen me first. I see him when he's already nearly gone. There are better ways for a snake to live. I've seen it in the unhindered foraging of Daisy and Boldy before they have seen me. When they see my presence as harm, I see my presence as unfair, unjust.

It is difficult to think through, more rigorously, what 'justice' might mean for Daisy and His Majesty. Justice is an idea subject to all the 'subjectivity of moral hierarchies' (Chen, 2016, p. 4). My subjective idea of justice is that it 'values life without reducing value to utility' (Rose, 2007, p. 18). This is a justice best understood as part of the 'posthuman alliances' characterised in Serpil Opperman's new materialism, alliances which contain within them a 'radical critique of the anthropocentric premises of traditional humanism' (2013, p. 28). In this way, my work to imagine greater freedom in these snakes' choices is part of a broader shift towards multispecies justice.

One practical expression of multispecies justice is multispecies spatial planning, which balances human development and economic activities within ecological boundaries (Fieuw et al., 2022, p. 5). Critical animal scholar Yamini Narayanan and herpetologist Sumanth Bindumadhav describe what this might look like for snakes, envisaging an innovative 'widening of inclusivity' that they characterise as 'posthuman cosmopolitanism' (2019, p. 408). I see this kind of multispecies planning as 'just' because it explicitly includes snakes and all other members of society, envisioning all people, human and nonhuman, living together in cosmopolitan ways that transcend harmful and humanist othering.

This post-anthropocentric society is not only just, but necessary. Most animals (except privileged humans) are struggling to thrive as seasons blur into greater climatic extremities. In this fraught context, multispecies justice becomes, as political theorist Alasdair Cochrane puts it so compellingly, an 'integral and ineliminable aspect of social justice' (2018, p. 142). All this world's creatures are facing a jittery future, and the more strategically inclusive human/nonhuman relations become, the more chance there is of a greater majority of peoples enjoying the unhampered right to live to the point of self-actualisation. As it is for humans, it is more difficult for snakes to creatively live to their full potential when spring becomes summer and autumn becomes winter. Bare living itself becomes difficult.

Yet still, there is a contradiction here, in putting forward ideas of post-anthropocentric justice, when law is inevitably bound to Western administrative frameworks. To speak of justice, with its underpinning assumptions of moral significance, creates constraint. In the face of this limitation, I turn to what anthropologist Anna Tsing calls 'passionate immersion,' a love unbounded by species, driven by curiosity, focused on reciprocity without demands (2011, p. 201).[10] A passionate curiosity about snakes opens my view towards living with snakes in more just *and* hospitable ways.

Compassionate Avoidance and Conflict Protection

In this continent known as Australia, snakes have long been understood as divine. In Australia there are many First Nation cosmologies that privilege snakes as creators who are not only present in snake bodies but also present in rainbows and rivers. Snakes bring life to this world as water does.[11] I have a non-Indigenous ancestry which ensures that my efforts to find new relations with the snakes who live with me will never be commensurate with older snake/human interactions on this land. As Indigenous scholar Carl Te Hira Mika makes clear, the ground I stand '*upon*' is the ground that the First Nations peoples of this land stand '*within*'; the best I can do is acknowledge the 'mystery' in play here (2017, pp. 100–101, Mika's italics). To expect anything more as I live with Daisy and His Majesty would be appropriative foolery. It seems more appropriate to seek the 'world-disclosing' love that comes with an 'unsettling practice of wonder,' so beautifully described by political scientist Lida Maxwell (2017, p. 685). As Maxwell points out, as a human body I am oriented to the world around me as surely as there are affinities between bees and flowers, birds and trees. Snakes have been part of the biodiversity of this place, which includes humans living here, since time immemorial. Snakes and humans belong together.

Any belongingness I might feel, however, is, necessarily, a long way distant from the deeply storied kinship between the old local communities and snakes in this place. However, these First Nation knowledges do contextualise my reach towards new freedoms for the snakes who are ancestral locals in this place. For Māori education scholar Te Kawehau Hoskins and his Pākehā collaborator Alison Jones, First Nation knowledges have an openness to a fluidity and uncertainty that is part of researching the world (2017, p. 54). This method of living in curious relationality contrasts strongly with Western ideas of a teleologically inevitable species collapse. However, my awareness of such 'methods' of seeing the world, as Hoskins and Jones make clear, means only that I can 'reach toward something that exceeds language: an attitude, a sympathy, a feeling, an openness' (p. 53) – a reach *toward*, rather than a reach that can be fully achieved. I have no ancestral memory in my blood to make kin of the snakes that live here. All I can hope for is a lifelong experience of learning and connecting that 'reaches toward' the possibilities of a stronger affinity with snakes.[12]

Therefore, my best learning work may be to speculate a space for human/snake freedoms that allows me to live in this place with meaningful 'relational validity' (Tuck & McKenzie, 2015, p. 636). Such relationality is determined by the history of the place where I live, known by invaders/settlers as the Stony Rises, known by the Eastern Maar Nation as Djagurd Wurrong Country and known by many in both communities as Snake Country. I must reach beyond all the colonialist empire-building and invader/settler assumptions of belongingness that surround me, to the knowledges that are layered through the composting dirt, fungi-wired into the trees, and laced through the air by birds and insects. The snakes around here have relational validities that only involve me at the margins. As Daisy and His Majesty demonstrate, although perhaps less so Boldy, snakes can live quite closely together, even when they hunt the same prey.[13] I have not seen Daisy and His Majesty interact, but perhaps they do. My not-knowing is part of my work to unhamper their lives.

My learning is helped by appreciating old precedents of humans living reverently with snakes. Seeing the snakes here as sacred creatures pushes against my formative years of socially inscribed practices of hatred. There is joy in learning this new relational mode.[14] However, while it is at last starting to feel like a divine visitation when I see Daisy or His Majesty, I may not be as reverent as other people are in this world, if either or both came into the house. In some parts of Nigeria, snakes choosing to enter a home are considered a blessing.[15] If I saw, inside my home, a sparkly female tiger snake over half as long as I am tall, or a very long male creamy-green tiger snake with a body thicker than my forearm, I don't know if I'd be venerating their holy presence. It would be hard enough to wait until they chose to leave in their own time.[16] This summer my beloved and I watched Boldy wind her way along the concrete in front of the shed, a place where we relax and work. Great relief when she slid to the garden on the north side, rather than wending her way inside with us. It's a poor devotee who wants to avoid a god. As it turned out, that side of the shed was where she died, perhaps willed by us, if not on our watch.

This is my limitation. I like having a tightly enclosed large wooden box to go to at night to sleep. I keep the doors shut through the day and night, and try to keep frogs, mice, slaters and other tasty snake snacks from making their way inside. It is a selfish refuge-making that curtails Daisy's and His Majesty's choices. As social geographer Krithika Srinivasan argues, humans need to move 'away from logics and practices of protection-sacrifice, and towards the redistribution of the risks of earthly living in more equitable directions' (2022, p. 353). She is right. My desire for a snake-proof home speaks to my desire to outlive my animality. I am reducing risk to myself (protection-sacrifice), without accounting for how this might lessen the well-being of Daisy and His Majesty. Srinivasan specifically critiques humans living in homes that operate as fortresses. My wire doors; my fly-wired windows; they speak of poor multispecies spatial planning.[17] Srinivasan suggests the way to avoid the hurts created by such self-absorption is through a 'reanimalisation of human wellbeing' that involves learning to live in ways that nurture the long-term future of all, rather than circumscribing the lives of others for one's own life (p. 361). This is a posthuman cosmopolitanism that is beautifully illustrated by Sumanth Bindumadhav's story of an 'old woman' who lets 'several snakes' come in and out of the house she lives in 'regularly': she chooses to 'leave the door open and after a while, they find their way out' (Narayanan & Bindumadhav, 2019, p. 407). There is a similar story from around here told by my uncle (see also Pyke, 2019, p. 225). An old man, who lived close by when I was young, shared his house with a tiger snake, made comfortable by broken windows, loose floorboards and doors that didn't shut. When the snake was slain by an unthinking visitor playing hero to this local hermit, my neighbour berated him for his thoughtlessness. Who was going to kill the mice now?

I feel so fear-bound, compared to these two old wise ones, but perhaps even some of the First Humans of this place shared my tendency for avoidance. My local Landcare group writes that the 'traditional owners [of Eastern Maar] liked to camp with she-oaks' a tree with pine-like leaves but also with flowers, because it was more likely there would be 'no snakes' (Heytesbury, 2023). A closed house is not an open stand of she-oaks, but knowing there are, most likely, no snakes close to me makes it easier for me to sleep.[18] I'm not planning to move out of this weatherboard enclosure any time soon, even as I agree with Srinivasan's philosophy. I hover between reverence and avoidance.

All that fear of death, of my animality. There are risks in living with snakes, even with a snake-proofed house. Daisy is well over a metre in length and His Majesty is bigger and thicker by quite a way. If I was bitten by either of them I would get very sick and could die. However, as Srinivasan goes on to argue, there is nothing out of order in this. Dying is not an act of war, and fearing death undermines my commitment to sociable post-anthropocentric living. Besides, if there is plenty of room, the hazard factor is reduced. As a global research team focused on snake envenoming note, snake bite is a 'public health problem' correlated with population density and poverty (Malhotra et al., 2021, n.p.).[19] I live in an unpopulated rural area so the snakes living here generally have space enough to make a choice to stay, or strike or move away, depending on the choices I make with my body. The fact that the health care system is of a good standard also reduces my risk of death. Once I walk outside the house, in the words of political scientist Dinesh Wadiwel, there is always potential for a 'frictional tussle of forces' between these snakes and me (2016, p. 201). However, from what humans know of snakes, it seems unlikely either Daisy or His Majesty would choose to bite me without provocation. Poison is a snake's tool for prey, and I am far too big for food. If they were to bite me, they would be left without adequate venom to kill and eat. Their wellbeing would be compromised. If I am bitten, this means I have not cared enough to keep my eyes wide open. The risk lies with me.

I am doing my best to mitigate the risk I present to Daisy and His Majesty. Communications scholar Núria Almiron argues that without a commitment to do something to improve interspecies relations, avoidance is a performative response (2021, p. 62). As unnecessary venom depletion is the worst thing for a snake who needs their poison to live well, the last thing I should do is create unnecessary conflict. I need to avoid presenting a threat. What I need to reach for might be compassionate avoidance.

I am only human, and as social scientist Catherine Hill has outlined, human ideas of peaceful coexistence refuse the fluid dynamics involved in active encounter and relationality. I tend to dream and rush. In the fallible world Hill describes, conflict is inevitable. But I can do my best to fail less. Good planning can create better multispecies justice. An interdisciplinary study from Sri Lanka has found that the incidence of snakebite is best predicted when the 'spatio-temporal synchronicity in both snake and farmer behaviours' are taken into account (Goldstein et al., 2021, p. 14). Sri Lanka is as different a place to Eastern Maar Country, as a rice paddy is different to a rural patch of farm-damaged land, but the principles of being aware of time and space, as outlined by these scholars, are instructive. The wellbeing of Daisy and His Majesty requires me to plan. Zoologist Eyal Goldstein and his co-researchers suggest social and physical infrastructure can help avoid unnecessary conflict. Closed doors perhaps can be part of multispecies spatial planning, even if she-oaks do a better job. An ethological study in India suggests heavy rubber boots, torches, and mosquito nets are also helpful in reducing conflict (Malhotra et al., 2021). Dressing differently might also be part of my infrastructural strategy. Biological scientists Whitaker and Shine (1999) have found Australian brown snakes are more likely to avoid humans in clothes that are dark or light enough to contrast with their surroundings. Perhaps light brown overalls for winter and spring, and a dark green pair for summer and autumn.

This kind of immersive work, making practical plans to create a more relational world, a better, if not perfect place for snakes to live, involves a way of thinking through equivalences that is different to Western ideas about human exceptionalism. As political theorist Mathias Thaler argues, in his outline of non-escapist desire that can help create a future that is just for all who have a place in what is to come, the 'intellectual legacies' surrounding the idea of multispecies justice require my considerations to be part of the 'complex processes of problematizing the present' (2022, pp. 268, 263). I need to get over my desire to protect myself, as Srinivasan advises, while reaching towards doing nothing at all that might cause Daisy or His Majesty to waste their venom on my inedible body.

These responsive and hospitable adaptations involve hampering my life to give Daisy and His Majesty more choice. As the weather warms up in spring, and snakes come out looking for a feed, moving at the slow pace their metabolism has given them over winter, all that sugar to get through the cold, I follow the zoological advice to use a torch when it's dark, wear rubber boots outside, and walk very slowly if I can't see what's underfoot. In summer and autumn I do the same, even if Daisy and His Majesty are by then going to be quicker than a wink to disappear. In winter I do the same, because snakes can get hungry in the middle of the hibernation season. I wake up in the middle of the night hungry too, sometimes.

Conclusion

As I reach towards more relationally valid responses to the snakes in my vicinity, my multispecies spatial planning becomes more intimate. What do Daisy and His Majesty do in their work to keep this world alive? How do they meet their responsibility to the dirt, the air, the trees, the grasses? And how do I assist them to do their work as best they can? Do I leave the holes in the crumbling concrete path that Daisy knows backwards, despite the trip hazard they present? Do I keep the long grass and leave the felled wood on Habitat Hill, despite the fire hazard this presents? And that metre-high dry dockweed that Daisy surfed as if it was a wave, that weedy eyesore perhaps should stay in place as well. I need to learn Daisy's and His Majesty's routines, without scaring them, to get better at getting out of their way.[20] The only way of doing this is by maintaining a careful slow walk around the house to make sure I don't give them a fright. I have to work harder at not getting caught up in my rush.

In my fallible ways, I will reach towards this compassionate avoidance, and attend to the possibilities of responsive hospitality, in ways that are more than a refusal to face morbidity. The light I carry, the boots and overalls I wear, the pace of my feet, the attention of my eyes, all these choices diffuse the potential for a conflict that would hamper these two snakes' ability to live towards a self-actualisation that is not mine to know. In that spirit, in the name of their divine bodies, not the names I put upon them in my all-too-human way, I pray that they keep their venom for their prey. May their will, much more than mine, be done.

Notes

1 Naming the neighbouring snakes helps my relational position, but it is a fraught anthropomorphism which is accompanied by the equally questionable suggestion that one of my neighbours is male and the other is female. Identifying a snake's sex involves an invasive examination. The general rule that male tiger snakes have larger heads and bodies and female tiger snakes have a longer thinner tail has shaped my naming of Daisy and His Majesty, but this creates a potential gendering unsupported by the multifactorial ethological research required (Meynell & Lopez, 2021). I could use 'they' and refer to Daze or Majesty, but names and pronouns stick, even when they should not.

2 I use scare quotes here to make it clear I eschew this term for the reasons outlined by William Cronon: 'wilderness' is a contested term for the work it does to suggest a (non-existent) utopia untouched by human hands. This fantasy accelerates dystopian damage to non-'wild' habitats. However, some Indigenous communities in the continent known now as Australia use the word 'wild' for Country where the governance of the first humans of that place has been stymied through white invasion and possession. Such 'wild' Country is unwell, sick, it 'has gone bad' (Pickerill, 2008, p. 98). In that sense, the farming land around here *is* a wilderness.

3 In Australia, wildlife protection legislation is implemented state by state. According to the Victorian Government's Conservation Regulator, under Section 43(1) of the Wildlife Act 1975, anyone caught killing a snake without authorisation faces a maximum penalty of 50 penalty units (in the 2023–24 financial year, one penalty unit was $192.31, so this equates to $9615.50) or six months imprisonment or both the fine and imprisonment (2023). This fine is waived if the killing is an act of self-defence.

4 Conservation is touted as a core value for many zoos now, although there is much more talk than money involved (Arcari, 2022). Enclosures are deemed necessary, nonetheless, for the visitors who offer donations in a hope to fund rescue and habitat for animals 'in the wild.' Such efforts, which reinforce the ideal of a 'wild' no-go-zone that limit multispecies living, are not good for anyone. Ideas of 'wild life' justify the containment of humans and nonhumans alike into a limited life with reduced personhood. For Giorgio Agamben, this oppression of the living potential of others is a 'naked life' (2000, p. 17). While Agamben's project does not explicitly speak to the nonhuman, his delineation of a productive 'space beyond' speaks to separations between humans and all other animals (Wadiwel, 2015, p. 79).

5 Unless scientific endeavours prove otherwise, it seems sensible to assume snakes have the personhood established in animal behavioural science and critical animal studies. The limitations in herpetological findings to date are human limitations, not the limitations of the animals under study (or not under study).

6 Extending Western ideas of personhood to all other animals is just a start. It is convenient to use taxonomic territories, such as 'snake,' and 'human,' but such terms create boundaries that would better be undone. Such divisions are an ontological violence that

erases the shared animality that creates shared relations in the making and remaking of this world.

7 These categories are used in the animal welfare sector to regulate human care of nonhuman animals.

8 There are interstices between these categories, but safety and security, in most situations, is a precondition for what is sometimes described in shorthand as wellbeing and happiness, or, the equally unmeasurable idea of self-actualisation (Ott, 2020, p. 40). These categories relate to relationships with others, esteem and a sense of psychological growth.

9 I make this differentiation alongside Helen Proctor (2012), who describes cognition as involving mental action or the perception, processing and storing of information, and sentience as involving feelings, states of being, and sensations. All the same, currently, humans who choose to share their homes with snakes are less likely to see them as cognisant family members or sentient communicative people, even as they speak to the grace and beauty of these companion animals (Azevedo et al., 2022).

10 This seems to be a key characteristic of the field of 'multispecies studies' outlined by Thom Van Dooren and his colleagues, a mode of thinking which has sympathies with other ontological frameworks including those of some Indigenous communities.

11 I have not yet heard a creation story of snakes from the Eastern Maar Country where I live, but the Mayapa U Budj Bim creation story is beautiful (Gunditj Mirring, 2020). Here I am thinking particularly of Alexis Wright's wonderful depictions of holy snakes as 'respected and trusted citizens,' as outlined in my reading of *Carpenteria*, alongside other First Peoples' literary works that involve the depiction of snake/human relations (2017, p. 227).

12 The ongoing generosity of the First Nations in this colonised land is led, where I live, by the Gunditjmara community, who, as part of their ongoing work to find ways to live generatively with invader/settlers, are opening the Budj Bim Cultural Landscape to people outside community who are willing to be part of their strategic plans to enhance and revitalise local ecologies.

13 According to one group of ethologists, reptiles (these scholars studied lizards) who share habitat with others of the same species can have reduced stress levels (Londono et al., 2018).

14 For example, I have been confabulating a snake church, with the help of kookaburras, who call me into this playful worship space, formed so I might get better at letting snakes be (2022). This is a cultural repositioning but not, I hope, a cultural appropriation. My less histrionic responses make sense in the philosophy of tendencies, attractions, and repulsions that inform the affect philosophy of Brian Massumi. As Massumi describes it, affect comes before emotion, it is 'the tendencies' or the 'pastnesses opening onto a future, but with no present to speak of' (1995, p. 91). By changing my past ideations, new responses can take place. The repulsion of fear shifts to a magnetism that is closer to love.

15 In some parts of Nigeria hereditary priests mediate between gods and goddesses indwelling in the organic forms of both animal and non-animal beings. For those following these traditional beliefs, the python is 'the god of wisdom, earthly bliss and benefaction' (Rim-Rukeh et al., 2013, p. 429). At the same time, there are 20,000 snakebites a year in Nigeria, resulting in 2,000 deaths and 1,700 amputations (Adebowale-Tambe, 2021). The incidence of snakebites in Africa is much higher than elsewhere in the world (10 for every 10,000 people compared to 1 for every 10,000 people in Australia), largely due to different population densities and different farming practices (Chippaux, 2009, p. 198).

16 George Pointon (2022), a primary school teacher who writes about his students on social media, reports that five-year-old Zahra proposed the following business proposition: 'The Cat Shop: A Shop for Cats.' There were rules. 'No humans allowed in store. Cats

will be fitted with a contactless card on their collar. They basically come in, take what they want and owners will get automatically charged.' Pointon writes, 'this is so genius it borders on the line of evil.' I think it's so genius it goes a step further than the 'tolerance and acceptance' that generally is associated with co-existence (Glikman et al., 2021). Were it not for the owners and the collars, this would be unhampered hospitality, where all humans are carers and there is no sense of possession. Yet I'm not ready to fully apply Zahra's business model with its insistence on a house with open doors, where Daisy or His Majesty could 'basically come in, take what they want.' Especially if their choice to stay outlasted my ability to stay awake.

17 Nonhuman housing can also impact on the choices for snakes. For example, if I chose to 'keep' or 'rescue' chickens, I would need chicken wire to keep them safe from predators. However, a chicken wire enclosure can be fatal for snakes. A dead tiger snake was found caught in a roll of chicken wire in my shed when I first moved back home and started cleaning up. That small head must have gone in first, the larger body wedged in, and there they were, unable to back up. A fence can be fatal for snakes, even if it's not set up to trap, enclose, or contain them.

18 Both of these approaches are different to those taken by the farmer featured in Aesop's tale of the Farmer and the Viper. He revived a snake to the point of self-defence and was promptly killed.

19 The work being done in those areas where snake bites are a health risk is centred around conflict prevention. Eradication is not encouraged, but the removal of snakes from places of conflict is supported.

20 I am only moderately optimistic I can do this. To become aware of a different species' 'phenomenal lifeworlds, biotic affordances, capacities, and needs' takes time (Celermajer, 2020, p. 479). To make this more difficult, most herpetological findings are limited to observing snakes in captivity. This is why the properties of non-captured snake venom are still largely unknown (McCleary et al., 2016). Similarly, while some predictions can be made about snakes' temporal patterns, the behaviour of a snake appears to be quite variable (Parkin et al., 2020). Foraging routines and other behaviours are particularly unpredictable for snakes forced out of their own habitat (Narayanan & Bindumadhav, 2019, p. 5). Preferred habitats are better known. Snakes do like an old tree with good hollows (Shelton et al., 2020). So do I.

References

Adebowale-Tambe, N. (2021). Nigeria Records 2,000 Deaths from Snake Bites Every Year – Minister. *Premium Times*, September 21. www.premiumtimesng.com/news/top-news/485929-nigeria-records-2000-deaths-from-snake-bites-every-year-minister.html

Agamben, G. (2000). *Means Without End: Notes on Politics*. University of Minnesota Press.

Almiron, N. (2021). *Like an Animal: Critical Animal Studies: Approaches to Borders, Displacement, and Othering*. Brill. https://doi.org/10.1163/9789004440654_003

Andrews, K. M., & Gibbons, J. W. (2005). How Do Highways Influence Snake Movement? Behavioral Responses to Roads and Vehicles. *Copeia, 4*, 772–782. www.jstor.org/stable/4098651.

Arcari, P. (2022). To Hell in a Gift Basket: The Deceptions and Dangers of Corporate Conservation and Philanthrocapitalism. *Medium*, February 18. https://medium.com/@parcari/to-hell-in-a-gift-basket-the-deceptions-and-dangers-of-corporate-conservation-and-2eaee8093951.

Azevedo, A., Guimarães, L., Ferraz, J., Whiting, M., & Magalhães-Sant'Ana, M. (2022). Understanding the Human – Reptile Bond: An Exploratory Mixed-Methods Study. *Anthrozoos, 35*(6), 755–772.

Baxter-Gilbert, J. H., Riley, J. L., Lesbarr'eres, D., & Litzgus, J. D. (2015). Mitigating Reptile Road Mortality: Fence Failures Compromise Ecopassage Effectiveness. *PloS One*, *10*(3), e0120537. https://doi.org/10.1371/journal.pone.0120537

Blackwell, B. F., DeVault, T. L., Fernandez-Juricic, E., Gese, E. M., Gilbert-Norton, L., & Breck, S. W. (2016). No Single Solution: Application of Behavioural Principles in Mitigating Human-Wildlife Conflict. *Animal Behavior*, *120*, 245–254. https://doi.org/10.1016/j.anbehav.2016.07.013.

Brooks-Pribac, T. (2021). *Enter the Animal: Cross Species Perspectives on Grief and Spirituality*. Sydney University Press.

Butler, H., Malone, B., & Clemann, N. (2005). The Effects of Translocation on the Spatial Ecology of Tiger Snakes (*Notechis scutatus*) in a Suburban Landscape. *Wildlife Research*. *32*(2), 165–171. https://doi.org/10.1071/WR04020.

Celermajer, D., Chatterjee, S., Cochrane, A., Fishel, S., Neimanis, A., O'Brien, A., Reid, S., & Srinivasan, K. (2020). Justice Through a Multispecies Lens. *Contemporary Political Theory*, *19*(3), 475–512. https://doi.org/10.1057/s41296-020-00386-5.

Chippaux, J. P. (2009). Incidence mondiale et prise en charge des envenimations ophidiennes et scorpioniques [Global Incidence of Snake and Scorpion Envenoming]. *Medecine Sciences: M/S*, *2*, 197–200. https://doi.org/10.1051/medsci/2009252197. PMID: 19239853.

Cochrane, A. (2018). *Sentientist Politics: A Theory of Global Inter-Species Justice*. Oxford University Press.

Cornelis, J., Parkin, T., & Bateman, P. (2021). Killing Them Softly: A Review on Snake Translocation and an Australian Case Study. *Herpetological Journal*, *31*, 118–131. https://doi.org/10.33256/31.3.118131.

Cronon, W. (1996). The Trouble with Wilderness: Or, Getting Back to the Wrong Nature. *Environmental History*, *1*(1), 7–28. www.jstor.org/stable/3985059.

Cudworth, E. (2005). *Developing Ecofeminist Theory: The Complexity of Difference*. Palgrave Macmillan.

Dillon, R., Boyle, S., Litzgus, J., & Lesbarreres, D. (2020). Build It and Some Will Use It: A Test of Road Ecopassages for Eastern Gartersnakes. *Journal of Herpetology*, *54*(1), 19–23. https://doi.org/10.1670/18–163

Fieuw, W., Foth, M., & Caldwell, G. A. (2022). Towards a More-than-Human Approach to Smart and Sustainable Urban Development: Designing for Multispecies Justice. *Sustainability*, *14*(2), 948. https://doi.org/10.3390/su14020948

Garcia, J. (2023). Not So Fast . . . Smoko Rudely Interrupted by Deadly Intruder. *Brisbane Times*, February 5. www.brisbanetimes.com.au/national/queensland/not-so-fast-smoko-rudely-interrupted-by-deadly-intruder-20230205-p5chzk.html.

Gigliotti. C. (2022). *The Creative Lives of Animals*. New York University Press. nyupress.org/9781479815463/the-creative-lives-of-animals/

Glikman, J. A., Frank, B., Ruppert, K. A., Knox, J., Sponarski, C. C., Metcalf, E. C., Metcalf, A. L., & Marchini, S. (2021). Coexisting with Different Human-Wildlife Coexistence Perspectives. *Frontiers in Conservation Science*, *2*, n.p. https://doi.org/10.3389/fcosc.2021.703174

Goldstein, E., Erinjery, J. J., Martin, G., Kasturiratne, A., Ediriweera, D. S., de Silva, H. J., Diggle, P., Lalloo, D. G., Murray, K. A., & Iwamura, T. (2021). Integrating Human Behavior and Snake Ecology with Agent-Based Models to Predict Snakebite in High Risk Landscapes. *PLoS Neglected Tropical Diseases*, *15*(1), e0009047 https://doi.org/10.1371/journal.pntd.0009047.

Gunditj Mirring. (2020). *Mayapa U Budj Bim* [animation]. Monash University. https://vimeo.com/408223290?embedded=true&source=video_title&owner=21986079

Heytesbury District Landcare Network. (2023). *Great Discussion About Understory Plants* [status update], February 13. www.facebook.com/HeytesburyDistrictLandcare Network./.

Hill, C. M. (2021). Conflict Is Integral to Human-Wildlife Coexistence. *Frontiers in Conservation Science, 2*, n.p. https://doi.org/10.3389/fcosc.2021.734314

Hoskins, T. K., & Jones, A. (2017). Non-Human Others and Kaupapa Māori Research. In T. K. Hoskins & A. Jones (Eds.), *Critical Conversations in Kaupapa Māori* 48–50. Huia Publications.

Londono, C., Bartolome, A., Carazo, P., & Font, E. (2018). Chemosensory Enrichment as a Simple and Effective Way to Improve the Welfare of Captive Lizards. *Ethology, 124*, 674–683. https://doi.org/10.1111/eth.12800

Malhotra, A., Wüster, W., Owens, J. B., Hodges, C. W., Jesudasan, A., Gnaneswar, C., Kartik, A., Christopher, P., Louies, J., Naik, H., Santra, V., Kuttalam, S. R., Attre, S., Sasa, M., Bravo-Vega, C., & Murray, K. A. (2021). Promoting Co-Existence Between Humans and Venomous Snakes Through Increasing the Herpetological Knowledge Base. *Toxicon: X, 12*(100081), 1–18. https://doi.org/10.1016/j.toxcx.2021.100081.

Massumi, B. (1995). The Autonomy of Affect. *Cultural Critique, 31*, Autumn, 83–109.

Massumi, B. (2014). *What Animals Teach Us About Politics*. Duke University Press.

Maxwell, L. (2017). Queer/Love/Bird Extinction: Rachel Carson's *Silent Spring* as a Work of Love. *Political Theory, 45*(5), 682–704. https://doi.org/10.1177/0090591717712024

McCausland, C. (2014). The Five Freedoms of Animal Welfare Are Rights. *Journal of Agricultural and Environmental Ethics, 27*, 649–662. https://doi.org/10.1007/s10806-013-9483-6

McCleary, R. J. R., Sridharan, S. Dunstan, N. L., Mirtschin, P. J., & Kini, R. M. (2016). Proteomic Comparisons of Venoms of Long-Term Captive and Recently Wild-Caught Eastern Brown Snakes (*Pseudonaja textilis*) Indicate Venom Does Not Change Due to Captivity. *Journal of Proteomics, 144*, 51–62. https://doi.org/10.1016/j.jprot.2016.05.027

Meynell, L., & Lopez, A. (2021). Gendering Animals. *Synthese: An International Journal for Epistemology, Methodology and Philosophy of Science, 199*, 4287–4311. https://doi.org/10.1007/s11229-020-02979-4.

Mika, C. T. H. (2017). The Uncertain Kaupapa in Kaupapa Māori. In T. K. Hoskins & A. Jones (Eds.), *Critical Conversations in Kaupapa Māori* (pp. 100–110). Huia Publications.

Nagabaskaran, N., Oliver, H. P., Burman, O. H. P., Hoehfurtner, T., & Wilkinson, A. (2021). Environmental Enrichment Impacts Discrimination Between Familiar and Unfamiliar Human Odours in Snakes (*Pantherophis guttata*). *Applied Animal Behaviour Science, 237*(105278), 1–5. https://doi.org/10.1016/j.applanim.2021.105278

Narayanan, Y., & Bindumadhav, S. (2019). 'Posthuman Cosmopolitanism' for the Anthropocene in India: Urbanism and Human-Snake Relations in the Kali Yuga. *Geoforum, 106*, 402–410. https://doi.org/10.1016/j.geoforum.2018.04.020.

Opperman, S. (2013). Feminist Ecocriticism: A Posthumanist Direction in Ecocritical Trajectory. In G. Gaard, S. Estok, & S. Oppermann (Eds.), *International Perspectives in Feminist Ecocriticism: Making a Difference* 19–36. Taylor & Francis Group. https://ebookcentral.proquest.com/lib/unimelb/detail.action?docID=1211708.

Ott, J. (2020). *Beyond Economics*. Palgrave Macmillan. https://doi.org/10.1007/978-3-030-56600-5_3

Parkin, T., Jolly, C. J., de Laive, A., & von Takach, B. (2020). Snakes on an Urban Plain: Temporal Patterns of Snake Activity and Human – Snake Conflict in Darwin, Australia. *Austral Ecology, 46*(3), 449–462. https://doi.org/10.1111/aec.12990

Pickerill, J. (2008). From Wilderness to Wild Country: The Power of Language in Environmental Campaigns in Australia. *Environmental Politics, 17*(1), 95–104. www.jstor.org/stable/40338997.

Plumwood, V. (2000). Integrating Ethical Frameworks for Animals, Humans, and Nature: A Critical Feminist Eco-Socialist Analysis. *Ethics and the Environment, 5*(2), 285–322.

Pointon, G. (2022). Zarah [Tweet], June 24. *Twitter*. https://twitter.com/GeorgePointon_/status/154028937232240640

Proctor, H. (2012). Animal Sentience: Where Are We and Where Are We Heading? *Animals*, *2*(4), 628–639. https://doi.org/10.3390/ani2040628.

Pyke, S. (2017). Citizen Snake: Uncoiling Human Bindings for Life. In A. Malinowska & M. Gratzke (Eds.), *The Materiality of Love: Essays on Affection and Cultural Practice* 223–236. Routledge. www.taylorfrancis.com/chapters/edit/10.4324/9781315228631-16/citizen-snake-susan-pyke

Pyke, S. H. (2022). Snake Church. *Animal Studies Journal*, *11*, 102–120. https://doi.org/10.14453/asj/v11i1.5

Rim-Rukeh, A., Irerhievwie, G., & Agbozu, I. E. (2013). Traditional Beliefs and Conservation of Natural Resources: Evidences from Selected Communities in Delta State, Nigeria. *International Journal of Biodiversity and Conservation*, *5*(7), 426–432. https://doi.org/10.5897/IJBC2013.0576.

Rose, D. B. (2007). Justice and Longing. In E. Potter, A. Mackinnon, S. McKenzie, & J. McKay (Eds.), *Fresh Water: New Perspectives on Water in Australia* (pp. 8–20). Melbourne University Press.

Rose, D. B. (2011). *Wild Dog Dreaming: Love and Extinction*. University of Virginia Press.

Shelton, M. B., Phillips, S. S., & Goldingay, R. L. (2020). Habitat Requirements of an Arboreal Australian Snake (*Hoplocephalus bitorquatus*) Are Influenced by Hollow Abundance in Living Trees. *Forest Ecology and Management*, *455*(117675), 1–9. https://doi.org/10.1016/j.foreco.2019.117675

Srinivasan, K. (2022). Re-Animalising Wellbeing: Multispecies Justice After Development. *Sociological Review*, *70*(2), 352–366. https://doi.org/10.1177/00380261221084781.

Thaler, M. (2022). What If: Multispecies Justice as the Expression of Utopian Desire. *Environmental Politics*, *31*(2), 258–276. https://doi.org/10.1080/09644016.2021.1899683

Tsing, A. L. (2011). Arts of Inclusion, or, How to Love a Mushroom. *Australian Humanities Review*, *50*, 5–22.

Tuck, E., & McKenzie, M. (2015). Relational Validity and the "Where" of Inquiry: Place and Land in Qualitative Research. *Qualitative Inquiry*, *21*(7), 633–638.

Van Dooren, T., Kirksey, E., & Münster, U. (2016). Multispecies Studies: Cultivating Arts of Attentiveness. *Environmental Humanities*, *8*(1), 1–23. https://doi.org/10.1215/22011919-3527695.

Victorian Government. (2023). Personal Communication, September 4. Conservation Regulator, Department of Energy, Environment and Climate Action.

Wadiwel, D. (2015). *The War Against Animals*. Brill.

Wadiwel, D. (2016). Do Fish Resist? *Cultural Studies Review*, *22*(1), 196–242. https://search.informit.org/doi/10.3316/informit.947695681113380

Whitaker, P., & Shine, R. (1999). Responses of Free-Ranging Brownsnakes (*Pseudonaja textilis*: Elapidae) to Encounters with Humans. *Wildlife Research*, *26*(5), 689–704. https://doi.org/10.1071/WR98042

Chapter 14

A New Pedagogy of Sharing Multispecies Sentience

Coexisting in Spaces of Love and Compassion

Jennifer Rebecca Schauer

> Love is of all passions the strongest, for it attacks simultaneously the head, the heart and the senses.
>
> ~Lao Tzu

I am heart centered in a multispecies liberationist pedagogy. It is through this that I believe love and compassion are the ways to sovereignty. This writing showcases my first-hand knowledge and observations based on my work with students. I teach about nature and animals founded in Environmental Sociology and Human Animal Studies, yet so much beyond Western Hegemonic Science and Scholarship informs my pedagogy and inspires me to co-create a space of learning that is intentionally full of hope. I believe my work with students is unique, novel, and exciting, and offers key insights into the promise of a new pedagogy, and, in that, holds space for transformation.

Pedagogy is becoming more open to qualities that have been historically under-represented in academia, including compassion (Barrable, 2019) and love (Loreman, 2011). There is no doubt scholarship has been shaped by qualities of ration, logic, and reason, which have been aligned more closely with men and the masculine forces of our world. Scholars have quite solidly critiqued such a narrow lens as an interpretation of the world that excludes many voices (Engelsman, 1989; Smith, 2021; Curato, 2013), and I would add many beings. Compassion and love are universal, yet within a hegemonic and Western construction such qualities are nested as feminine in nature (Gabrielson, 2021 citing Plumwood, 2002). Yet the ways in which we know the world and our experiences in it are much beyond the thinking mind. Feminist scholarship encases a body of work around embodied experiences and related epistemologies (Haraway, 1988, 1991). Corporeal feminist scholar Grosz (1987) writes of the body as a relational vessel of capacities to imprint. And arriving at the cusp of a New Era, we have emerging research on heart intelligence (Sands, 2022; Gillani & Ferdoos, 2022) providing a new understanding that mind, body, and spirit are all connected. The heart is emotional (Gillani & Ferdoos, 2022; Sands, 2022), intuitive (Sands, 2022), rational, and spiritual (Gillani & Ferdoos,

DOI: 10.4324/9781003366706-19

2022). Thus, decisions are made not simply by the brain but rather by the connection of the brain with the heart (Surel, 2014).

Certainly, scholarship supports the body, heart, and mind connection. Perhaps not the first, yet in 2004, Steiner wrote on emotional literacy as a "love-centered emotional intelligence" (xi), which is where we develop our intuitive compass (xii). John Muir Laws (2016, p. 45), too, writes of emotional intelligence as a practice of awareness and attention to the natural world, and how it dances with our own self-awareness. Other scholars remind us of our multiplicities. For example, Paulo Freire, a forefather of the critical pedagogy movement (Darder, 2017) as well as philosophers Gilles Deleuze and Félix Guattari (2022) in their work on *Becoming Animal* focus on the dimensionality of humans. Such scholars, I believe, extend traditional and orthodox frames of ontological understandings and conventional epistemologies in the West to expand into other parts of our human essence. It is here, I suggest, that inclusion of compassion and love in pedagogy invites in a move toward equanimity in ourselves and in our world. It is a calling to reunite with a part of human nature that is inherent in our very essence. It is our very humanness that allows us to connect with nonhuman species in sentient ways, in ways in which we know, but perhaps have forgotten.

Theoretical Foundation

Scholarship on Love

A New Pedagogy of Sharing Multispecies Sentience aligns with Loreman's (2011) call for a radical shift to bring love into pedagogy. Loreman (2011) writes that love and pedagogy are bound to one another in effective teaching and learning. Love is a power within teaching (Darder, 2017). This love is innate in us all yet needs practice to translate into action, into the being of each and every student. Steiner's (2003) emotional literacy involves having the capacity to receive love and also the capability to love another. Nelson and Durham (2023, p. 30) write of Interspecies Love that "is fostered in shared moments of vulnerability and beholdenness, smallness, and humility. Upending human-centered notions of domestication and cultivation, interspecies love is nourished by practices that slow things down, make us pause to look around, to *notice.*" In a sharing of multispecies sentience, I know that multispecies love is also unconditional, and those who calibrate with this vibration, know too, in their heart of compassion. An unconditional love is one that cannot be judged or determined, rather it is a knowing and a wisdom. From my own intuitive wisdom, the teacher with this kind of human spirit imbues not love and compassion alone but also peace and harmony. I see these as the alchemy for a New Pedagogy of Sharing Multispecies Sentience, and certainly foundational for a teaching methodology with an intention to liberate nonhuman species.

Ray (1982, p. 299) writes that benevolence forms the foundation for many of the world's religions. For example, ahimsa is nonharm and extends to animals in

Jainism, where "the Jains of India . . . avoid harming even microbes and insects." Buddhism supports loving kindness toward all the world, including nonhuman species (Waldau, 2006). It has been acknowledged that some Hebrews (Elkayam, 2014) and various sects of Sufism (McLeod, 2018) have fundamental love for nonhuman life, and espouse practices in alignment with virtuous regard to life beyond humans. The unconditional love, compassionate love, and eternal love I speak of is distinct from discussions around loving some nonhuman species and not others (Joy, 2020). It is a universal love, not culturally or species bound. Religious traditions remind us of a commonality in ideas around universal love, and that, while the notion of love is not tethered to religion, there is continuity in love for nonhuman life in ancient ways of understanding the world and in non-Western cultures.

Scholarship on Compassion

Compassion is an expression of love, and an expression of the heart. Barrable (2016, p. 1) created a learning paradigm of connectivity with cultivating compassion as one of four foundational points of reference along with sustained contact with the natural world, engagement with nature's beauty, and mindfulness. Buddhism's compassion "encompasses all who suffer" and therefore is a way to circumvent suffering of all beings. Drawing on Buddhist stories and parables in the classroom, Adarkar and Keiser encourage "awakening or reinforcing compassion and mindfulness in teachers, students, and administrators" (2007, p. 246). This can then guide desires and actions "to do the least harm possible to any living being in whatever circumstance" (Phelps, 2004, p. xv).

Importantly then, similar to love, compassion toward nonhuman animals is practiced all over the world not only through Buddhist understandings but also by Jains, some sects of Hebrew, Sufi, and Krishna Consciousness traditions.[1] Yet, compassion is not exclusive to religious traditions. Many compassionate animal people are atheists or at least agnostic so we must invite in all who practice and follow compassionate and loving ways to coexist with the life world. That said, I find these to be imperative traditions, cultures, and religiosities for study in the New Pedagogy of Sharing Multispecies Sentience, one that embraces love and compassion and also, embodied wisdom, toward nonhuman species. Thus, notwithstanding other groups and practices, including veganism and other spiritual practices, that claim no affiliation to institutional religion, I offer here that religious traditions provide wisdom for generating varying forms of peacefully coexisting in the world with all life.

Beyond pedagogy, scholars, especially in conservation, are speaking to the necessity of compassion with nonhuman species (Johnson et al., 2019; Wallach et al., 2018). Compassionate conservation and multispecies justice are two emerging fields in this area concerned with challenging the normative harms done to nonhuman species (Santiago-Ávila & Lynn, 2020).

Rooting A New Pedagogy of Sharing Multispecies Sentience

Scholarship certainly shares a disruptive pedagogical intent with my proposed New Pedagogy of Sharing Multispecies Sentience. Yet, I distinguish this Pedagogy from adjacent formulations, which have recently gained momentum.

Wild Pedagogies reexamines our relations with the natural world, and all life, rethinking concepts of wilderness, nature, and freedom, while simultaneously challenging conformity within pedagogical structure (Jickling, 2018). Of the six touchstones in *Wild Pedagogies*, the life world, including all species and nature, are co-teachers (Morse et al., 2018). A New Pedagogy of Sharing Multispecies Sentience is similar to *Wild Pedagogies* to the extent that it relinquishes pedagogical authority and invites in wisdom beyond the written and spoken word. Love and compassion are the strongholds of A New Pedagogy of Sharing Multispecies Sentience, which invite more of a sitting with such qualities, as well as more conversation and reflection around what they might mean in a world of peaceful interspecies coexistence.

Similar to theorizations of decolonization within multispecies pedagogies, a New Pedagogy of Sharing Multispecies Sentience aims to situate all external knowledge in humility. For example, similar to Fawcett and Johnson's (2019) work in experiential environmental education, my pedagogical philosophy expands into Indigenous and total liberation pedagogies, critical animal studies, environmental ethics, and phenomenological understandings, as well as decolonial work embedded in indigenous ontologies of living among other species (Taylor et al., 2021). In this way, it situates external knowledge in scholarship as disruptive to conventional modes of scholarship.

Scholars are also offering critical animal pedagogies as alternatives to conventional teaching and learning frames with animals (Dinker & Pederson, 2016). Pedersen's (2023) work seeks to reshape higher education and recognizes that entering spaces of discomfort is natural to the reflective practices of studying relationships with other species. Similarly, a New Pedagogy of Shared Multispecies Sentience invites students to write personal narratives about their lived, embodied, and felt experiences with animals drawing on course literature to support their reflective expressions. Taylor and Pacini-Ketchabaw (2015) describe vulnerability in pedagogy as a space to move through in order to co-learn with other species and reconsider our own place in the world. In this way, vulnerability is a necessary space to traverse in order to recalibrate toward an alignment with nonhuman species and the natural world.

There is alignment across these alternative pedagogies in terms of disrupting pedagogical structures and conventions, and shifting course materials to voices that emerge from worlds where other than human species are seen, felt and heard. Yet, the primary space of distinction for a New Pedagogy of Shared Multispecies Sentience is the powerful grounding in the intuitive, embodied wisdom of each individual student. While there is some alignment with Critical Animal Pedagogies, the New Pedagogy I offer seeks even further to break down hierarchies by challenging the notion of external knowledge as the only source of wisdom. As Taylor and colleagues (2021) describe, it is a pedagogy that sits with the life world. Most distinctively, this New Pedagogy of Sharing Multispecies Sentience is grounded in love and compassion.

A New Pedagogy of Sharing Multispecies Sentience

Pedagogical Approach, Four Pillar Framework

The New Pedagogy of Sharing Multispecies Sentience comprises four fundamental pillars. The first pillar is intuition and logic – understood as embodied wisdom – which are both invited wisdoms in the pedagogical lineage. Second, empowerment is unveiled inside of each learner as they allow their own intuition and felt wisdom to emerge in conversation with the logic and reason of the external knowledges presented in the learning experience.[2] This empowerment in allowing the students to be co-creators of knowledge within the classroom, within academic institutions dissolves tyranny. Third, an affinity and becoming allow for fluidity, as each learner and teacher have their own unique and authentic blueprint and are both radically accepted as part of the community, as well as irreplaceable in human constitution.[3] Such a celebration of differences shifts into a space of nonjudgment, which is the final and fourth pillar. This is a space of moving beyond the rational mind, of thoughts and opinions to feel compassion for another human, even if their conditioning has taught them to differ in their words, and actions, and in their compassion for self and other species.

In the New Pedagogy of Sharing Multispecies Sentience, I invite students to cultivate new ways of thinking that involve their heart spaces, where the mind of constant thoughts merges with the heart of feelings, and to write and reflect from this space of their lived, felt, and embodied understandings. This is not a space of opinion and judgement involving hierarchies and categories but a space of interconnected feelings between the students as sentient beings, and nonhuman animals as sentient beings. The semester opens with several ways to cultivate sentient awareness including mindfulness practices with meditation and breath; sitting on the Earth in nature; and readings on sentience, emotions, and culture in other species. To navigate our own sentience, I draw on Muir Laws (2016), who guides us to sit in our own consciousness, which is nested in the natural world and all life beings. To see and feel animals as sentient, I use excerpts from the work of de Waal (2019)[4] and Bentham (2022) as philosophical underpinnings. With reference to sentience, de Waal (2019) highlights the consciousness of all species, while Bentham (2022) reminds us all species suffer, it is sentience that binds us.

Multispecies shared sentience is feeling ourselves as sentient. To align with this intention, students practice sentience understandings in written reflections and interspecies field interactions with wild and free animals, where they are invited to simultaneously feel themselves and another species who is near them, however fleeting, in an outdoor environment. This culminates with an essay written in an empowered first-person voice, where students are invited to interweave their own heartfelt journey in conversation with scholarship, empirical evidence, social theories, and our ontological understandings, which are external perhaps to our felt embodied wisdom, yet shape and reshape our social reality. The grading rubric grants value to the sentient portion of the lived, felt, and embodied experience with animals. The essay narratives are reflections and heartfelt journeys, where students' personal sentience understandings, intuition, and wisdom are equal to the scholarly readings.

A New Pedagogy of Sharing Multispecies Sentience, Lived, Felt, and Embodied Experiences

The following sections draw on my work teaching an Animals and Society course to approximately 250 students attending a minority-serving public university in California during Fall 2023 and Winter and Spring 2024.[5] The course opens with a unit exploring multispecies field observations, interconnectivity (Kimmerer, 2013), and sentience (de Waal, 2019; Muir Laws, 2016), and includes field observations and personal essay narratives. Unit 2 covers oppression, capitalism, and colonialism, and encourages students to reimagine the dominant paradigm. The final unit offers students the opportunity to co-create an intentional interspecies community[6] of peaceful coexistence with nonhuman animals – free-living, commensal, domesticated, captive, and companions. Most fundamentally, the foundation of my sharing sentience pedagogy is this, *I invite students to write about their felt, lived, and embodied experiences with nonhuman species. This is how we know other species, and how we know ourselves with other species.*

Particularly in their collaborative projects and final reflections, which both arrive as opportunities at the end of the semester to demonstrate thematic learning expressions in art, and in writing, students repeatedly demonstrate, in creative and compassionate ways, how they have awakened to new ways of being in the world with animals, ways demonstrating connection, respect, and massive transformations in learning. The following examples showcase such pinnacle learning expressions. Pseudonyms have been used in place of students' real names, in some cases. In other cases, students wished to have their names credited with their learning expressions.

Sentience and Co-creating a New World

One student, Takuma Funakoshi, when sharing his field observation with the class, orchestrated the phrase "practicing sentience" as a way to coexist with wild animals and all sentient species. Such a learning expression is alchemically embedded in my emerging research on Sharing Sentience (Schauer, 2020a, 2020b, 2021; Schauer et al., 2024). It is what I feel, think, and see as a professor, and yet curiously, it is nowhere structurally in my course.

Echoing the feelings of many students in the class, the work depicted in Figure 14.1 entitled "We Are All One," co-created by Melanie Mendoza and Seungha Seo, expresses the love, affinity, becoming, and interconnectedness that are seen as key to a betterment of the lives of animals.

Another provocative final semester project was created by an individual student whose learning trajectory is expressed in Figures 14.2 to 14.5.

Beginning with the pyramid in Figure 14.2, Kasey Nguyen writes,

The inspiration piece represents a chain of what I believe to be the relationship between humans and animals. The top being humans who use all beings as a means of development, next being predators and exotic animals as humans have different views and are most likely to protect them over the next category, being animals who are known to be mistreated and slaughtered for food, and the final . . . plants. This has inspired me to delve into the different themes and meanings that can be pieced together through this chart.

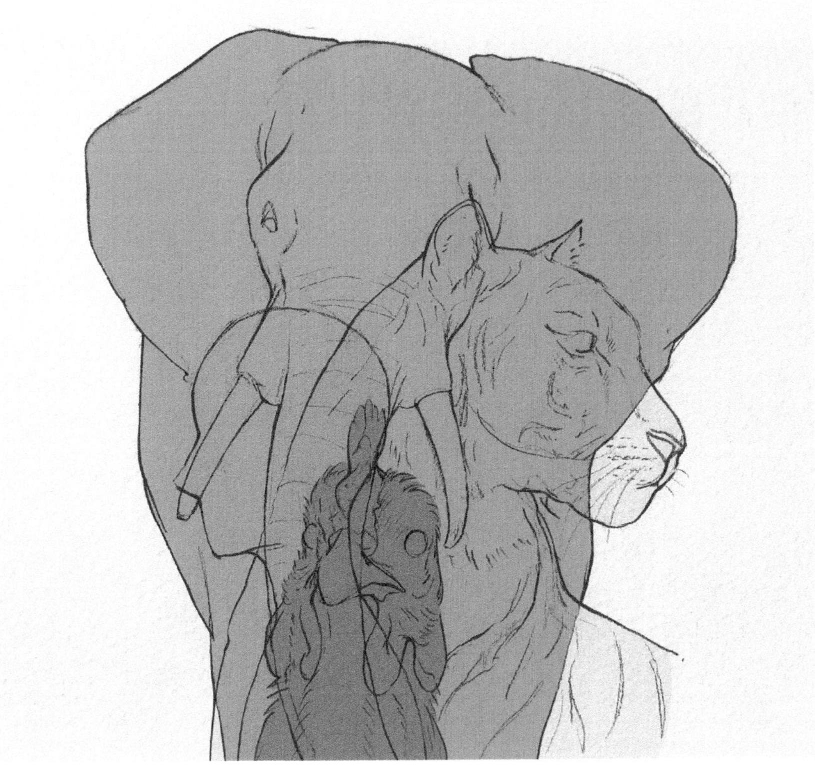

Figure 14.1 Creation. Melanie and Seungha.

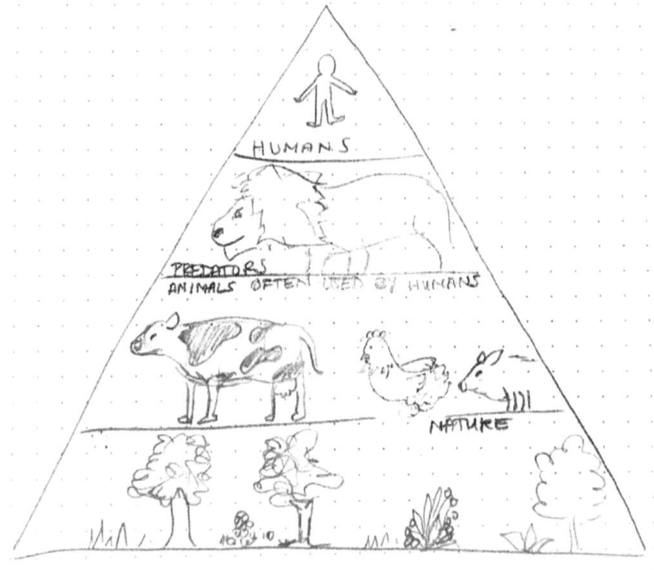

Figure 14.2 Inspiration. Kasey Nguyen's Final Project.

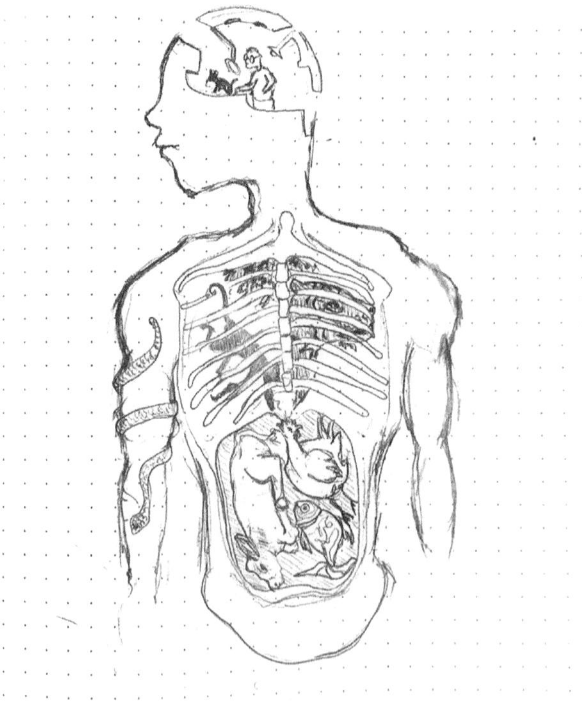

Figure 14.3 Preparation. Kasey's Final Project.

Figure 14.4 Preparation. Kasey's Final Project.

In her next figure (Figure 14.3), Kasey represents an intricate web of human centrism and domination. Reflecting her move through anthropocentric understandings of the world and one of the course themes, Kasey depicts the suffering of animals within the lived and felt body of the human – their suffering is inevitably part of, and not separate from. This corporeal understanding is also affinity and becoming, one of the four pillars of this New Pedagogy.

This quote demonstrates her understanding of affinity and becoming where lines are blurred. Furthermore, her journey within the semester then saw animals as sentient persons, Figure 14.4, which is what she moved through, to arrive within coexistence, Figure 14.5.

The culmination of Kasey's learning expression was her vision of a new world (Figure 14.5). She writes,

> My creation piece expands on the themes of a healthy coexistence between humans and animals, avoiding animals as a food source which is represented with fruits as restaurants that the animals are freely able to enjoy comfortably at their own will.

Kasey writes of her learning experience throughout the semester as leading her to a "recognition of animals as persons who are non-human, that have sentience and personhood all while experiencing the same world and emotions as humans do" (Figure 14.4).

Figure 14.5 Creation. Kasey's Final Project.

The creation piece was titled "Representing a World of Coexistence with Animals Granting Them Rights to Their Own Existence." Peaceful coexistence is possible with the four pillars founded in love and compassion.

Sharing Multispecies Sentience: Love, Compassion, and Inspiration to Become

Sharing Multispecies sentience, love, compassion, and becoming have emerged as key themes in the final course reflections, and field observations. Love and compassion are the theoretical underpinnings of A New Pedagogy of Sharing Multispecies Sentience, where becoming is one of the four foundational pillars.

Sentience. Students recognize their own sentience and the sentience in another, this is the sharing of a multispecies sentience. As Swetha Nair writes, "I learned about each individual having their own souls that are clearly sentient and noticeable. They [two squirrels] interacted with me without human words but still spoke clearly." Takuma's field observation with a crow founded a becoming through sentience, he writes,

I started making more and more connections with the crow that made me realize that him and I are one in the same. Both the crow and I were traveling from afar with our friends, and we both had decided to rest at the same area. We both decided to rest because we both got tired, we both chose to come to this specific rest area and were fond of our environment. I started to observe our surroundings and saw the big trees that stood above us. While the crow and I compared are different sizes and different species of animal, it doesn't make a difference when we are both underneath the same big tree. To the tree, he and I are both very small and that is something the crow and I can relate to. So why must I have the authority over him, the crow, when he belongs underneath this tree, this same environment, just like myself. Who is to decide the difference between human and animal, and what is to determine our relationship as fellow sentient beings? And the answer is, no one. No one is in nature to determine this. This was the conclusion I had come to. . . . The crow and I are one in the same underneath the tree. The only difference that separates him and I are our appearances. But we both exemplify sentience, and we both suffer. Suffering is about sentience, and whether we can suffer as existing beings or not is the question that should be raised when seeking superiority. This small encounter with the crow has been an immeasurable learning journey for me and it makes a great impact on my understanding of what my individual behavior as a human should be when sharing this Earth.

Love. Love was also a learning expression. Nelson and Durham (2023) suggest love is nourished and practiced when we pause to see the world around us, the living animals. Many students commented on the meditative quality of their interspecies field observations[7] and of how they felt deeply connected with animals during these moments of stillness. Such an opportunity allows students to practice sitting with other species. From here love is born. As Radda Triova writes in her final reflection,

> The experience of sitting with animals has really been special to me and is something I will carry with me for my entire life . . . what has truly resonated with me is the openness to love. The capacity to cry for beings that may seem so different from myself or form such a bond with a being that protecting their rights becomes essential to me is most meaningful. To me, this is what learning is.

Radda's pause and stillness to sit with wild animals allowed for an "openness to love," which aligned with the capacity to bond, protect, and find meaning in other species.

For other students, wild species become more similar to animals who are loved, such as companions. As Nadia writes in her final reflection,

> From deepening my love for animals, to changing my daily routines, this class has left a mark on me that will not leave my side. . . . From learning about applying the same love I have for Grace [her beloved cat who had

passed away the year before], to other animals in the wild, it made me so fascinated behind the idea of love itself. . . . I learned about the personalities, emotions, experiences, and lifestyles that all animals possess. . . . With your help and allowing me to have a voice, the love I have for animals are being expressed to all my loved ones around me and will also be taught to my future children.

Nadia expresses her new love for wild animals, which is being shared with her loved ones, and her future children. This love is supported as Nadia feels empowered to share her voice with others. Ricardo Esparza, another student, also feels empowered by love and writes of his love for animals and the changes that ensued from such love,

Loving animals and starting to learn more about veganism . . . has illuminated the interconnectedness of all living beings. . . . My love for animals provides me with motivation to question the status situation. It drives me on to support animal welfare organizations, eat and live ethically, and participate in discussions that highlight how urgent it is for things to change . . . love is not a passive force, rather it is an active force that inspires advocacy, compassion, and group efforts to bring about a world in which animals receive the dignity and kindness that they are due.

Here Ricardo discusses his love as active, one that inspires change in the way he lives in the world with animals. Within this kind of love, compassion is an active agent of change toward extending kindness and dignity toward animals.

Compassion. In pedagogy, compassion has been discussed as actively doing the least harm to other living beings (Phelps, 2004). Such an awareness was developed in my classes where students were intentionally seeking ways to avoid harm and suffering toward animals. As Anthony Zesati writes,

I started correcting my niece and nephews about killing spiders. . . . I can tell that this was the first time they have heard this or thought about it. . . . Spiritually I even feel awakened. I will continue to push myself and the younger generations to cherish every life and to help spread awareness and compassion to one day lead to an interspecies coexisting society.

For Anthony, the shift to end the killing of certain species of animals, such as spiders, extended to teaching the same to his niece and nephews. This was active compassion and awakened dimensions of spirituality for him to see anew. Similarly, Nadia discussed in class sending intentions and prayers to wild animals who she saw suffering. This is compassion.

Compassion was also commonly discussed around veganism and plant-based eating, which is one theme I use in the class to disempower capitalism and empower our own possibilities around assuaging the suffering of animals globally. Amarbir Singh writes of being on a journey toward compassionate living,

> The adoption of a plant-based diet, explored extensively in our course, has been another transformative aspect of my journey . . . this dietary shift has become an embodiment of my commitment to peaceful coexistence . . . the Animals and Society course has been a profound and deeply enriching experience that transcends academic learning. It has guided me towards a more intentional, compassionate, and interconnected way of living.

To Amarbir, the shift to plant-based foods and dietary choices is embodied. What is good for him is good for earth and is good for animals. This is the message of my teachings. Amarbir's daily affirmation is extraordinary intention and compassion and a connected way to live our lives. Compassion is action.

Love and compassion hold all such qualities. As Darder (2017) writes, love is power in pedagogy, and Loreman (2011) similarly argues that love and pedagogy are not separate. Such empirical evidence here is demonstrative of this – that we cannot teach without love, when we root into the core of our humanness, intentions, and voices, we become agents of change.

Inspiration to Become. The becoming theme was expressed Takuma's field observation with a Crow yet is also present in individual final reflections as I showcase here. Such a quality is within the Affinity and Becoming pillar of this New Pedagogy. It broadly displays a transformation of which the student is intently and powerfully aware. Becoming animal is multidimensional, blurs boundaries, and is fluid (Deleuze & Guattari, 2022). As Radda Triova writes,

> The emotions such as love, fear, pain and joy that I believe define human existence is [sic] everything that any animal will feel in their lifetime. . . . I could no longer separate myself from the butterfly I spotted on my walk. She and I have both experienced both the beauty and hardships of change.

The Animals and Society course invites students to undergo transformation, and students feel inspired to become, knowing this is a space to enter and undergo a metamorphosis in order to transmute qualities that no longer serve their life journey. As Catalina, wrote, "This journey has been both enlightening and transformative marking a significant chapter in my academic and personal growth." There is a poignancy of intention that is invited in the course, where love and compassion exist in connection with scholarship, embodied wisdom, empowerment, affinity and becoming and nonjudgement. Such qualities are powerful markers of change, as Swetha Nair wrote, "Through the course of this class, I have felt my world view change."

When students are invited to join a journey about animals that also invites in a journey of self-discovery, they have the opportunity to see the world in a way that aligns with a part of them that they may just be beginning to know. This is why students such as Brian Castruita write, "Thank you so much. . . . It's one of the most impactful classes I have had that changed me personally and my views on animals." Students are grateful to see the world in a way that aligns with who they wish to be, rather than who they were conditioned to be. This becomes possible, allowing for empowerment through embodied wisdom, in a space that is both non-judgemental and invites in affinity and becoming. Affinity and becoming are then deeply contemplated, personal transformative spaces, which emerge as profound ideologies around powerful shifts in our world, as Melanie Mendoza writes in her final reflection,

> If we as people start to understand animals are living breathing souls, then the idea that they should not be slaughtered would be an obvious right. They, in reality, should have autonomy over their bodies, just as it is a right we grant ourselves.

My Sentience Collective

My sentience collective is an expansion of such classroom learning expressions to connect deeply and profoundly with animals. I formed a collaborative made up of myself, the mentor, and eight undergraduate research fellows. We met weekly to discuss our manuscript projects, which jointly had a theme of sharing sentience with animals. Through this work I mentored and supported students to write and publish work on animal sentience in peer-reviewed journals (Patterson, 2021; Schauer et al., 2024). The undergraduate research fellows were funded through the University, and we supported one another in ways of peer review, sharing resources, and working toward a common understanding of empowering student voices in academia. Several more manuscripts are at different stages of review as a research collective, and although I am now at another University, the Sentience Collective continues among new student researchers, lives in the hearts of those who came before, and in our continued interactions and experiences with other species and Earth Mother. The value of a New Pedagogy of Sharing Multispecies Sentience is that it liberates the learner. Undergraduate research is the stronghold for this New Pedagogy of Sharing Multispecies Sentience as it cultivates and co-creates the space for the next generation of scholars, activists, and ambassadors for other species to form the lived, felt, and embodied sentient capacities that will herald the New Era of interspecies peaceful coexistence.

Conclusion

I make a call for radical transformation alongside other scholars who are revolutionizing changes in pedagogy (Herbst, 2023; Loreman, 2011, Fawcett & Johnson, 2019; Dinker & Pedersen, 2016; Morse et al., 2018; Pedersen, 2023; Nocella et al., 2019). I foresee A New Pedagogy of Sharing Multispecies Sentience implemented globally. This is different from alternative pedagogies mentioned earlier in that it is a pedagogy which invites in love and compassion so we may align with a natural quality of our humanness. The connection of love and compassion with nonhuman species and self allows for a transformative experience that fosters new ways of relating with animals, seeing them as sentient beings with whom we share the Earth. *The deep space this holds is to move beyond the thinking mind, to allow the wisdom of the body, heart and spirit to become fully present to our connection to animals.*

The first iteration of a New Pedagogy of Sharing Multispecies Sentience has liberationist undertones, yet it is rooted in four mutually dependent fundamental pillars, and theoretically underpinned by compassion, love, and the creation of new knowledge.[8] The marriage of intuition and embodied wisdom with logic, a deeply sensed empowerment, and a space of affinity to become are the bones that hold together this living and breathing way to teach and learn. More structure will perhaps be needed to fully realize this New Pedagogy and to develop it as a foundation with breath. This then is a summons to all those seeking to disrupt old pedagogies and expand into new ways – a call to join forces in co-creating a New Pedagogy that holds and sees nonhuman species, and is embodied in wisdom, love, and compassion. I position myself here, as one among so many, to come forward, at the brink of a new Era that is emerging from the ashes of Tyranny, to rise above and to teach with more than our minds. We are meant to teach with our hearts, our souls, and our embodied wisdom, all of this in unison, in oneness. Oneness with ourselves, with our student learners, in oneness with nonhuman species, and in oneness with Earth Mother.

My contributions are three-fold. First, pedagogies exist within our capacity to love, to love ourselves, other species, Earth Mother, and one another. This is our truth. Second, I integrate into my classes studies of indigeneity through traditional ecological knowledge into my classes, and Eastern cultural understandings that have a deep connection with Earth and animals, such as Taoism, Jainism, Hinduism, and Buddhism. These voices have been excluded from animal liberation conversations for far too long. Finally, students are co-creating the future. It is us, now, who can guide them toward liberation for all, with skills and capacities to live in peaceful coexistence. This New Pedagogy of Sharing Multispecies Sentience tells the story of connection and the story is told best through our own authentic felt, embodied, and lived experiences.

Notes

1 I view Buddhism, Jainism, some sects of Hebrew, Sufi and other traditions which seek to live in harmony with the nonhuman life world, as bearers of knowledge and wisdom which support a deep understanding for how cultures are living and coexisting in peaceful ways with nonhuman species.
2 Such a unification of intuition and empowerment in the felt, lived, and experienced body is distinctly different from scholarly synthesis and analysis, which may often feel external and unattainable, and remains hierarchical.
3 In this work the human constitution is the physical body and its functions, as well as the mind, heart and spirit.

4 I use excerpts with de Waal (2019) only to emphasize the ethological scholarship on consciousness and sentience within other species.
5 Institutional Review Board for Human Subject Research approval was granted in April 2024.
6 The intentional interspecies community is presented in *Just Wondering* (Nor et al., 2021), a short film I show students to invite understandings of living in a world where animals are granted rights and part of a larger community of living beings. This work is founded in Zoopolis from Donaldson and Kymlicka (2011).
7 Students are to conduct 10 Interspecies Field Observations throughout the semester, where they sit in a natural outdoor setting and see wild and free animals. They are to take an image, journal, and they are invited to write about how not only they see the animal, yet also how the animal is observing them. Sentience, emotions, culture, and consciousness are the themes that are the focus of the student narrative within each of the 10 experiences.
8 This extends to Earth Mother, and all Animate Life, yet for now, I focus on Nonhuman Animals here, because I love them deeply.

References

Adarkar, A., & Keiser, D. L. (2007). The Buddha in the Classroom: Toward a Critical Spiritual Pedagogy. *Journal of Transformative Education, 5*(3), 246–261.

Barrable, A. (2019). Refocusing Environmental Education in the Early Years: A Brief Introduction to a Pedagogy for Connection. *Education Sciences, 9*(61), 1–8.

Bentham, J. (2022 [1789]). Principles of Morals and Legislation. In L. Kalof & A. Fitzgerald (Eds.), *The Animals Reader: The Essential Classic and Contemporary Writings* (2nd ed., pp. 16–17). Routledge.

Curato, N. (2013). A Sociological Reading of Classical Sociological Theory. *Philippine Sociological Review, 61*(2), 265–287.

Darder, A. (2017). *Reinventing Paulo Freire: A Pedagogy of Love.* Routledge.

Deleuze, G., & Guattari, F. (2022 [1987]). Becoming Animal. In L. Kalof & A. Fitzgerald (Eds.), *The Animals Reader: The Essential Classic and Contemporary Writings* (2nd ed., pp. 63–77). Routledge.

de Waal, F. (2019). *Mama's Last Hug: Animal Emotions and What They Tell Us About Ourselves.* WW Norton & Company.

Dinker, K. G., & Pedersen, H. (2016). Critical Animal Pedagogies: Re-Learning Our Relations with Animal Others. In H. E. Lees & N. Noddings (Eds.), *The Palgrave International Handbook of Alternative Education* (pp. 415–430). Palgrave Macmillan.

Donaldson, S., & Kymlicka, W. (2011). *Zoopolis: A Political Theory of Animal Rights.* Oxford University Press.

Elkayam, S. (2014). "Food for Peace": The Vegan Religion of the Hebrews of Jerusalem. *Idea. Studia nad strukturą i rozwojem pojęć filozoficznych, XXVI,* 317–340.

Engelsman, J. C. (1989). Rediscovering the Feminine Principle. In S. J. Nicholson (Ed.), *The Goddess Re-Awakening: The Feminine Principle Today* (pp. 100–107). Quest Books.

Fawcett, L., & Johnson, M. (2019). Coexisting Entities in Multispecies Worlds: Arts-Based Methodologies for Decolonial Pedagogies. In T. Lloro-Bidart & V. S. Banschbach (Eds.), *Animals in Environmental Education: Interdisciplinary Approaches to Curriculum and Pedagogy* (pp. 175–193). Palgrave Macmillan.

Gabrielson, T. (2021). The Nature of Gender. In J. Cohen & S. Foote (Eds.), *The Cambridge Companion to Environmental Humanities* (pp. 56–69). Cambridge University Press.

Gillani, I., & Ferdoos, A. (2022). Heart Intelligence from Quran and Science: An Integrated Approach for the Evolution of Enlightened Self in the Society. *International Research Journal on Islamic Studies (IRJIS), 4*(1), 1–17.

Grosz, E. (1987). Notes Towards a Corporeal Feminism. *AustralianFeministStudies*, *2*(5), 1–16.

Haraway, D. (1988). Situated Knowledges: The Science Question in Feminism and the Privilege of Partial Perspective. *Feminist Studies*, *14*(3), 575–599.

Haraway, D. (1991 [1985]). *Simians, Cyborgs and Women*. Routledge.

Herbst, J. H. (2023). Current and Future Potentials of Liberation Pedagogies: A Discussion of Paulo Freire's, Augusto Boal's, and Johannes A. van der Ven's Approaches. *Religions*, *14*(2), 145.

Jickling, B. (2018). Why Wild Pedagogies? In B. Jickling, S. Blenkinsop, N. Timmerman, & S-S. M. De Danann (Eds.), *Wild Pedagogies* (pp. 1–22). Palgrave Macmillan.

Johnson, P. J., Adams, V. M., Armstrong, D. P., Baker, S. E., Biggs, D., Boitani, L., Cotterill, A., Dale, E., O'Donnell, H., Douglas, D. J. T., Droge, E., Ewen, J. G., Feber, R. E., Genovesi, P., Hambler, C., Harmsen, B. J., Harrington, L. A., Hinks, A., Hughes, J., . . . Dickman, A. (2019). Consequences Matter: Compassion in Conservation Means Caring for Individuals, Populations and Species. *Animals*, *9*(12), 1115.

Joy, M. (2020). *Why We Love Dogs, Eat Pigs, and Wear Cows: An Introduction to Carnism*. Red Wheel.

Kimmerer, R. W. (2013). *Braiding Sweetgrass: Indigenous Wisdom, Scientific Knowledge and the Teaching of Plants*. Milkweed Editions.

Laws, J. M. (2016). *The Laws Guide to Nature Drawing and Journaling*. HeyDay Books.

Loreman, T. (2011). *Love as Pedagogy*. Springer Science & Business Media.

McLeod, D. (2018). *A Path of Justice and Compassion: About Vegan Muslims and Islamic Veganism*. University of Otago.

Morse, M., Jickling, B., & Quay, J. (2018). Rethinking Relationships Through Education: Wild Pedagogies in Practice. *Journal of Outdoor and Environmental Education*, *21*, 241–254.

Nor, A., Martelli, M., & Mimosa, M. (2021). *A Political Theory of Animals Rights After Sue Donaldson and Will Kymlicka* [Film]. Just Wondering Project. www.justwondering.io/a-political-theory-of-animal-rights/

Nelson, P., & Durham, B. S. (2023). Desire, Interspecies Love, and Becoming-Animal: Reading "The Overstory" in Social Studies Education. *Journal of Curriculum Theorizing*, *38*(1), 23–45.

Nocella II, A. J., Drew, C., George, A. E., Ketenci, S., Lupinacci, J., Purdy, I., & Schatz, J. L. (2019). *Education for Total Liberation: Critical Animal Pedagogy and Teaching Against Speciesism*. Peter Lang.

Patterson, C. (2021). Humans and Nonhumans: Coexistence Continuum and Approaches for Working Toward Shared Sentience. *Student Journal for Vegan Sociology*, *1*, 58–73.

Pedersen, H. (2023). Post-Anthropocentric Pedagogies: Purposes, Practices, and Insights for Higher Education. *Teaching in Higher Education*, 1–15. https://www.tandfonline.com/doi/pdf/10.1080/13562517.2023.2222087

Phelps, N. (2004). *The Great Compassion: Buddhism and Animal Rights*. Lantern Books.

Plutarch, M. (2022). The Eating of Flesh. In L. Kalof & A. Fitzgerald (Eds.), *The Animals Reader: The Essential Classic and Contemporary Writings* (2nd ed., pp. 234–238). Routledge.

Ray, J. J. (1982). Love of Animals and Love of People. *The Journal of Social Psychology*, *116*(2), 299–300.

Sands, B. E. (2022). Overview of the Heart's Intelligence: A Dynamic Perspective into the World of Energetic Wellness. *Journal of American Holistic Vet Medicine Association*, *68*, 22–30.

Santiago-Ávila, F. J., & Lynn, W. S. (2020). Bridging Compassion and Justice in Conservation Ethics. *Biological Conservation*, *248*, 108648.

Schauer, J. R. (2020a). *Shared Sentience*. Invited Presentation for Greener Lib. Boston College.

Schauer, J. R. (2020b). *Shared Sentience: Other Species and Humans*. Lecture Fall Semester for Human Wildlife Conflict and Coexistence. Boston College.

Schauer, J. R. (2021). *Sharing Sentience: Other Species and Humans.* Presentation at the Animals and Society section of the Annual Meeting of the American Sociological Association, August.

Schauer, J. R., Walsh, E., & Patterson, C. (2024). Human Wildlife Coexistence: A Shark Centric Perspective. *Trace Journal of Human Animal Studies*, *10*, 204–218.

Smith, B. J. (2021). Sufism and the Sacred Feminine in Lombok, Indonesia: Situating Spirit Queen Dewi Anjani and Female Saints in Nahdlatul Wathan. *Religions*, *12*(8), 1–22.

Surel, D. (2011). Speaking from the Heart. *EdgeScience*, *6*, January–March, 5–8.

Steiner, C. (2003). *Emotional Literacy: Intelligence with a Heart*. Personhood Press.

Taylor, A., & Pacini-Ketchabaw, V. (2015). Learning with Children, Ants, and Worms in the Anthropocene: Towards a Common World Pedagogy of Multispecies Vulnerability. *Pedagogy, Culture & Society*, *23*(4), 507–529.

Taylor, A., Zakharova, T., & Cullen, M. (2021). Common Worlding Pedagogies: Opening Up to Learning with Worlds. *Journal of Childhood Studies*, *46*(4), 74–88.

Treves, A., Santiago-Ávila, F. J., & Lynn, W. S. (2019). Just Preservation. *Biological Conservation*, *229*, 134–141.

Wallach, A. D., Bekoff, M., Batavia, C., Nelson, M. P., & Ramp, D. (2018). Summoning Compassion to Address the Challenges of Conservation. *Conservation Biology*, *32*(6), 1255–1265.

Afterword

A Methodological Side-note

Disrupting the Animal-Industrial Complex: A Counter-map to Guide Heterotopic Interventions and Radical Imaginaries

Paula Arcari

In my analyses and writing to date about anthropocentrism, animal oppression, and animal advocacy, I have been keen to emphasise the scope of practices involving the use and abuse of animals that constitute the Animal-Industrial Complex (A-IC). This is pertinent to the current volume because the multiplication of present heterotopia, the radical imagination of more liberatory nonhuman futures, and the work of shattering orders must similarly encompass all nonhumans who are restricted, coerced, controlled, caught, entrapped, enslaved, bred, modified, violated, and killed at the will and behest of humans. As Angie Pepper (2017) notes, and underscoring the need for an integrated *and* intersectional perspective, the violence done to animals is part of a global institutional order that perpetuates climate change, natural disasters, human conflicts, the trade in animal bodies, and habitat destruction.[1]

A comprehensive approach is required for the comprehensive disruption and dissolution of anthropocentric orders. I therefore offer here a way of thinking about and visualising the A-IC that could usefully inform frameworks for research, advocacy, and intervention aimed at challenging its social, economic, and political hegemony.

In his rearticulation of Barbara Noske's (1989) concept of the A-IC, Richard Twine recognises the range of relations, infrastructures, competencies, and understandings that comprise and shape it, and which had tended to be overlooked up to his writing. Hence, he defines the A-IC as:

> a partly opaque and multiple set of networks and relationships between the corporate (agricultural) sector, governments, and public and private science. With economic, cultural, social and affective dimensions it encompasses an extensive range of practices, technologies, images, identities and markets.

> (2012, p. 23)

The A-IC as a stand-alone concept was, and still is, predominantly deployed in relation to animal agriculture. However, Twine emphasises that "a working definition of the A-IC must not be confined to this domain" (ibid., 23) and proceeds to describe various components of what is essentially a comprehensive counter-map of animal exploitation that encompasses all forms and dimensions of animal use and reframes them in ways that can help "challenge the dominant power structures

DOI: 10.4324/9781003366706-20

and . . . articulate progressive, alternative, and, sometimes, radical interests" (Kindynis, 2014, pp. 227–228).

In the interests of clarifying and popularising this holistic conception, Table BM1.1 provides a non-exhaustive list of practices directly involving animals and the related practices they rely on and/or support, primarily in terms of the supply and movement of animals and/or their body parts.[2] For the sake of legibility, the animals involved in each practice are not listed, but the intended scope is inclusive. For example, farming for protein includes traditional 'farm' animals, as well as farmed fish, insects, and wildlife, while bullfighting includes horses.

Table BM1.1 A non-exhaustive list of practices involving animals and the co-constitutive practices to which they can be linked.

Practice	*Linked products, practices, and industries*
Farming (animal protein)	Leather, wool, feathers, zoos, mobile animal exhibits, circuses, rodeos, testing, pharmaceuticals, medicines.
Farming (leather, fur, wool, silk, down)	Farming (animal protein), fashion, and other goods.
Hunting, poaching, culling and/or 'pest' control	Testing, zoos, aquaria and other 'encounters,' mobile animal exhibits, pet industry, meat industry*, production of fur & leather, tourism.
The wildlife trade	Hunting, fishing, meat industry, zoos, circuses, animal exhibits and 'encounters', pet industry.
Fishing (commercial & sport)	Food industry, pharmaceuticals, medicines.
Horse riding, horse racing, harness racing, dressage, show jumping, carriage rides	Circuses, rodeos, testing, meat industry, pharmaceuticals, leather.
Greyhound racing, hare coursing	Testing, meat industry, zoos.
Rodeos	Meat industry, horse riding.
Bullfighting	Meat industry, horse riding.
Falconry	Poaching, pet industry, zoos and exhibits, meat industry.
Zoos & aquaria, mobile animal exhibits, animal rides & swims, other 'encounters'	Hunting, poaching, testing, pet industry, meat industry, circuses.
Circuses	Hunting, poaching, zoos, testing, pet industry, meat industry.
Street events (e.g., Pamplona, Siena)	Horse riding, meat industry.
Cultural & religious festivals	Hunting, farming (animal protein).
Endurance races (e.g., Iditarod, Mongol Derby)	Pet industry, horse riding.
Animal testing	Hunting, poaching, farming (animal protein), zoos, pet industry.
Pharmaceuticals & medicines	Farming (animal protein), fishing, hunting, poaching.
Pet ownership	Hunting, poaching, zoos, testing, racing, fighting.

* Hunting is related to the meat industry not only with respect to the consumption of hunted animals but also because the protection of farmers' interests (i.e., maximisation of profit) is used to justify routine (legal and illegal) culls of alleged threats, including coyotes, wolves, dingoes, eagles, kangaroos, and other animals.

This provides a basis from which to start conceptually and spatially mapping the co-constitutive and co-dependent relations between practices involving other animals. Building on the work of Twine and others, Figure BM1.1 shows one way this might be approached. Simplified nodes on an interconnected global network conceptually represent each dimension of the A-IC. In the same way that the multiple lenses of an insect's compound eye render an image of the whole, the entirety of the animal oppression matrix is constituted by the sum of its parts (O'Sullivan, 2020). By zooming in (Nicolini, 2010) on the wildlife trade as one element of this matrix, the practices (and elements thereof) relevant to that industry, and the pathways connecting them, can start to be unpacked, again focusing in BM1.1 on the supply and movement of animals. Cross-sectional layers could be used to zoom out again to foreground different dimensions of overlapping networks depending on the focus (e.g., physical infrastructures, forms/structures of knowledge, social and symbolic meanings) (Shove et al., 2012). Shared techniques of oppression (Arcari, 2022) could also be represented in this counter map.

Another example of counter-mapping is The Farm Transparency Map[3] from Aussie Farms, which offers a strictly spatial approach. Described as an interactive map of animal exploitation facilities in Australia, this online project displays the location of sites associated with specific categories of use (food, pets, wildlife, entertainment, clothing, and science) and the types of nonhumans used. Based on significant industry backlash generated by this and similar projects, Barnes and White (2020)

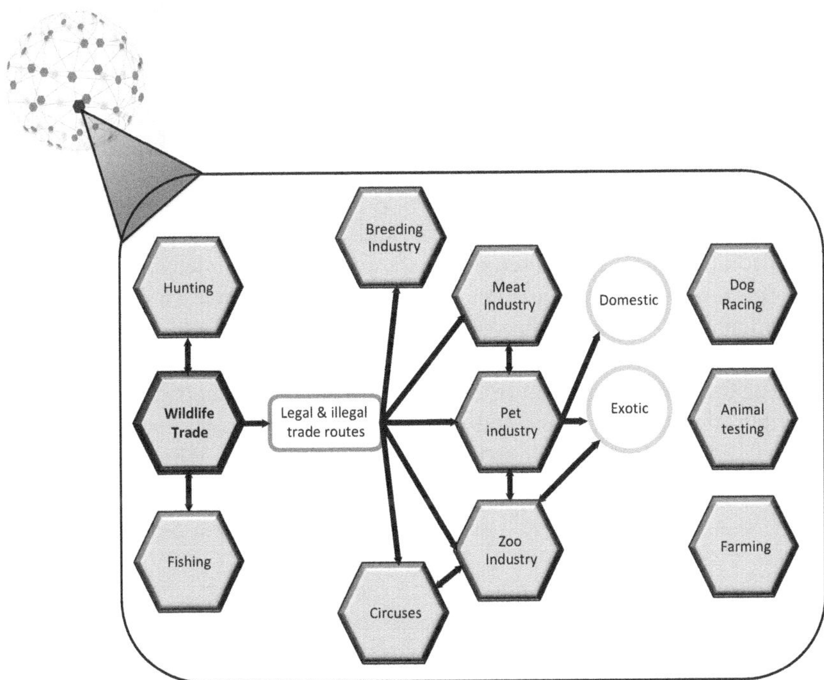

Figure BM1.1 A conceptual map of the wildlife trade and its shared, co-constitutive practices (global vector adapted from macrovector at Freepik).

use it to demonstrate the power of counter-maps to challenge hegemonic industries and views. As yet, the Farm Transparency Map does not make connections between the different industries and there is no overarching narrative or representation that situates each use category within a unified complex. However, the map is a dynamic work-in-progress and represents a heterotopic site of disruption that could be part of ongoing and collective efforts to emphasise the full scope of the A-IC.

Representing the "abstract machines" (Pedersen, 2019, p. 25) of the A-IC in a way that usefully and succinctly conveys their scope and scale could constitute a valuable tool for scholars, advocates, activists, and their combinations – a map on which all associated practices may be located and contextualised. Indeed, Twine contends, "It is impossible to envision a transition to more convivial human-animal relations without a wholesale challenge to the animal-industrial complex" (Twine, 2013, p. 91), and Giraud similarly argues that the entire "matrix of oppressive relations" needs to be challenged before fundamental social change can be achieved (2013, pp. 110–111). A focus on isolated parts of the A-IC and/or certain practices is not sufficient.

With a comprehensive and dynamic model of practices that comprise the A-IC, it becomes possible to consider where else the disordered ways of thinking, feeling, speaking, and acting explored in this volume's chapters could be applied. Which practices are so far under-represented in analyses? Which are largely excluded, understood as less problematic or even benign based on incomplete narratives and the smoothing of violences? The contextualisation offered through the compound lens of the A-IC also points to the potential for disruptive counter-sites and radical reimaginings to apply themselves to vertical deep dives into the machinations of one set of arrangements (such as the wildlife trade) or horizontal, cross-practice analyses of shared mechanisms and techniques of power (e.g., breeding) across several arrangements.

Ideologies that support the use and abuse of nonhuman animals have certainly not expired, but they should be approaching their terminal stages. Broader recognition of this must be built not only on the urgency of mitigating climate change (Burke et al., 2023), biodiversity loss (Henry et al., 2019), and environmental degradation (Scarborough et al., 2023), or the associated benefits to human health (Springmann et al., 2016) and global food security (Ritchie, 2021) but also on the fact that other animals are individual subjectivities with inherent rights whose sovereign lives and communities humans are not entitled to assimilate in pursuit of anthropocentric, capitalocentric, human supremacist, or any other ends.

A global, collective embrace of this fact will require, among other things, the multiplication of heterotopia and radical imaginaries directed at liberating other animals, and other humans, from the life-negating confines of the obstinate human–animal duality. Hopefully, this volume can serve to advance this mission.

Notes

1 A euphemism for the potential and actual extermination of animals (Brooks, 2021, p. 132).
2 The range of industries that use animal products in their manufacturing processes are too numerous and therefore excluded.
3 See: https://map.farmtransparency.org/

References

Arcari, P. (2022). (Animal) Oppression: Responding to Questions of Efficacy and (Il)Legitimacy in Animal Advocacy with a New Collective Action/Master Frame. *Animal Studies Journal, 11*(2), 68–108.

Barnes, A., & White, R. (2020). Mapping Emotions: Exploring the Impact of the Aussie Farms Map. *Journal of Contemporary Criminal Justice, 36*(3), 303–326.

Brooks Pribac, T. (2021). *Enter the Animal: Cross-species Perspectives on Grief and Spirituality*. Sydney University Press.

Burke, D. T., Hynds, P., & Anushree, P. (2023). Quantifying Farm-to-Fork Greenhouse Gas Emissions for Five Dietary Patterns Across Europe and North America: A Pooled Analysis from 2009 to 2020. *Resources, Environment and Sustainability, 12*, 100108.

Giraud, E. (2013). 'Beasts of Burden': Productive Tensions Between Haraway and Radical Animal Rights Activism. *Culture, Theory and Critique, 54*(1), 102–120.

Henry, R. C., Alexander, P., Rabin, S., Anthoni, P., Rounsevell, M. D. A., & Arneth, A. (2019). The Role of Global Dietary Transitions for Safeguarding Biodiversity. *Global Environmental Change, 58*, 101956.

Kindynis, T. (2014). Ripping up the Map: Criminology and Cartography Reconsidered. *British Journal of Criminology, 54*(2), 222–243.

Nicolini, D. (2010). Zooming In and Out: Studying Practices by Switching Theoretical Lenses and Trailing Connections. *Organization Studies, 30*(12), 1391–1418.

Noske, B. (1989). *Humans and Other Animals: Beyond the Boundaries of Anthropology*. Pluto Press.

O'Sullivan, V. (2020). Non-Human Animal Trauma During the Pandemic. *Postdigital Science and Education, 2*(3), 588–596.

Pedersen, H. (2019). *Schizoanalysis and Animal Science Education*. Bloomsbury Academic.

Pepper, A. (2017). Justice for Animals in a Globalising World. In A. Woodhall & G. G. da Trindade (Eds.), *Ethical and Political Approaches to Nonhuman Animal Issues* (pp. 149–176). Palgrave Macmillan.

Ritchie, H. (2021). If the World Adopted a Plant-Based Diet, We Would Reduce Global Agricultural Land Use from 4 to 1 Billion Hectares. *OurWorldInData.org*. https://ourworldindata.org/land-use-diets

Scarborough, P., Clark, M., Cobiac, L., Papier, K., Knuppel, A., Lynch, J., Harrington, R., Key, T., & Springmann, M. (2023). Vegans, Vegetarians, Fish-Eaters and Meat-Eaters in the UK Show Discrepant Environmental Impacts. *Nat Food, 4*(7), 565–574.

Shove, E., Pantzar, M., & Watson, M. (2012). *The Dynamics of Social Practice: Everyday Life and How It Changes*. SAGE Publications Ltd.

Springmann, M., Godfray, H. C., Rayner, M., & Scarborough, P. (2016). Analysis and Valuation of the Health and Climate Change Cobenefits of Dietary Change. *Proceedings of the National Academy of Sciences (PNAS), 113*(15), 4146–4151.

Twine, R. (2012). Revealing the 'Animal-Industrial Complex' – a Concept & Method for Critical Animal Studies. *Journal for Critical Animal Studies, 10*(1), 12–39.

Twine, R. (2013). Addressing the Animal – Industrial Complex. In R. Corbey & A. Lanjouw (Eds.), *The Politics of Species: Reshaping our Relationships with Other Animals* (pp. 77–91). Cambridge University Press.

Index

For Product Safety Concerns and Information please contact our
EU representative GPSR@taylorandfrancis.com Taylor & Francis
Verlag GmbH, Kaufingerstraße 24, 80331 München, Germany